Springer Proceedings in Mathematics & Statistics

Volume 303

D1795684

Springer Proceedings in Mathematics & Statistics

This book series features volumes composed of selected contributions from workshops and conferences in all areas of current research in mathematics and statistics, including operation research and optimization. In addition to an overall evaluation of the interest, scientific quality, and timeliness of each proposal at the hands of the publisher, individual contributions are all refereed to the high quality standards of leading journals in the field. Thus, this series provides the research community with well-edited, authoritative reports on developments in the most exciting areas of mathematical and statistical research today.

More information about this series at http://www.springer.com/series/10533

Praveen Agarwal · Dumitru Baleanu ·
YangQuan Chen · Shaher Momani ·
José António Tenreiro Machado
Editors

Fractional Calculus

ICFDA 2018, Amman, Jordan, July 16–18

 Springer

Editors
Praveen Agarwal
Department of Mathematics
Anand International College of Engineering
Jaipur, Rajasthan, India

Harish-Chandra Research Institute (HRI)
Allahabad, India

International Centre for Basic
and Applied Sciences
Jaipur, India

YangQuan Chen
School of Engineering (MESA Lab)
University of California, Merced
Merced, CA, USA

José António Tenreiro Machado
Institute of Engineering
Polytechnic Institute of Porto
Porto, Portugal

Dumitru Baleanu
Department of Mathematics
and Computer Science
Çankaya University
Ankara, Turkey

Shaher Momani
Department of Mathematics
The University of Jordan
Amman, Jordan

ISSN 2194-1009   ISSN 2194-1017   (electronic)
Springer Proceedings in Mathematics & Statistics
ISBN 978-981-15-0432-7   ISBN 978-981-15-0430-3   (eBook)
https://doi.org/10.1007/978-981-15-0430-3

Mathematics Subject Classification (2010): 26-XX, 33-XX, 35-XX, 39-XX, 41-XX, 45-XX, 65-XX, 68-XX

This Springer imprint is published by the registered company Springer Nature Singapore Pte Ltd.
The registered company address is: 152 Beach Road, #21-01/04 Gateway East, Singapore 189721, Singapore

Preface

These are the keynote and invited talks of the *International Conference on Fractional Differentiation and its Applications* (ICFDA-2018), which was held at the Amman Marriot Hotel, Sheissani in Amman, The Hashemite Kingdom of Jordan from July 16 to July 18, 2018. The ICFDA18 is a specialized conference on fractional-order calculus and its applications. It is a generalization of the integer-order ones. The fractional-order differentiation of arbitrary orders takes into account the memory effect of most systems. The order of the derivatives may also be variable, distributed, or complex. Recently, fractional-order calculus became a more accurate tool to describe systems in various fields in mathematics, biology, chemistry, medicine, mechanics, electricity, control theory, economics, and signal and image processing.

For this edition, we were happy to have 23 invited speakers who gave talks on a subject for which they are internationally known experts. Thirteen of these talks are collected in this volume. Throughout this book, the fractional calculus concepts have been explained very carefully in the simplest possible terms, and illustrated by a number of complete solved examples. This book contains some theorems and their proofs.

The book is organized as follows. In chapter "Closed-Form Discretization of Fractional-Order Differential and Integral Operators", a closed-form concretization of fractional-order differential or integral Laplace operators is introduced. The proposed method depends on extracting the necessary phase requirements from the phase diagram. The magnitude frequency response follows directly due to the symmetry of the poles and zeros of the finite z-transfer function. Unlike the continued fraction expansion technique, or the infinite impulse response of second-order IIR-type filters, the proposed technique generalizes the Tustin operator to derive a first-, second-, third-, and fourth-order discrete-time operators (DTO) that were stable and of minimum phase. The proposed method depends only on the order of the Laplace operator. The resulted discrete-time operators enjoy flat phase response over a wide range of discrete-time frequency spectrum. The closed-form DTO enables one to identify the stability regions of fractional-order discrete-time systems or even to design discrete-time-fractional-order $PI^\lambda D^\mu$ controllers.

The effectiveness of this work was demonstrated via several numerical simulations. In chapter "On Fractional-Order Characteristics of Vegetable Tissues and Edible Drinks", we are concerned about frequency response techniques to characterize vegetable tissues and edible drinks. In the first phase, the impedance of the distinct samples is measured and fractional-order models are applied to the resulting data. In the second phase, hierarchical clustering and multidimensional scaling tools are adopted for comparing and visualizing the similarities between the specimen.

In chapter "Some Relations Between Bounded Below Elliptic Operators and Stochastic Analysis", we apply Malliavin calculus tools to the case of a bounded below elliptic right-invariant pseudo-differential operators on a Lie group. We give examples of bounded below pseudo-differential elliptic operators on $\mathbb{R}^d$ by using the theory of the Poisson process and the Garding inequality. In the two cases, there are no stochastic processes because the considered semi-groups do not preserve positivity. In chapter "Discrete Geometrical Invariants: How to Differentiate the Pattern Sequences from the Tested Ones?" based on the new method (defined below as the discrete geometrical invariants—DGI(s)), one can show that it enables to differentiate the statistical differences between random sequences that can be presented in the form of 2D curves. We generalized and considered the Weierstrass–Mandelbrot function and found the desired invariant of the fourth order that connects the WM-functions with different fractal dimensions. Besides, we consider an example based on real experimental data. A high correlation of the statistically significant parameters of the DGI obtained from the measured data (associated with reflection optical spectra of olive oil) with the sample temperature is shown. This new methodology opens wide practical applications in the differentiation of the hidden interconnections between measured by the environment and external factors.

In chapter "Nonlocal Conditions for Semi-linear Fractional Differential Equations with Hilfer Derivative", we study the existence of solutions and some topological proprieties of solution sets for nonlocal semi-linear fractional differential equations of Hilfer type in Banach space by using noncompact measure method in the weighted space of continuous functions. The main result is illustrated with the aid of an example. In chapter "Offshore Wind System in the Way of Energy 4.0: Ride Through Fault Aided by Fractional PI Control and VRFB", we present a simulation about a study to improve the ability of an offshore wind system to recover from a fault due to a rectifier converter malfunction. The system comprises: a semi-submersible platform; a variable speed wind turbine; a synchronous generator with permanent magnets; a five-level multiple point diode clamped converter; a fractional PI controller using the Carlson approximation. Recovery is improved by shielding the DC link of the converter during the fault using as further equipment a redox vanadium flow battery, aiding the system operation as desired in the scope of Energy 4.0. Contributions are given for: (i) the fault influence on the behavior of voltages and currents in the capacitor bank of the DC link; (ii) the drivetrain modeling of the floating platform by a three-mass modeling; (iii) the vanadium flow battery integration in the system.

In chapter "Soft Numerical Algorithm with Convergence Analysis for Time-Fractional Partial IDEs Constrained by Neumann Conditions", a soft numerical algorithm is proposed and analyzed to fitted analytical solutions of PIDEs with appropriate initial and Neumann conditions in Sobolev space. Meanwhile, the solutions are represented in series form with accurately computable components. By truncating the n-term approximate solutions of analytical solutions, the solution methodology is discussed for both linear and nonlinear problems based on the nonhomogeneous term. Analysis of convergence and smoothness is given under certain assumptions to show the theoretical structures of the method. Dynamic features of the approximate solutions are studied through an illustrated example. The yield of numerical results indicates the accuracy, clarity, and effectiveness of the proposed algorithm as well as provide a proper methodology in handling such fractional issues. Chapter "Approximation of Fractional-Order Operators" deals with the several comparisons in the time response and Bode results between four well-known methods; Oustaloup's method, Matsuda's method, AbdelAty's method, and El-Khazali's method are made for the rational approximation of fractional-order operator (fractional Laplace operator). The various methods along with their advantages and limitations are described in this chapter. Simulation results are shown for different orders of the fractional operator. It has been shown in several numerical examples that the El-Khazali's method is very successful in comparison with Oustaloup's, Matsuda's, and AbdelAty's methods.

In chapter "Multistep Approach for Nonlinear Fractional Bloch System Using Adomian Decomposition Techniques", we discuss a superb multistep approach, based on the Adomian decomposition method (ADM), which is successfully implemented for solving nonlinear fractional Bolch system over a vast interval, numerically. This approach is demonstrated by studying the dynamical behavior of the fractional Bolch equations (FBEs) at different values of fractional order α in the sense of Caputo concept over a sequence of the considerable domain. Further, the numerical comparison between the proposed approach and implicit Runge–Kutta method is discussed by providing an illustrated example. The gained results reveal that the MADM is a systematic technique in obtaining a feasible solution for many nonlinear systems of fractional order arising in natural sciences.

The chapter "Simulation of the Space–Time-Fractional Ultrasound Waves with Attenuation in Fractal Media" deals with the simulation of the space–time-fractional ultrasound waves with attenuation in fractal media. In chapter "Certain Properties of Konhauser Polynomial via Generalized Mittag-Leffler Function", we establish several new properties of generalized Mittag-Leffler function via Konhauser polynomials. Properties like mixed recurrence relations, differential equations, pure recurrence relations, finite summation formulae, and Laplace transform have been obtained. In chapter "An Effective Numerical Technique Based on the Tau Method for the Eigenvalue Problems", we consider the (presumably new) effective numerical scheme based on the Legendre polynomials for approximate solution of eigenvalue problems. First, a new operational matrix, which can be represented by sparse matrix is defined by using the Tau method and orthogonal functions. Sparse data is by nature more compressed and thus require significantly less storage.

A comparison of the results for some examples reveals that the presented method is convenient and effective, also we consider the problem of column buckling to show the validity of the proposed method. Finally, in chapter "On Hermite–Hadamard-Type Inequalities for Coordinated Convex Mappings Utilizing Generalized Fractional Integrals", we obtain the Hermite–Hadamard-type inequalities for coordinated convex function via generalized fractional integrals, which generalize some important fractional integrals such as the Riemann–Liouville fractional integrals, the Hadamard fractional integrals, and Katugampola fractional integrals. The results given in this chapter provide a generalization of several inequalities obtained in earlier studies.

Jaipur, India   Praveen Agarwal
Ankara, Turkey   Dumitru Baleanu
Merced, USA   YangQuan Chen
Amman, Jordan   Shaher Momani
Porto, Portugal   José António Tenreiro Machado

Contents

Closed-Form Discretization of Fractional-Order Differential and Integral Operators

Reyad El-Khazali and J. A. Tenreiro Machado

Abstract This paper introduces a closed-form discretization of fractional-order differential or integral Laplace operators. The proposed method depends on extracting the necessary phase requirements from the phase diagram. The magnitude frequency response follows directly due to the symmetry of the poles and zeros of the finite z-transfer function. Unlike the continued fraction expansion technique, or the infinite impulse response of second-order IIR-type filters, the proposed technique generalizes the Tustin operator to derive a first-, second-, third-, and fourth-order discrete-time operators (DTO) that are stable and of minimum phase. The proposed method depends only on the order of the Laplace operator. The resulted discrete-time operators enjoy flat-phase response over a wide range of discrete-time frequency spectrum. The closed-form DTO enables one to identify the stability regions of fractional-order discrete-time systems or even to design discrete-time fractional-order $PI^\lambda D^\mu$ controllers. The effectiveness of this work is demonstrated via several numerical simulations.

Keywords Fractional calculus · Transfer function · Discrete-time operator · Discrete-time integro-differential operators · Frequency response

1 Introduction

Fractional calculus is a generalization of the integer-order one. Most practical systems exhibit fractional-order dynamics, which could be of real or complex values. Fractional-order systems enjoy the hereditary effect that is approximated by infinite-dimensional models [8, 20]. It is used in many fields such as in economy, physics,

R. El-Khazali (✉)
ECE Department, Khalifa University, Abu Dhabi, United Arab Emirates
e-mail: reyad.elkhazali@ku.ac.ae

J. A. T. Machado
Department of Electrical Engineering, Institute of Engineering, Polytechnic of Porto, Porto, Portugal
e-mail: jtm@isep.ipp.pt

© Springer Nature Singapore Pte Ltd. 2019
P. Agarwal et al. (eds.), *Fractional Calculus*, Springer Proceedings
in Mathematics & Statistics 303, https://doi.org/10.1007/978-981-15-0430-3_1

biology, chemistry, medicine, social sciences, and engineering. To analyze fractional-order systems, one has to look for finite-dimensional and realizable models that approximate such systems [11, 12, 19, 21–25].

The use of microprocessors nowadays are necessary for signal processing and system analysis. Thus, a straightforward method is required to discretize a continuous-time fractional-order system into a discrete-time one. This can be accomplished by discretizing the fractional-order Laplacian operator s^α and replacing it with a finite-order DTO. In general, there are two methods that are used to discretize s^α; i.e., a direct and an indirect one. In the indirect discretization method, a rational continuous-time operator (CTO) is first obtained and then discretized using techniques such as the bilinear transformation, the Al-Alaoui operator, the Euler's backward method, or the stable Simpson's method [1–3]. The direct method, however, allows one to generate discrete-time operators that converts a continuous-time operator (CTO) into a DTO [4, 5, 17].

The indirect discretization method is achieved in two steps; the first one is to approximate the Laplacian operator s^α by a rational transfer function in the s-domain, which is then simplified using the continued fraction expansion (CFE), and the second step is to discretize the expanded form using either the bilinear transformation, Simpson's method, Euler's method, or a linear combination of them or other existing forms [6, 24]. It is important to realize that the CFE method could yield an unstable non-minimum phase discrete-time operator. An alternative approach to the CFE was discussed in [19], where infinite impulse response (IIR) autoregressive moving-average (ARMA) models are used to develop DTO operators, which may result in developing higher order approximation. Notice that the Al-Alaoui operator is obtained as a linear combination of the trapezoidal and the rectangular integration rules [2, 14, 15, 26]. The interpolation and inversion processes may induce, in some cases, unstable fractional-order operators.

This work introduces a straightforward discretization direct method to discretize continuous differential and/or integral operators. It can be considered as a dynamic (or adaptive) discretization technique, where the poles and the zeros of the generated z-transfer function are all located inside the unit disc and their values depend only on the fractional-order α. The proposed method yields finite-order DTO that exhibits a competitive frequency response to higher order operators developed in [2, 6, 16].

The paper is organized as follows. Section 2 summarizes some preliminary concepts and background. Section 3 introduces the main results of first-, second-, third-, and fourth-order operators, while Sect. 4 summarizes the numerical simulation and a comparison between different operators. Section 5 outlines the main conclusions.

2 Preliminary Concepts and Background

The general fractional-order differential (integral) operator is denoted by $_a D_t^{\pm\alpha}$ ($_a I_t^\alpha$), respectively [18], where a and t represent the starting time and $\alpha \in \mathbb{R}$ is the order of the operator. For example, if one wishes to implement a discrete-time fractional-order controller, then it is necessary to look for a stable non-minimum phase DTO operator of low order. The design and implementation of fractional-order discrete-time controllers cannot accommodate higher order operators since this will increase the complexity of the controlled system, and could yield unstable ones. Therefore, the proposed technique provides a competitive DTO that benefits from the IIR structure of such operators; i.e., a second-order DTO is competitive to that of a ninth-order one introduced in [6, 19, 20].

As mentioned in Sect. 1, the indirect discretization method starts by developing a rational finite-order transfer functions, that is, $s^{\pm\alpha} \approx \frac{N(s,\alpha)}{D(s,\alpha)}$ [10, 13, 21], and it is followed by using any existing discretization technique, or a linear combination of such methods. For example, the Al-Alaoui discrete-time integral operator is simply a linear interpolation of the backward rectangular rule and the trapezoidal rule, namely $H(z) = aH_{Rect}(z) + (1-a)H_{Trap}(z)$, where $0 < a < 1$ [1–3]. A similar approach was used to derive a hybrid digital integrator using a linear combination of Trapezoidal and Simpson integrator [6, 10]. Such interpolation reduces the frequency warping over a limited frequency band, and their phase frequency response is not constant. For comparison, Fig. 1 displays the frequency response of the Tustin operator, $s = H(z) = \frac{2}{T}\frac{1-z^{-1}}{1+z^{-1}}$, Al-Alaoui operator, $s = H(z) = \frac{8}{7T}\frac{1-z^{-1}}{1+\frac{1}{7}z^{-1}}$, and Chen discrete-time operator [5]. Another discrete-time operator that approximates an integer-order integrator was also introduced in [6] and given here for completeness:

$$H(z) = \frac{6(z^2 - 1)}{T(3-a)(z+p_1)(z+p_2)}, \tag{1a}$$

$$p_1 = \frac{3+a+2\sqrt{3a}}{3-a}, \tag{1b}$$

$$p_2 = \frac{3+a-2\sqrt{3a}}{3-a}, \tag{1c}$$

where T is the sampling time and $0 < a < 1$ is a scaling factor. Equation (1) can then be used to generate several quadratic forms that discretize $s^{\pm 1}$.

Figure 1 shows the frequency response of the aforementioned three DTO operators that approximate s^1 for $T = 0.001$. The magnitude response of Tustin operator exhibits large errors at both ends of the frequency spectrum. The magnitude response of the Al-Alaoui operator, however, is almost identical to that of the Tustin operator at low frequency, but provides a better response at high frequency. Moreover, it yields a linear phase response due to the asymmetric pole-zero location, while the hybrid ninth-order operator reported in [6] yields a perfect phase behavior. However, one cannot afford this size of an operator since a discrete-time fractional-order phase-locked loop, for example, will be modeled by an 18th-order discrete-time z-transfer function.

Since the goal is to look for a closed-form discrete-time model for $s^{\pm\alpha}$, the direct approach is adopted here to develop a straightforward discretization method.

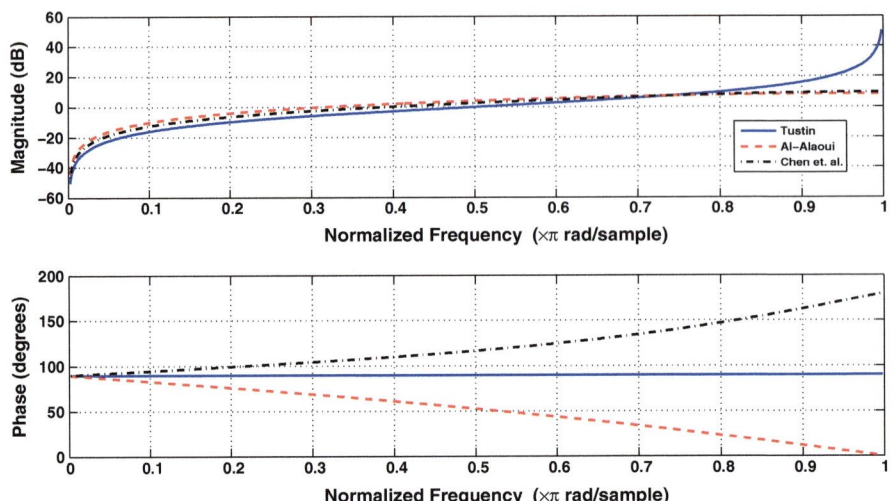

Fig. 1 Frequency response of Tustin, Al-Alaoui, and the DTO of Eq. (1) for $a = 1$

In all direct methods, the continuous frequency operator is replaced by a generating function, that is, $s^{\pm\alpha} = \left(\omega\left(z^{-1}\right)\right)^{\pm\alpha}$. To gain more insight, one may start with the Grünwald–Letnikov (GL) definition of the fractional-order differential (integral) operator [7, 13, 14, 21, 22]:

$$_a D_t^{\pm\alpha} f(t) = \lim_{h \to 0} \frac{1}{h^{\pm\alpha}} \sum_{j=0}^{\infty} C_j^{\pm\alpha} f((t-j)h). \tag{2}$$

where

$$C_j^{\pm\alpha} = (-1)^j \binom{\pm\alpha}{j} = \left(1 - \frac{1 \pm \alpha}{j}\right) C_{j-1}^{\pm\alpha}, \ j = 1, \ldots, n, \tag{3a}$$

$$C_0^{\pm\alpha} = 1. \tag{3b}$$

Taking the $\mathcal{Z}$-transform of (2) and using the short memory principle [14], the following generating function may discretize $s^{\pm\alpha}$:

$$\left(\omega\left(z^{-1}\right)\right)^{\pm\alpha} = T^{\mp\alpha} \left(\sum_{j=0}^{\left[\frac{L}{T}\right]} C_j^{\pm\alpha} z^{-j}\right), \tag{4}$$

where $T = h$ is the sampling time, and $\frac{L}{T} = \left[\frac{nh-a}{h}\right]$ is an increasing memory size $L - nh - a$.

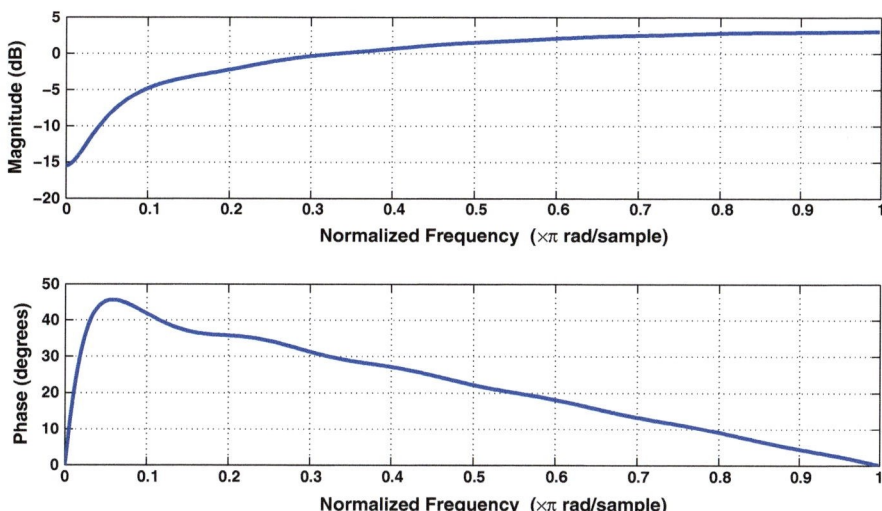

Fig. 2 Frequency response of a FIR-type discrete-time differentiator for $s^{0.5}$ with $T = 0.001$ s, and $L = 0.011$

Equation (4) defines a transfer function of a finite impulse response (FIR) discrete-time system of $s^{\pm\alpha}$. The memory size, L, determines the accuracy of the approximation. Hence, a compromise has to be made between the accuracy and the size of the operator. Figure 2 shows the frequency response of (4) for $\alpha = 0.5$, $T = 0.001$, and $L = 11$. Clearly, the phase diagram is close to the expected angle of $\frac{\pi}{4}$ over a very narrow frequency band $\omega \in (0.06, 0.08)$ rad/s, which may not be suitable for realization techniques.

Obviously, in spite of its large size, the frequency response of the FIR discrete-time form of Eq. (4) does not provide the expected constant phase response. Therefore, an alternative discrete-time IIR-type rational z-transfer function, of lower size than the FIR form, to discretize $s^{\pm\alpha}$ will be the choice to overcome such problem.

Since, the CFE approach does not always yield a minimum phase and stable system, or a flat-phase response [2, 6, 7, 11, 13], a compromise has to be made between the size of the expansion and the type of the generating functions used for approximation. The following generating functions can be used to discretize $s^{\pm\alpha}$ and replace it with DTO operators [5, 6, 14, 17]:

(a) Backward-Euler method: $\left(\omega\left(z^{-1}\right)\right)^{\pm\alpha} = \left(\frac{1-z^{-1}}{T}\right)^{\pm\alpha}$

(b) Trapezoidal (Tustin) discretization rule: $\left(\omega\left(z^{-1}\right)\right)^{\pm\alpha} = \left(\frac{2}{T}\frac{1-z^{-1}}{1+z^{-1}}\right)^{\pm\alpha}$

(c) Al-Alaoui Operator: $\left(\omega\left(z^{-1}\right)\right)^{\pm\alpha} = \left(\frac{8}{7T}\frac{1-z^{-1}}{1+\frac{1}{7}z^{-1}}\right)^{\pm\alpha}$

(d) A Hybrid interpolation of Simpson and Trapezoidal discrete-time integrators:

$$H(z) = aH_S(z) + (1-a)H_T(z), \ 0 < a < 1, \tag{5}$$

where $H_S(z) = \frac{T}{3}\frac{1+4z^{-1}+z^{-2}}{1-z^{-2}}$ and $H_T(z) = \frac{T}{2}\frac{1+z^{-1}}{1-z^{-1}}$.

The interpolation in (5) represents a generalization of the first three methods. Since the magnitude frequency response of the integer-order integrator, s^{-1}, lies between the Simpson rule and that of the Trapezoidal discrete-time integrator [2, 3], the linear combination in (5) for $0 < a < 1$ can be used to generate a typical IIR-type discrete-time operator as follows [6]:

$$\left(\omega\left(z^{-1}\right)\right)^{\pm\alpha} = k_0 \left(\frac{1-z^{-2}}{\left(1+bz^{-1}\right)^2}\right)^{\alpha}, \tag{6}$$

where $\alpha \in [0, 1]$, $k_0 = \left(\frac{6z_2}{T(3-a)}\right)^{\alpha}$ and $b = z_2 = \frac{3+a-2\sqrt{3a}}{3-a}$.

Several transfer functions of different sizes can be obtained to approximate $\left(\omega\left(z^{-1}\right)\right)^{\pm\alpha}$. For example, when $\alpha = 0.5$ and $T = 0.001$, Eq. (6), yields the following z-transfer functions, $G_{(n,a)}(z)$, that discretize $s^{0.5}$, where n and a represent the order and the weighting factor of the approximation, respectively [6]:

$$G_{(2,0.5)}\left(z^{-1}\right) = \frac{127 + 41.26z^{-1} - 112.6z^{-2}}{4 + 2.98z^{-1} - z^{-2}}, \tag{7a}$$

$$G_{(3,0.5)}\left(z^{-1}\right) = \frac{1501 - 503.6z^{-1} - 1298z^{-2} + 446.5z^{-3}}{47.26 + 4z^{-1} - 23.63z^{-2} - z^{-3}}, \tag{7b}$$

$$G_{(4,0.5)}\left(z^{-1}\right) = \frac{508.1 - 1501z^{-1} - 4.478z^{-2} + 1298z^{-3} - 382.9z^{-4}}{16 - 40.54z^{-1} - 12z^{-2} + 20.27z^{-3} + z^{-4}}. \tag{7c}$$

Figure 3 shows the frequency response of (7) for $\omega \in (-\pi, \pi)$. The magnitude frequency response of the second-order approximation yields a warping effect at high frequency, while the phase diagram of the three forms exhibit a decreasing phase value over most of the spectrum.

Remark 1 The approximation given by (7c), reported in [6], represents an unstable non-minimum phase DTO since it has a pole and a zero outside the unit circle at $p = 2.6298$, and $z = 2.6328$, respectively. Even though $p \approx z$, that almost cancel each other, implementing such an operator would cause system instability. Furthermore, according to [6], one must improve the phase performance of $G_{(4,0.5)}(z)$ by cascading a causal lead compensator $z^{0.5} = \frac{z^{-0.5}}{z-1}$, which requires the implementation of a fractional-order sampler.

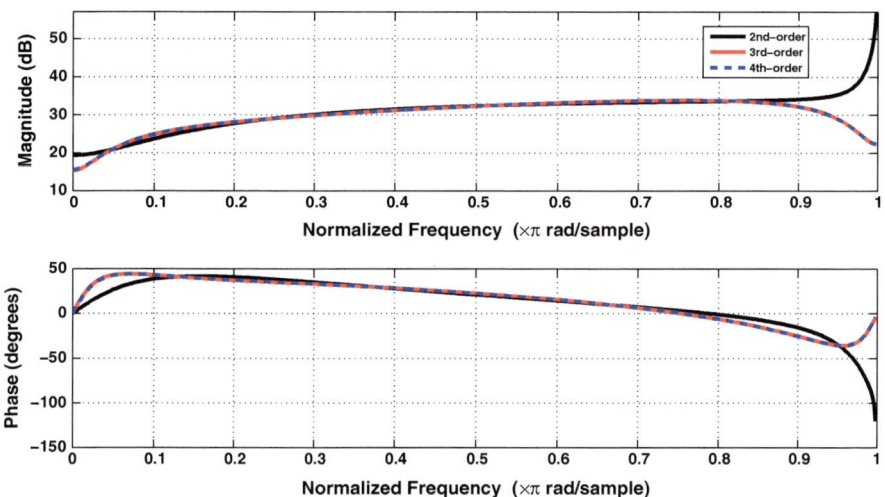

Fig. 3 Frequency response of $s^{0.5}$ using (7a), (7b), (7c)

3 New Fractional-Order Discrete-Time Operators

As discussed in Sect. 2, the discretization technique of generating functions using the CFE yields high order and an unstable non-minimum phase discrete-time approximation. The aim of this work is to avoid such subtleties by developing an adaptive closed-form DTO that effectively discretizes the fractional-order operators, $s^{\pm\alpha}$, which only depend on its order $\pm\alpha$. Furthermore, one can also define the stability region of the discrete form of $s^{\pm\alpha}$.

3.1 First-Order Operators

The following first-order operator based on a closed-form solution was first introduced in [8, 9]. It represents an approximation of a first-order discrete-time differential operator (DTDO), where its reciprocal also defines a discrete-time integral operator (DTIO):

$$s^{\pm\alpha} \approx H_{1_K}(z) = \left(\frac{2}{T}\right)^{\pm\alpha} \frac{z \mp z_1(\alpha)}{z \mp p_1(\alpha)}, \tag{8}$$

where

$$z_1(\alpha) = -p_1(\alpha) = \frac{1}{\tan\left((2-\alpha)\frac{\pi}{4}\right)}, \quad 0 < \alpha < 1, \tag{9}$$

and where $z_1(\alpha) = -p_1(\alpha) \in \mathbb{R}$.

Obviously, for $0 < \alpha < 1$, $|z_1(\alpha)| = |p_1(\alpha)| < 1$ are located inside the unit circle.

3.2 Second-Order Operator

The second-order discrete-time operator was also introduced in [8, 9]. It yields a normalized biquadratic discrete-time transfer function that approximates $s^{\pm\alpha}$ and is given by (Fig. 4):

$$s^{\pm\alpha} \approx H_{2_K}(z) = \left(\frac{2}{T}\right)^{\pm\alpha} \frac{(z \mp z_1(\alpha))(z \mp z_2(\alpha))}{(z \mp p_1(\alpha))(z \mp p_2(\alpha))}, \tag{10}$$

where

$$z_1(\alpha) = \frac{\eta_2 - 2 + \sqrt{5\eta_2^2 + 4}}{2\eta_2}, \quad \eta_2 = \tan\left(\alpha\frac{\pi}{4}\right), \tag{11}$$

and

$$\begin{cases} z_2(\alpha) = z_1(\alpha) - 1 \\ p_1(\alpha) = -z_2(\alpha) \\ p_2(\alpha) = -z_1(\alpha) \end{cases} . \tag{12}$$

Clearly, for large values of α, the first-order DTO yields a competitive frequency response to that of the second-order DTO as shown in Fig. 5.

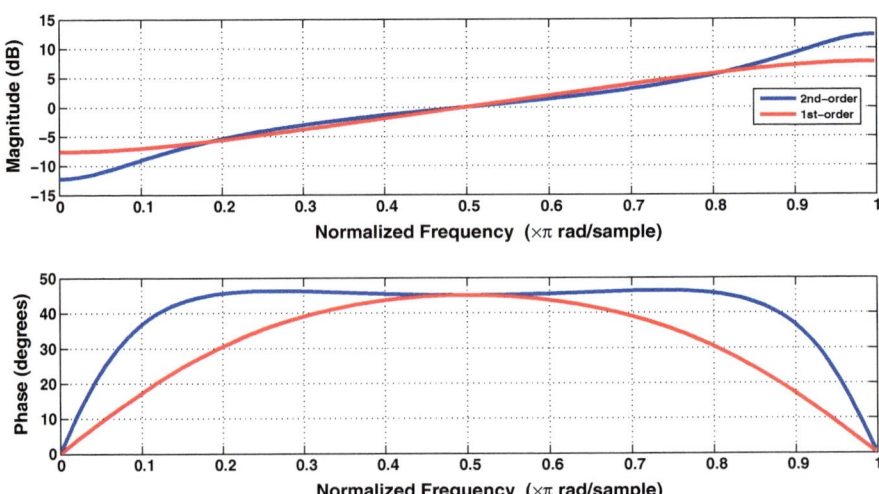

Fig. 4 Frequency response of discrete-time first- and second-order operators for $s^{0.5}$

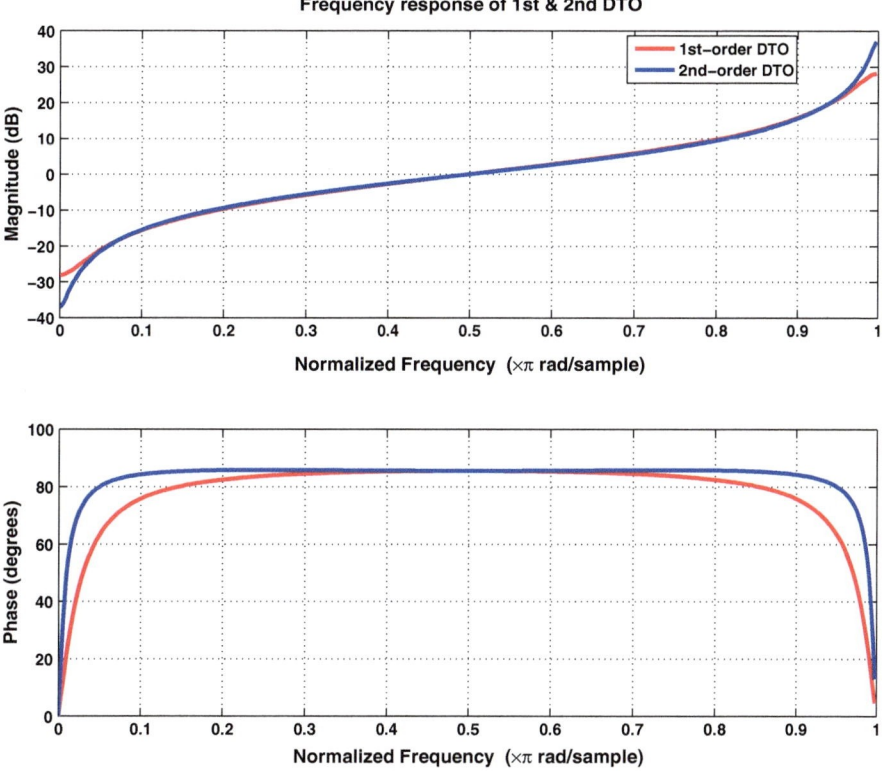

Fig. 5 Frequency response of (8) and (10) for $s^{0.95}$

3.3 Third-Order Operator

The third-order operator is developed to improve the accuracy of the discrete-time approximation over a wider frequency range. Similar to (10), the third-order operator is given by

$$s^{\pm\alpha} \approx H_{3_K}(z) = \left(\frac{2}{T}\right)^{\pm\alpha} \frac{(z \mp z_1(\alpha))(z \mp z_2(\alpha))(z \mp z_3(\alpha))}{(z \mp p_1(\alpha))(z \mp p_2(\alpha))(z \mp p_3(\alpha))}, \quad (13)$$

where

$$\begin{cases} p_3(\alpha) = -z_1(\alpha) \\ z_2(\alpha) = 1 - z_1(\alpha) \\ p_2(\alpha) = -z_2(\alpha) \\ z_3(\alpha) = -p_1(\alpha) \end{cases} \quad (14)$$

The pole-zero map of (14) is shown in Fig. 6, which represents a distribution of alternating real poles and zeros.

Fig. 6 Pole-zero map of
third-order DTO operator

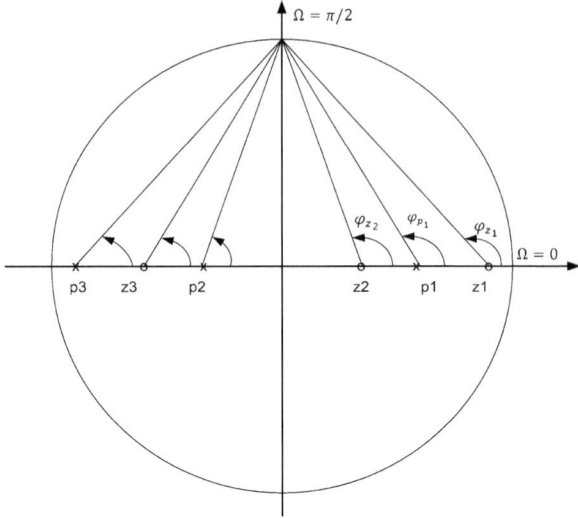

Due to the symmetry of the poles and zeros and since $z_i(\alpha) = -p_i(\alpha)$, $i = 1, 2, 3$, the phase requirement is assumed to meet the phase contribution of the fractional-order operator at the discrete-time frequency $\Omega = \alpha \frac{\pi}{2}$:

$$\left(\varphi_{z_1} + \varphi_{z_2} + \varphi_{z_3}\right) - \left(\varphi_{p_1} + \varphi_{p_2} + \varphi_{p_3}\right) = \alpha \frac{\pi}{2} \tag{15}$$

Substituting (14) into (15) yields

$$z_1 = \max\left(roots\left(z_1^2 - z_1 + q(\alpha)\right)\right), \tag{16}$$

where

$$q(\alpha) = \frac{2 - \alpha(1 + \eta_3)}{1 + \eta_3(1 - \alpha)} \tag{17}$$

and

$$\eta_3 = \tan\left(\alpha \frac{\pi}{4}\right). \tag{18}$$

Hence z_1 is found, the rest of poles and zeros are determined from (17) and (15). For example, for $\alpha = 0.5$ Eq. (19) gives $z_1 = 0.8425$, $z_2 = 0.1575$, and $z_3 = -0.5$, while $p_1 = -z_3$, $p_2 = -z_2$ and $p_3 = -z_1$. Therefore, the third-order DTO that discretizes $s^{0.5}$ for $T = 2$ is given by

$$s^{0.5} \approx H_{3_K}(z) = \frac{z^3 - 0.5z^2 - 0.3673z + 0.06635}{z^3 + 0.5z^2 - 0.3673z - 0.06635}. \tag{19}$$

Fig. 7 Pole-zero map of the fourth-order DTO operator

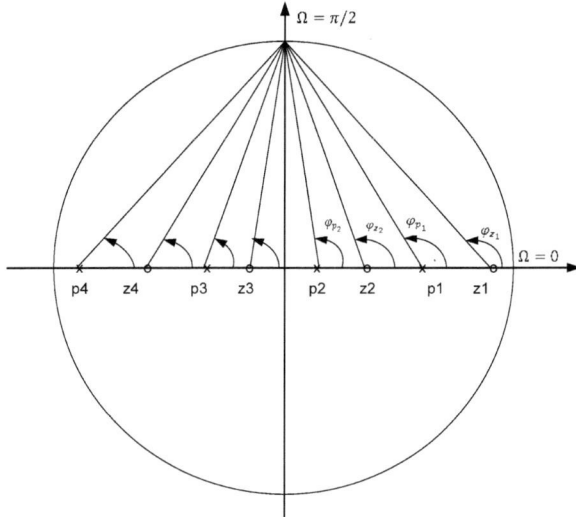

Remark 2 When $\alpha = 1$, then from (17), $q(1) = 0$, and Eq. (16) reduces to $z_1^2 - z_1 = 0$, which yields a nontrivial solution $z_1 = 1$, and the third-order DTO operator given by (13) and (14) for this case reduces to the well-known bilinear transformation $H_{3_K} = \frac{2}{T}\frac{1-z^{-1}}{1+z^{-1}}$.

3.4 Fourth-Order Operator

The fourth-order z-transfer function that discretizes the fractional-order operators is similarly developed as the previous three operators and given by the following finite z-transfer function:

$$s^{\pm\alpha} \approx H_{4_K}(z) = \left(\frac{2}{T}\right)^{\pm\alpha} \frac{(z \mp z_1(\alpha))(z \mp z_2(\alpha))(z \mp z_3(\alpha))(z \mp z_4(\alpha))}{(z \mp p_1(\alpha))(z \mp p_2(\alpha))(z \mp p_3(\alpha))(z \mp p_4(\alpha))},$$
(20)

where

$$\begin{cases} p_4(\alpha) = -z_1(\alpha) \\ p_2(\alpha) = 1 - z_1(\alpha) \\ z_3(\alpha) = -p_2(\alpha) \\ z_4(\alpha) = -p_1(\alpha) \\ p_3(\alpha) = -z_2(\alpha) \end{cases}$$
(21)

The pole-zero map of (21) is shown in Fig. 7, which also represents a distribution of alternating real poles and zeros of (20).

The phase contribution of (20) is given by

$$\left(\varphi_{z_1} + \varphi_{z_2} + \varphi_{z_3} + \varphi_{z_4}\right) - \left(\varphi_{p_1} + \varphi_{p_2} + \varphi_{p_3} + \varphi_{p_4}\right) = \alpha \frac{\pi}{2} \tag{22}$$

Since there is a symmetry between the poles and zeros as depicted in Fig. 7, one may focus on the phase contribution of the poles and zeros that lie on the positive real axis. By other words, from the symmetry, and without loss of generality, one may conclude from (22), that,

$$\left(\varphi_{z_1} + \varphi_{z_2}\right) - \left(\varphi_{p_1} + \varphi_{p_2}\right) = \alpha \frac{\pi}{4} \tag{23}$$

where

$$\varphi_{z_i} = \pi - \arctan\left(\frac{1}{z_i\left(\alpha\right)}\right), \; \varphi_{p_i} = \pi - \arctan\left(\frac{1}{p_i\left(\alpha\right)}\right), \; i = 1, 2. \tag{24}$$

Assumption 2 Let $p_1\left(\alpha\right)$ and $z_2\left(\alpha\right)$ lie in the geometric mean of their adjacent zeros and poles, respectively,

$$p_1\left(\alpha\right) = \sqrt{z_1\left(\alpha\right) z_2\left(\alpha\right)}, \tag{25a}$$

$$z_2\left(\alpha\right) = \sqrt{p_1\left(\alpha\right) p_2\left(\alpha\right)}. \tag{25b}$$

Substituting (24) and (25) into (23) yields the following nonlinear function in $z_1\left(\alpha\right)$:

$$
\begin{aligned}
f\left(z\right) &= \eta z_1^4 + 2\left(1 - \eta\right) z_1^3 - \left(\eta + 3\right) z_1^2 + \left(2\eta - 1\right) z_1 + \left(\eta + 1\right) \\
&+ \eta \left[z_1^{\frac{5}{3}}\left(1 - z_1\right)^{\frac{1}{3}} + z_1^{\frac{1}{3}}\left(1 - z_1\right)^{\frac{5}{3}} - z_1^{\frac{4}{3}}\left(1 - z_1\right)^{\frac{2}{3}} - z_1^{\frac{2}{3}}\left(1 - z_1\right)^{\frac{4}{3}}\right] \\
&+ z_1^{\frac{5}{3}}\left(1 - z_1\right)^{\frac{4}{3}} - z_1^{\frac{4}{3}}\left(1 - z_1\right)^{\frac{5}{3}} + z_1^{\frac{2}{3}}\left(1 - z_1\right)^{\frac{1}{3}} - z_1^{\frac{1}{3}}\left(1 - z_1\right)^{\frac{2}{3}} = 0,
\end{aligned}
\tag{26}
$$

where $\eta = \tan\left(\alpha \frac{\pi}{4}\right)$.

Obviously, the nonlinearities in (26) are due to placing the inner pole/zero at the geometric mean of its surrounding zeros/poles. Solving (26) numerically with an accuracy $|f\left(z\right)| < \epsilon$, for small $\epsilon > 0$ yields a desired solution $0 < z_1 < 1$. For $\alpha = 0.5$, the fourth- order operator that discretizes $s^{0.5}$ is found to be

$$s^{0.5} \approx H_{4k}\left(z\right) = \left(\frac{2}{T}\right)^{\alpha} \frac{z^4 - 0.5295z^3 - 0.3835z^2 + 0.05843z + 0.01218}{z^4 + 0.5295z^3 - 0.3835z^2 - 0.05843z + 0.01218}. \tag{27}$$

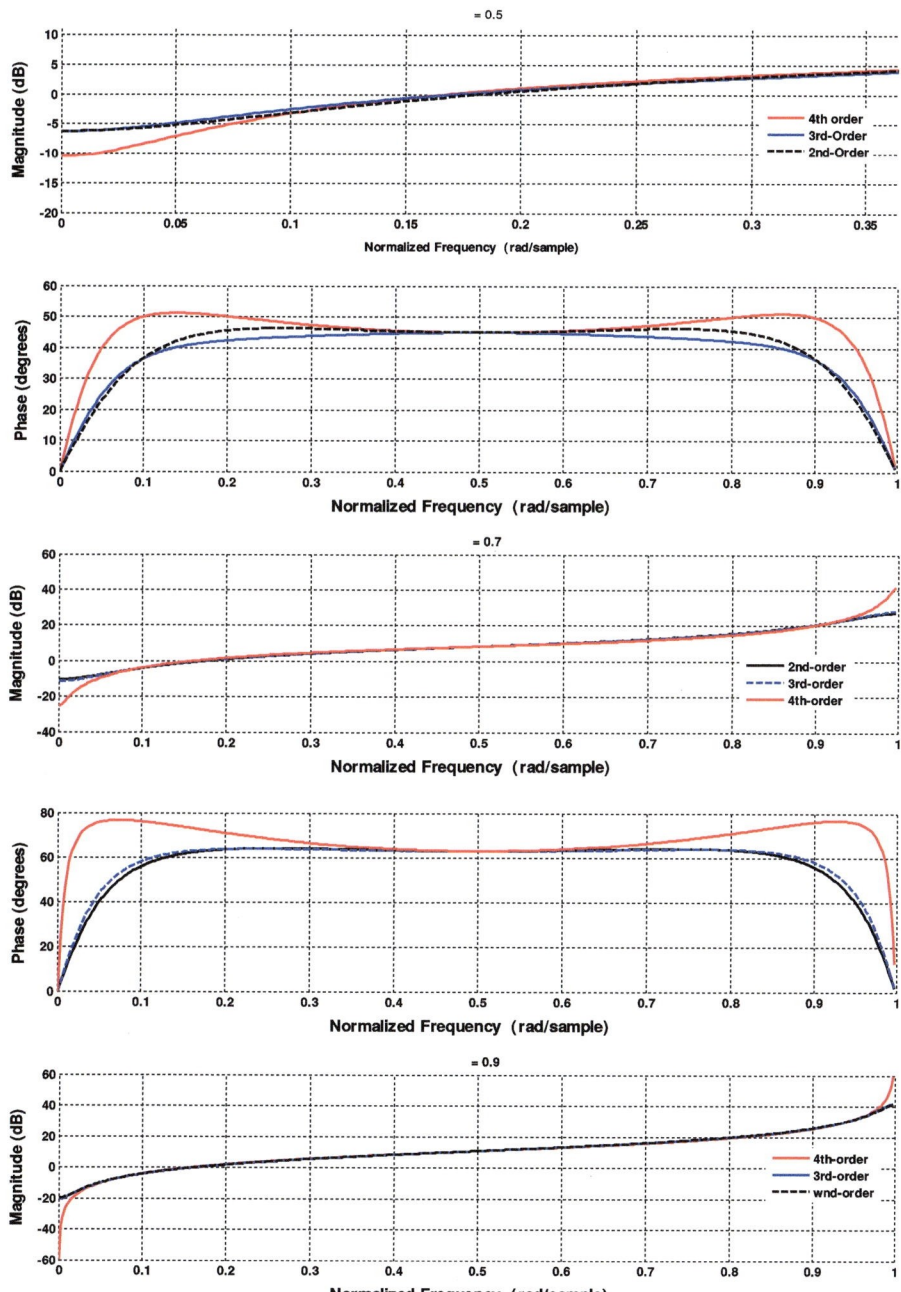

Fig. 8 Frequency response of the second-, third-, and fourth-order operators for $\alpha = 0.5$, $\alpha = 0.7$ and $\alpha = 0.9$

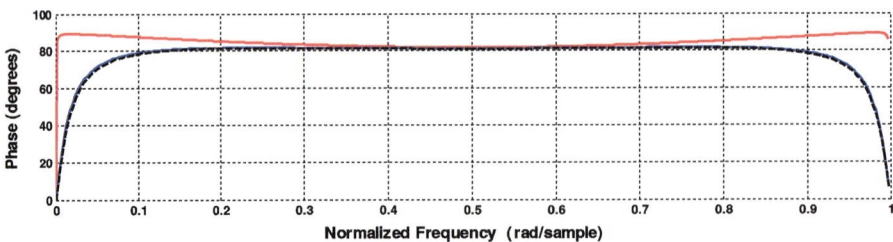

Fig. 8 (continued)

4 Numerical Simulation

Figure 8a–c shows the frequency response of the second-, third-, and the fourth-order operators for $\alpha = 0.5, \alpha = 0.7$, and $\alpha = 0.9$, respectively. As noted, the second-order operator is a good competitor to the third-order one, especially for $\alpha > 0.7$, while

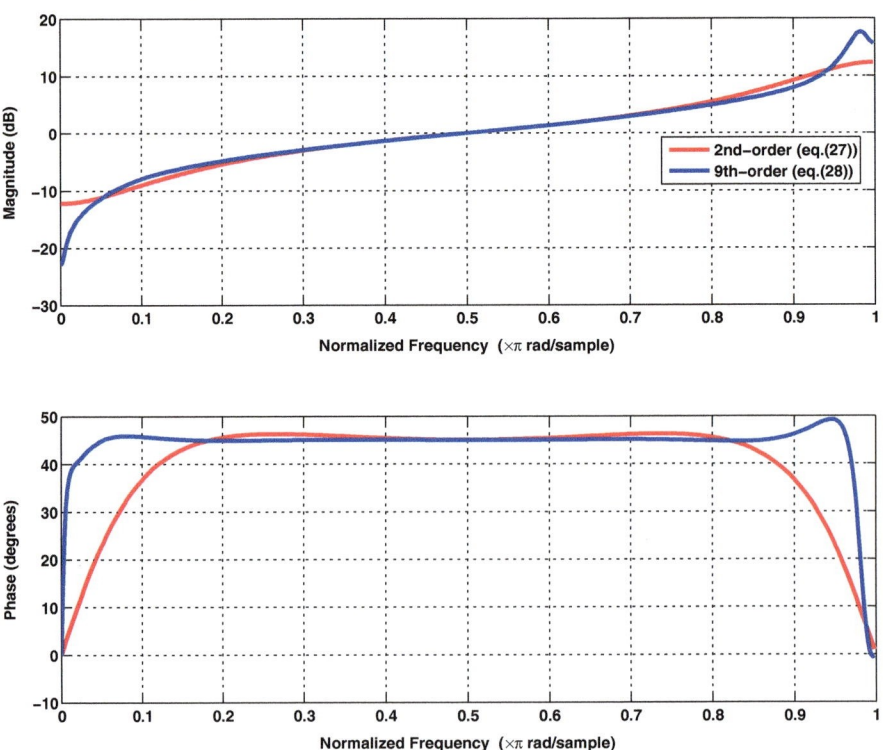

Fig. 9 Frequency response of (27) and (28) for $s^{0.5}$

the fourth-order operator exhibits a much better frequency response with a constant phase at the middle frequency with some overshoot at both ends of the spectrum.

To appreciate the proposed DTO, the frequency response of the second- order operator described by (10) is compared with other forms of DTO reported in [2, 6]. The case when $\alpha = 0.5$ for $T = 0.001$ is taken as a benchmark. Equations (10)–(12) then yield

$$s^{0.5} \approx \frac{44.7214 - 22.0313z^{-1} - 8.4670z^{-2}}{1.0 + 0.4926z^{-1} - 0.1893z^{-2}}. \tag{28}$$

The following ninth-order DTO that discretizes $s^{0.5}$ using the CFE and reported in [6] is investigated against the one given by (28)

$$G_9(z) = 44.72 \frac{z^9 - 0.5z^8 - 2z^7 + 0.875z^6 + 1.313z^5 - 0.4688z^4}{z^9 + 0.5z^8 - 2z^7 - 0.875z^6 + 1.313z^5 + 0.4688z^4} \cdots$$

$$\frac{-0.3125z^3 + 0.07813z^2 + 0.01953z - 0.001953}{-0.3125z^3 - 0.07813z^2 + 0.01953z - 0.001953}. \tag{29}$$

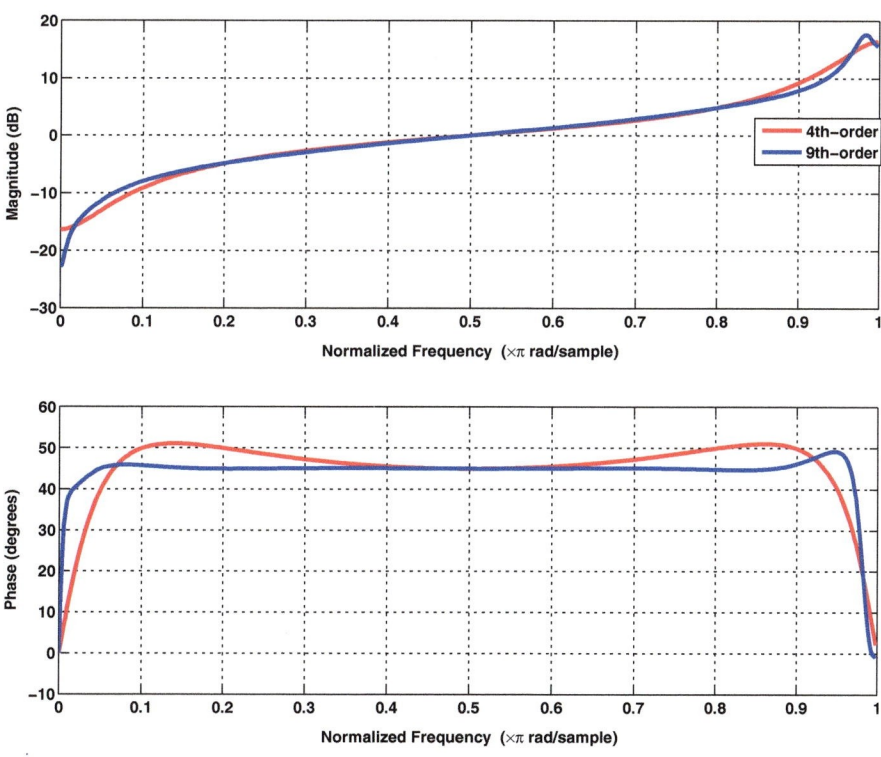

Fig. 10 Comparison between the approximations of (27) and (29) for $s^{0.5}$

Figure 9 displays the frequency response of both (28) and (29). Both forms exhibit similar magnitude and phase responses. However, the proposed second-order operator of (28) has a better magnitude response at low frequency than the ninth-order one given by (29), while the phase response of (29) at low frequency is better than that of (28). However, the significant improvement of (28) over the one in (29) is evident by the order reduction. For example, if one wishes to discretize a system with Laplace operators of two different orders, one would need an 18th-order model, while a second-order DTO given by (10) is faster and requires less hardware (Fig. 10).

The appealing factor in the proposed techniques lies in the fact that in all cases and for different fractional orders, the reciprocals of all proposed operators yield stable minimum phase discrete-time integrators.

5 Conclusion

A closed-form discrete-time first-, second-, third-, and fourth-order operators (DTO) are introduced to discretize the fractional-order Laplacian operator, $s^{\pm\alpha}$. Each operator is described by finite-dimensional rational z-transfer function. The discretization method is straightforward and depends on the order of the operator. The proposed method generates an adaptive, symmetrical real poles, and zeros that migrate to different locations inside the unit disc. The corresponding $z-$transfer functions represent stable non-minimum phase IIR-filters that exhibit constant phase and gain frequency responses over a wide frequency spectrum. As α approaches 1, the poles and zeros of all four operators converge to ± 1 and they reduce to the well-known discrete-time bilinear transformation. It is worth noting that the first and the second-order operators will be sufficient to discretize $s^{\pm\alpha}$ for high fractional orders; (say $0.8 \le \alpha \le 1$). The proposed DTO operators exhibit competitive frequency responses to those ones obtained by different discretization methods.

References

1. Al-Alaoui, M.A.: Novel digital integrator and differentiator. IEE Electron. Lett. **29**(4), 376–378 (1993)
2. Al-Alaoui, M.A.: Novel stable higher order s-to-z transforms. IEEE Trans. Circuit. Syst. I: Fundam. Theory Appl. **48**(11), 1326–1329 (2001)
3. Al-Alaoui, M.A.: Al-Alaoui operator and the α-approximation for discretization of analog system. FACTA Universitatis (NIS) **19**(1), 143–146 (2006)
4. Barbosa, R.S., Machado, J.A.T., Ferreira, I.M.: Pole-zero approximations of digital fractional-order integrators and diferentiators using modeling techniques. In: 16th IFAC World Congress. Prague, Czech Republic (2005)
5. Chen, Y., Moore, K.: Discretization schemes for fractional-order differentiators and integrators. IEEE Trans. Circuit. Syst.I: Fundam. Theory Appl. **49**(3), 363–367 (2002)
6. Chen, Y., Vinagre, B.M., Podlubny, I.: Continued fraction expansion to discretize fractional-order derivatives - an expository review. Nonlinear Dyn. **38**(1), 155–170 (2004)

7. Dorčák, L., Petráš, I., Terpák, J., Zborovjan, M.: Comparison of the method for discrete approximation of the fractional order operator. In: Proceedings of the International Carpathian Control Conference (ICCC'2003), pp. 851–856. High Tatras, Slovak Republic (2003)
8. El-Khazali, R.: Biquadratic approximation of fractional-order Laplacian operators. In: 2013 IEEE 56th International Midwest Symposium on Circuits and Systems(MWSCAS), pp. 69–72. Columbus, OH (2013)
9. El-Khazali, R.: Discretization of fractional-order differentiators and integrators. In: 19th IFAC World Congress, pp. 2016–2021. Cape Town, South Africa (2014)
10. Gupta, M., Yadav, R.: Design of improved fractional-order integrators using indirect discretization method. Int. J. Comput. Appl. **59**(14) (2012)
11. Kilbas, A., Srivastava, H., Trujillo, J.: Theory and Applications of Fractional Differential Equations, vol. 204. North-Holland Mathematics Studies, Elsevier, Amsterdam (2006)
12. Krishna, B.T.: Studies on fractional order differentiators and integrators: a survey. Signal Process. **59**(3), 386–426 (2011)
13. Lubich, C.: Discretized fractional calculus. SIAM J. Math. Anal. **17**(3), 704–719 (1986)
14. Machado, J.T.: Analysis and design of fractional-order digital control systems. Syst. Anal. Model. Simul. **27**(2–3), 107–122 (1997)
15. Machado, J.T.: Fractional-order derivative approximations in discrete-time control systems. Syst. Anal. Model. Simul. **34**, 419–434 (1999)
16. Matignon, D.: Stability results for fractional differential equations with applications to control processing. In: Computational Engineering in Systems Applications, vol. 2, pp. 963–968. Lille, France (1996)
17. Nie, B., Li, W., Ma, H., Wang, D., Liang, X.: Research of direct discretization method of fractional order differentiator/integrator based on rational function approximation. In: Tarn, T.J., Chen, S.B., Fang, G. (eds.) RoboticWelding, Intelligence and Automation. Lecture Notes in Electrical Engineering, pp. 479–485. Springer, Berlin, Heidelberg (2011)
18. Oldham, K., Spanier, J.: The Fractional Calculus: Theory and Application of Differentiation and Integration to Arbitrary Order. Academic Press, New York (1974)
19. Ortigueira, M.D., Serralheiro, A.J.: Pseudo-fractional ARMA modelling using a double Levinson recursion. IET Control Theory Appl. **1**(1), 173–178 (2007)
20. Oustaloup, A., Levron, F., Mathieu, B., Nanot, F.: Frequency-band complex noninteger differentiator: characterization and synthesis. IEEE Trans. Circuit. Syst. I: Fundam. Theory Appl. **47**(1), 25–39 (2000)
21. Podlubny, I.: The Laplace transform method for linear differential equations of the fractional order. Tech. Rep. UEF-02-9, Slovak Academy of Sciences Institute of Experimental Physics, Kosice, Slovakia (1994)
22. Podlubny, I.: Fractional differential equations, Volume 198: An Introduction to Fractional Derivatives, Fractional Differential Equations, to Methods of Their Solution. Mathematics in Science and Engineering. Academic Press, San Diego (1998)
23. Samko, S., Kilbas, A., Marichev, O.: Fractional Integrals and Derivatives and Some of Their Applications. Nauka i Tekhnika, Minsk (1987)
24. Siami, M., Tavazoei, M.S., Haeri, M.: Stability preservation analysis in direct discretization of fractional order transfer functions. Signal Process. **91**(3), 508–512 (2011)
25. Vinagre, B., Podlubny, I., Hernandez, A., Feliu, V.: Some approximations of fractional order operators used in control theory and applications. Fract. Calculus Appl. Anal. **3**(3), 231–248 (2000)
26. Vinagre,, B.M., Chen, Y.Q., Petras, I.: Two direct Tustin discretization methods for fractionalorder differentiator/integrator. J. Franklin Inst. **340**(5), 349–362 (2003)

On Fractional-Order Characteristics of Vegetable Tissues and Edible Drinks

J. A. Tenreiro Machado and António M. Lopes

Abstract This chapter uses frequency response techniques to characterize vegetable tissues and edible drinks. In the first phase, the impedance of the distinct samples is measured and fractional-order models are applied to the resulting data. In a second phase, hierarchical clustering and multidimensional scaling tools are adopted for comparing and visualizing the similarities between the specimens.

Keywords Frequency response · Fractional-order models · Clustering · Visualization

1 Introduction

The frequency response technique with electrical signals, often referred to as electrical impedance spectroscopy (EIS), measures the electrical impedance of a specimen across a given range of frequencies [5, 17, 22, 23, 25]. This technique has the advantage of being nondestructive, while avoiding complex and time-consuming experimental or laboratory procedures. The EIS has been widely used for studying vegetable tissues [4, 36], animal, and human samples [1, 13], beverages [30, 39], nonbiological materials [18, 37], and devices [2, 14].

This chapter addresses the application of the EIS for characterizing different products, namely plant leaves, vegetables, wine, and milk [22–24, 26]. In the first phase, the impedance $\mathbf{Z}(j\omega)$ is measured and fractional calculus (FC) is applied to model the samples with a reduced number of parameters. In a second phase, the EIS experimental data are processed by means of hierarchical clustering (HC) and

J. A. T. Machado (✉)
Department of Electrical Engineering, Institute of Engineering, Polytechnic of Porto, Rua Dr. António Bernardino de Almeida, 431, 4249–015 Porto, Portugal
e-mail: jtm@isep.ipp.pt

A. M. Lopes (✉)
Faculty of Engineering, UISPA–LAETA/INEGI, University of Porto, Rua Dr. Roberto Frias, 4200–465 Porto, Portugal
e-mail: aml@fe.up.pt

© Springer Nature Singapore Pte Ltd. 2019
P. Agarwal et al. (eds.), *Fractional Calculus*, Springer Proceedings in Mathematics & Statistics 303, https://doi.org/10.1007/978-981-15-0430-3_2

multidimensional scaling (MDS) algorithms for visualizing similarities between the specimen.

The chapter is organized as follows. Section 2 introduces the tools and methods adopted in the follow-up. Section 3 describes the impedance spectra, $\mathbf{Z}(j\omega)$, by means of FC models. Section 4 applies the MDS for clustering and visualizing similarities between the specimens. Finally, Sect. 5 draws the main conclusions.

2 Tools and Methods

2.1 The Canberra Distance

The Canberra distance was proposed, and later modified, by Lance and Williams [19, 20]. Given 2 points in a K-dimensional space, $X = (x_1, \ldots, x_K)$ and $Y = (y_1, \ldots, y_K)$, the Canberra distance between X and Y is given by

$$d_C(X, Y) = \sum_{k=1}^{K} \frac{|x_k - y_k|}{|x_k| + |y_k|}. \tag{1}$$

Equation (1) is a metric widely used for quantifying data scattered around an origin. The Canberra distance has several interesting properties, namely it is unitary when the arguments are symmetric, biased for measures around the origin, and highly sensitive for values close to zero.

2.2 Electrical Impedance Spectroscopy

In practical terms, the EIS method involves exciting a specimen with frequency-variable electric sinusoidal signals and registering the system response. The voltage $v(t)$ and current $i(t)$ across the specimen at steady state are sinusoidal functions of time given by

$$\begin{cases} v(t) = V \cos(\omega t + \theta_V) \\ i(t) = I \cos(\omega t + \theta_I) \end{cases}, \tag{2}$$

where $\{V, I\}$ are the amplitudes of the voltage and current, $\{\theta_V, \theta_I\}$ denote their phase shifts, $\omega = 2\pi f$ represents the angular frequency, and f is the frequency.

The voltage and current can be represented in the frequency domain by

$$\begin{cases} \mathbf{V}(j\omega) = V \cdot e^{j\theta_V} \\ \mathbf{I}(j\omega) = I \cdot e^{j\theta_I} \end{cases}, \tag{3}$$

where $j = \sqrt{-1}$. The experimental complex impedance $\mathbf{Z}_e(j\omega)$ is defined as the ratio of phasors:

$$\mathbf{Z}_e(j\omega) = \frac{\mathbf{V}(j\omega)}{\mathbf{I}(j\omega)} = \frac{V}{I} \cdot e^{j(\theta_V - \theta_I)} = |\mathbf{Z}_e(j\omega)| \cdot e^{j \arg [\mathbf{Z}_e(j\omega)]}. \tag{4}$$

Given an impedance spectrum $\mathbf{Z}_e(j\omega)$, it is often necessary to find a mathematical description, that is, a heuristic model, $\mathbf{Z}_m(j\omega)$, that fits well into the experimental data, and has a reduced number of parameters [28, 29].

Different empirical models in the scope of the dielectric relaxation phenomenon were proposed [12], namely the Debye, Cole-Cole, Cole-Davidson, and Havriliak-Negami models [7–10, 16, 21, 34, 35]:

$$\mathbf{Z}_D(j\omega) = \frac{1}{1 + j\omega\tau}, \tag{5}$$

$$\mathbf{Z}_{CC}(j\omega) = \frac{1}{1 + (j\omega\tau)^\alpha}, \tag{6}$$

$$\mathbf{Z}_{CD}(j\omega) = \frac{1}{(1 + j\omega\tau)^\beta}, \tag{7}$$

$$\mathbf{Z}_{HN}(j\omega) = \frac{1}{[1 + (j\omega\tau)^\alpha]^\beta}, \tag{8}$$

where $0 < \alpha, \beta \leq 1$, and τ denotes a relaxation time.

These empirical models are, in fact, particular cases of FC, and represent the fundamental bricks of any more complex fractional-order expressions, that may include further poles and zeros.

In this chapter, the Bode diagrams of the electrical impedance $\mathbf{Z}_e(j\omega)$ are approximated using fractional-order (FO) based models, while minimizing a fitness function, J, based on the Canberra distance [6] between the experimental, $\mathbf{Z}_e$, and model, $\mathbf{Z}_m$, impedances:

$$J = \frac{1}{L} \sum_{k=1}^{L} \left(\frac{|\Re[\mathbf{Z}_e(j\omega_k)] - \Re[\mathbf{Z}_m(j\omega_k)]|}{|\Re[\mathbf{Z}_e(j\omega_k)]| + |\Re[\mathbf{Z}_m(j\omega_k)]|} + \frac{|\Im[\mathbf{Z}_e(j\omega_k)] - \Im[\mathbf{Z}_m(j\omega_k)]|}{|\Im[\mathbf{Z}_e(j\omega_k)]| + |\Im[\mathbf{Z}_m(j\omega_k)]|} \right), \tag{9}$$

where L denotes the number of frequencies, ω_k, used for measuring the electrical impedance $\mathbf{Z}_e(j\omega)$, and $\Re(\cdot)$ and $\Im(\cdot)$ represent the real and imaginary parts of a complex number [22, 24].

The function, J, leads to good results because it calculates the ratio between the difference and the sum of two values. Therefore, it is possible to capture the relative error of the adjustment, avoiding saturation-like effects, that occurs when using the standard Euclidean norm due to the simultaneous presence of large and small values.

2.3 Experimental Setup for EIS Measurements

The diagram of Fig. 1 represents schematically the experimental arrangement adopted for the measurements [22–24, 26]. The specimens are connected in series with an adaptation resistance, $R_s = 15$ kΩ, for signal measurement, while yielding a good signal/noise ratio. A Hewlett Packard/Agilent 33220 A function generator applies a sinusoidal 5 V AC voltage to the circuit (i.e., a voltage divider). A Tektronix TDS 2002C two- channel oscilloscope measures the voltages $\mathbf{V}_{ab}$ and $\mathbf{V}_{cb}$. The impedance $\mathbf{Z}(j\omega)$ is obtained for the frequency range $2\pi \times 10 \leq \omega \leq 2\pi \times 10^5$ rad/s, at $L = 25$ logarithmically spaced points, using the expression:

$$\mathbf{Z}(j\omega) = R_s \cdot \left(\frac{\mathbf{V}_{ab}(j\omega)}{\mathbf{V}_{cb}(j\omega)} - 1 \right). \tag{10}$$

Several experimental tests demonstrated good stability in what concerns the oxidation of the copper electrodes, while different electrode geometries revealed a negligible influence on the results. Moreover, experiments with various amplitudes of the excitation signal showed good linearity, allowing data treatment using transfer function concepts.

2.4 Hierarchichal Clustering

Clustering is a data analysis technique [15] that groups similar items. In HC, two possible iterative strategies generate a hierarchy of clusters, namely the (i) agglomerative and the (ii) divisive clustering. With (i) each item starts in its own cluster and the algorithm merges the two most similar clusters until there is one single cluster. With (ii) all items start in a single cluster and the algorithm removes the outsiders from the least cohesive cluster, until each item is in its own cluster. In both cases, it is required a linkage criterion, that is a function of the distances between pairs of items, for quantifying the dissimilarity between clusters. For 2 clusters, R and S, the distance $d(x_R, x_S)$ between items $x_R \in R$ and $x_S \in S$ is based on metrics such as the maximum, minimum, and average linkages given by [3]

$$d_{max}(R, S) = \max_{x_R \in R, x_S \in S} d(x_R, x_S), \tag{11}$$

$$d_{min}(R, S) = \min_{x_R \in R, x_S \in S} d(x_R, x_S), \tag{12}$$

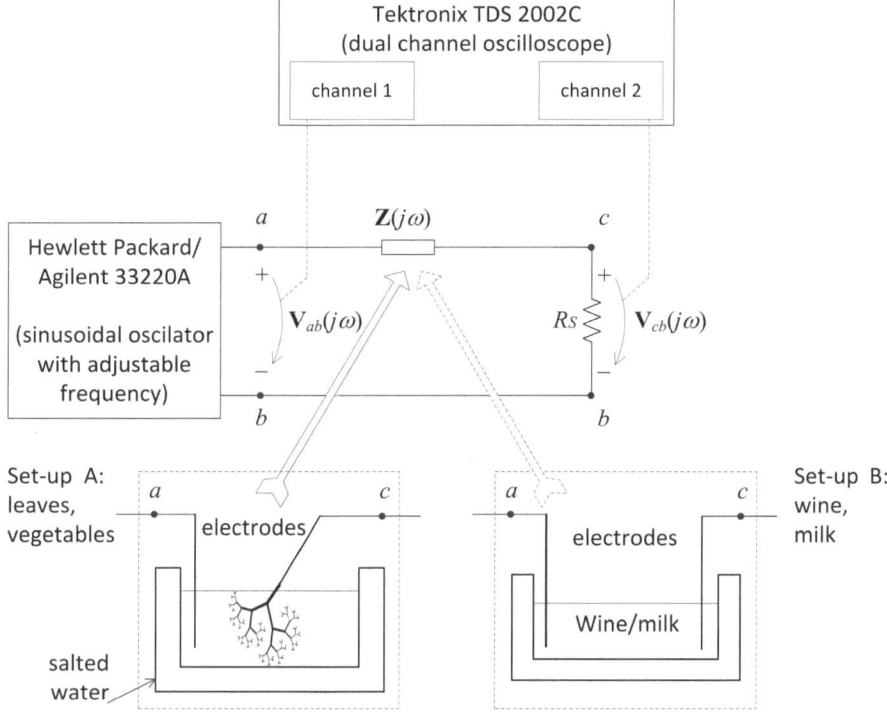

Fig. 1 Experimental EIS setup for measuring impedance $\mathbf{Z}(j\omega)$

$$d_{ave}(R, S) = \frac{1}{\| R \| \| S \|} \sum_{x_R \in R, x_S \in S} d(x_R, x_S). \tag{13}$$

After using one of the algorithms, the results of HC are presented in a graphical object such as a dendrogram or a hierarchical tree.

To assess the quality of the clustering, the cophenetic correlation (CC) coefficient is used [33]. The CC gives a measure of how well the generated graphical object preserves the original pairwise distances. If the clustering is successful, the links between items in the graphical object have a strong correlation with those in the original dataset. The closer the CC value to 1, the better the clustering result. The quality assessment is plotted in a Shepard diagram that compares the original and the cophenetic distances. A good clustering leads to a layout of points close to the 45-degree line.

2.5 Multidimensional Scaling

MDS is a computational technique for clustering and visualizing data [31]. In the first phase, given s items and a measure of dissimilarity, a $s \times s$ symmetric matrix, $\mathbf{C} = [c_{ij}]$, $(i, j) = 1, \ldots, s$, of item-to-item dissimilarities is calculated. The matrix $\mathbf{C}$ represents the input information for starting the MDS computational scheme. The MDS rational is to assign points for representing items in a multidimensional space and to try to reproduce the measured dissimilarities, c_{ij}. In a second phase, MDS evaluates different configurations for maximizing some fitness function, arriving at a set of point coordinates (and, therefore, to a symmetric matrix of distances $\mathbf{D} = [d_{ij}]$) with the reproduced dissimilarities that best approximates c_{ij}. A common fitness function for measuring the difference between c_{ij} and d_{ij} is the raw stress:

$$S = \left[d_{ij} - f(c_{ij}) \right]^2, \tag{14}$$

where $f(\cdot)$ indicates some types of transformation.

The MDS interpretation is based on the patterns of points that can be visualized in the generated map. Similar (dissimilar) objects are represented by points that are close to (far from) each other. So, the information retrieval is not based on the point coordinates, or the geometrical form of the clusters. Indeed, we can rotate, translate, or magnify the map (for better visualization) because the distances remain identical. The MDS axes have neither special meaning nor units.

The quality of the MDS mat can be assessed by means of the stress and Shepard plots. The stress plot represents S versus the number of dimensions m of the MDS map. The plot $S(m)$ is a monotonic decreasing chart and choosing the value of m is a compromise between achieving low values of S or m. Often the values $m = 2$ or $m = 3$ are adopted since they allow direct visualization. On the other hand, the Shepard diagram compares d_{ij} and c_{ij} for a particular value m. A narrow scatter around the 45-degree line represents a good fit between d_{ij} and c_{ij}.

3 Modeling Vegetable Tissues and Edible Drinks

3.1 EIS Analysis of Vegetable Tissues

A total of $N_l = 6$ angiosperm leaves and $N_v = 4$ vegetables are studied, as summarized in Tables 1 and 2, respectively [22, 23].

Each leaf (vegetable) is submerged in salted water, with except ion of its petiole (base). Two copper electrodes of 0.5 mm diameter connect the specimen to the measurement circuit. One electrode is inserted into the leaf petiole (vegetable base), aligned with its longitudinal axis, and the other one is placed in the water (see Fig. 1, setup A).

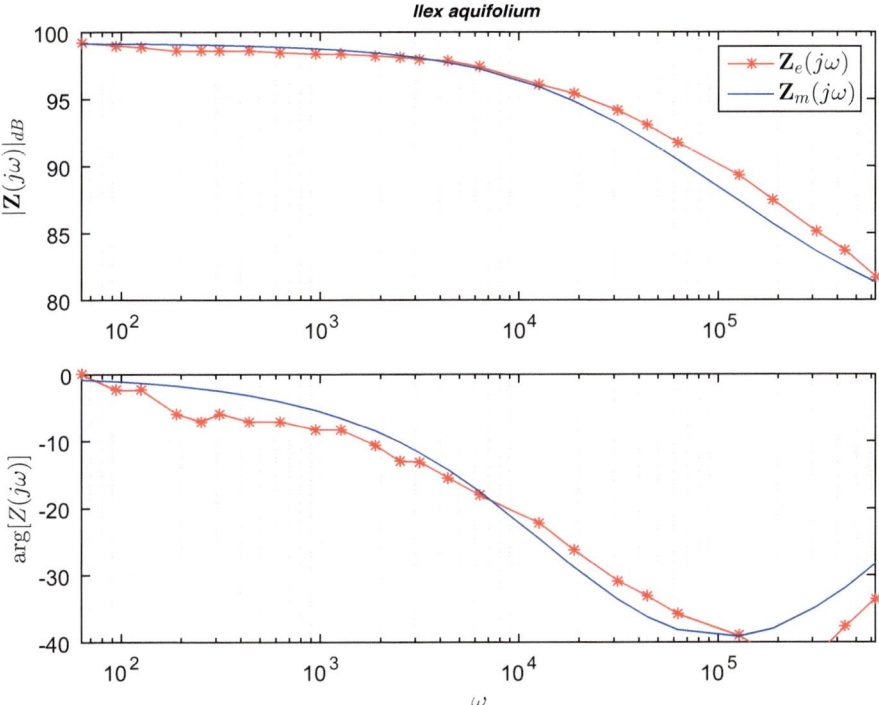

Fig. 2 The Bode diagram of the experimental, $\mathbf{Z}_e$, and model, $\mathbf{Z}_m$ impedance of the *Ilex aquifolium* (IA)

For the leaves, several numerical tests proved that the 4-parameter FO model

$$\mathbf{Z}_m(j\omega) = R + \frac{K}{1 + \left(\frac{j\omega}{p}\right)^\beta} \tag{15}$$

leads to a good approximation to the experimental data (see details in [22]). Figure 2 depicts the Bode diagram of the experimental, $\mathbf{Z}_e$, and model, $\mathbf{Z}_m$ impedance of the *Ilex aquifolium* (IA), illustrating the fit. Table 1 summarizes the values of the model parameters that approximate the spectra of all leaves.

For the vegetables, the following 5-parameter FO model is needed to fit the experimental data:

$$\mathbf{Z}_m(j\omega) = R + \frac{K}{\left[1 + \left(\frac{j\omega}{p}\right)^\alpha\right]^\beta}. \tag{16}$$

An example of the experimental and model Bode diagram is depicted in Fig. 3 for the *Cauliflower*. The model parameters of the four specimens are summarized in Table 2 (see details in [23]).

Table 1 Parameters of the FO-based model for $N_l = 6$ leaves

i	Species	Tag	R	K	p	α	J
1	*Citrus limon*	CL	8.9×10^3	5.6×10^4	2×10^3	0.59	0.035
2	*Ilex aquifolium*	IA	7.2×10^3	8.4×10^4	2.5×10^3	0.72	0.288
3	*Ficus elastica*	FE	7.5×10^3	7.8×10^4	6×10^2	0.55	0.338
4	*Hydrangea macrophylla*	HM	2×10^3	6.6×10^4	5×10^2	0.48	0.493
5	*Acacia dealbata*	AD	7×10^3	4.9×10^5	1.2×10^1	0.37	0.398
6	*Acer pseudoplatanus*	AC	1.5×10^4	5.5×10^5	1×10^2	0.47	0.581

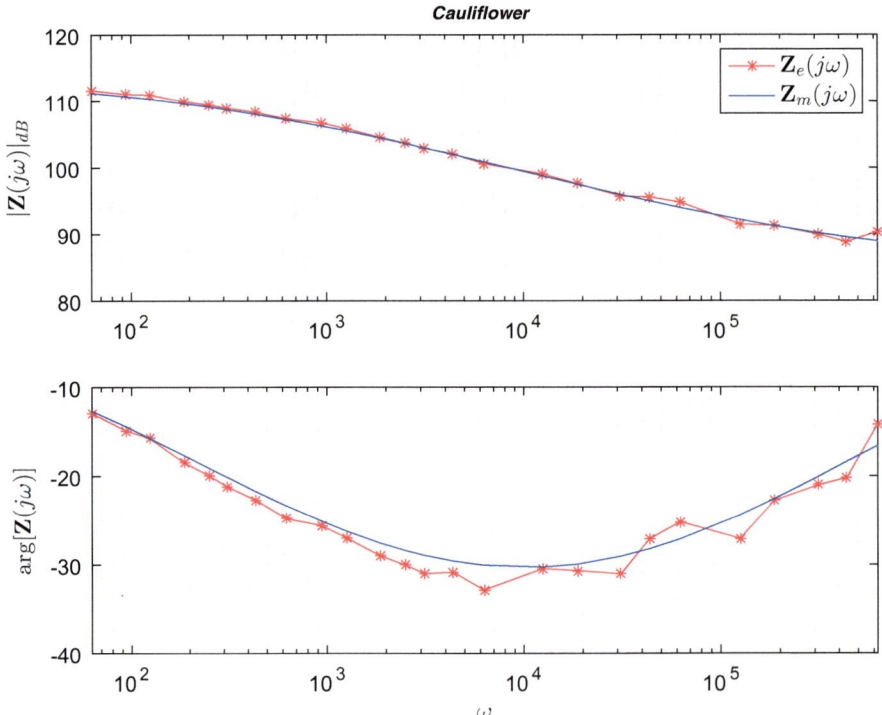

Fig. 3 The Bode diagram of the experimental, $\mathbf{Z}_e$, and model, $\mathbf{Z}_m$ impedances for the *Cauliflower*

Table 2 Parameters of the FO-based model for the $N_v = 4$ vegetables

i	Designation	Tag	R	K	p	α	β	J
1	*Cauliflower*	CF	1.8×10^4	4.8×10^5	5.5×10^1	0.470	1.046	0.0026
2	*Broccoli*	BR	0.1×10^4	4.0×10^5	5.5×10^1	0.859	0.897	0.0027
3	*Round cabbage*	RC	5.2×10^4	0.6×10^5	1.7×10^3	0.444	1.470	0.0002
4	*Brussels sprout*	BS	5.9×10^4	7.2×10^5	1.7×10^3	0.354	1.292	0.0008

Table 3 The set of $N_w = 16$ wine samples analyzed

i	Tag	Wine region	Wine style
1	W_1	Alentejo	White
2	W_2	Alentejo	White
3	W_3	Alentejo	White
4	W_4	Península de Setúbal	White
5	W_5	Tejo	White
6	W_6	Douro	White
7	W_7	Vinhos Verdes	Green white
8	W_8	Vinhos Verdes	Green white
9	R_1	Alentejo	Red
10	R_2	Alentejo	Red
11	R_3	Península de Setúbal	Red
12	R_4	Península de Setúbal	Red
13	R_5	Bairrada	Red
14	R_6	Douro	Red
15	R_7	Vinhos Verdes	Green red
16	R_8	Vinhos Verdes	Green red

These results demonstrate that FO empirical formulae constitute simple, yet reliable models to characterize vegetable structures.

3.2 EIS Analysis of Edible Drinks

In this Section, $N_w = 16$ wine types and $N_m = 12$ UHT milk varieties are studied. The wine set includes samples from distinct Portuguese regions [11], and involves a mix of ripe and green, both red and white, styles (Table 3). The milk set comprises samples from distinct brands, with different fat contents, and includes a mix of normal, reduced, and fortified milk varieties (Table 4).

The experimental setup for EIS measurements is depicted in setup B of Fig. 1. A parallelepipedic container with dimensions $(l \times w \times h) = (120 \times 100 \times 55)$ mm is filled with 200 ml of wine. Two 0.5 mm diameter copper electrodes connect the samples to the measurement circuit. The electrodes are immersed at 5 mm from the bottom of the container, and are placed diametrically opposed to each other.

For both wine and milk, several numerical tests revealed that a good fit between the experimental, $\mathbf{Z}_e$, and model, $\mathbf{Z}_m$, impedance occurs for the 6-parameter FO model:

$$\mathbf{Z}(j\omega) = K \cdot \frac{\left(1 + \frac{j\omega}{z_1}\right)^{\alpha_1} \cdot \left(1 + \frac{j\omega}{z_2}\right)^{\alpha_2}}{(j\omega)^{\beta}}. \tag{17}$$

Table 4 The set of $N_m = 12$ milk samples analyzed

i	Tag	Milk type
1	M_1	Skimmed
2	M_2	Skimmed
3	M_3	Skimmed
4	M_4	Semi-skimmed
5	M_5	Semi-skimmed
6	M_6	Semi-skimmed
7	M_7	Whole
8	M_8	Whole
9	M_9	Whole
10	M_{10}	Organic semi-skimmed
11	M_{11}	Reduced skimmed
12	M_{12}	Fortified skimmed

Table 5 Impedance parameters of the $N_w = 16$ wine samples

i	Tag	Impedance parameters						
		K	z_1	α_1	z_2	α_2	β	J
1	W_1	6700	1000	0.33	24×10^4	0.88	0.29	0.2298
2	W_2	5000	1000	0.32	19×10^4	0.87	0.32	0.2372
3	W_3	5800	1300	0.40	22×10^4	0.88	0.31	0.2678
4	W_4	7000	1000	0.35	22×10^4	0.86	0.31	0.2639
5	W_5	5500	1000	0.32	19×10^4	0.87	0.33	0.2885
6	W_6	6000	900	0.32	20×10^4	0.88	0.29	0.2888
7	W_7	7500	1600	0.40	15×10^4	0.75	0.36	0.2667
8	W_8	7500	1600	0.40	17×10^4	0.75	0.36	0.2339
9	R_1	5500	950	0.33	19×10^4	0.87	0.32	0.2859
10	R_2	6800	1100	0.33	23×10^4	0.92	0.33	0.2936
11	R_3	5000	1000	0.33	23×10^4	0.86	0.31	0.2731
12	R_4	6000	1000	0.34	22×10^4	0.88	0.33	0.2941
13	R_5	7500	1600	0.39	15×10^4	0.75	0.35	0.2644
14	R_6	5000	1000	0.32	20×10^4	0.88	0.32	0.2977
15	R_7	6500	1100	0.30	20×10^4	0.89	0.35	0.3651
16	R_8	7200	1700	0.41	19×10^4	0.88	0.38	0.2985

Figures 4 and 5 depict the Bode diagrams of $\mathbf{Z}_e$ and $\mathbf{Z}_m$ for the wine and milk samples W_5 and M_2, respectively, illustrating the adequacy of expression (17). Tables 5 and 6 summarize the impedance parameters for all wine and milk specimens (see details in [24, 26]).

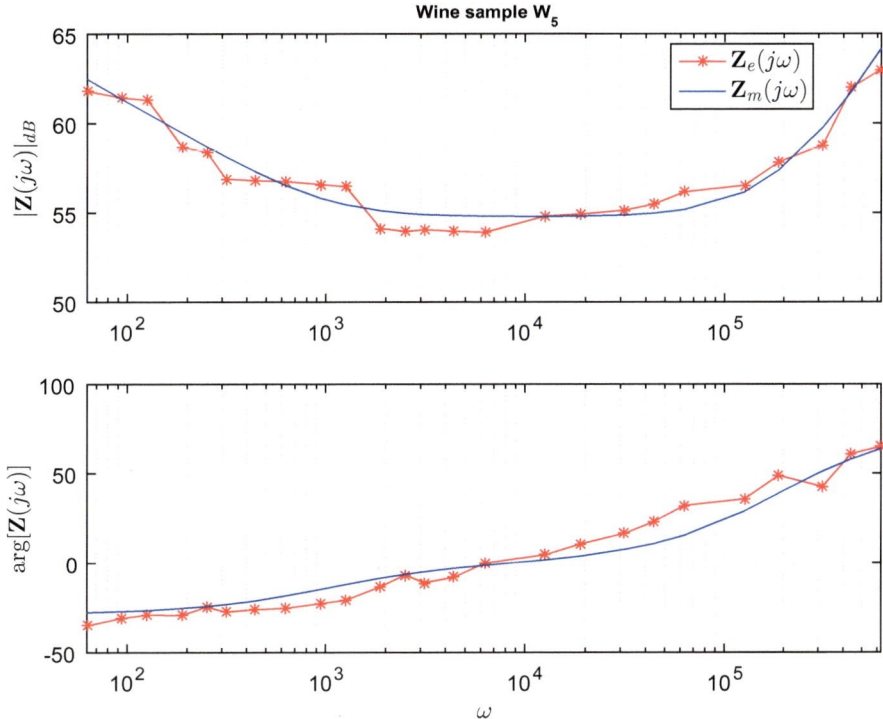

Fig. 4 The Bode diagram of the experimental, $\mathbf{Z}_e$, and model, $\mathbf{Z}_m$, impedances for the wine sample W_5

Table 6 Impedance parameters for the $N_m = 12$ milk samples

i	Tag	Impedance parameters						
		K	z_1	α_1	z_2	α_2	β	J
1	M_1	33325	5340	0.68	5606	0.57	0.7	0.735
2	M_2	36563	8840	0.67	8233	0.65	0.7	0.693
3	M_3	45133	3844	0.72	15813	0.64	0.72	0.71
4	M_4	29400	9880	0.67	8906	0.59	0.67	0.726
5	M_5	35150	4290	0.72	8190	0.56	0.73	0.772
6	M_6	34380	9172	0.76	7717	0.54	0.68	0.69
7	M_7	59463	2600	0.72	16225	0.68	0.8	0.691
8	M_8	35750	7150	0.66	8450	0.61	0.69	0.763
9	M_9	31859	4572	0.68	5999	0.62	0.76	0.71
10	M_{10}	32626	9667	0.76	6656	0.57	0.72	0.72
11	M_{11}	27300	3500	0.72	65000	0.53	0.64	0.717
12	M_{12}	34613	3500	0.85	65000	0.53	0.72	0.747

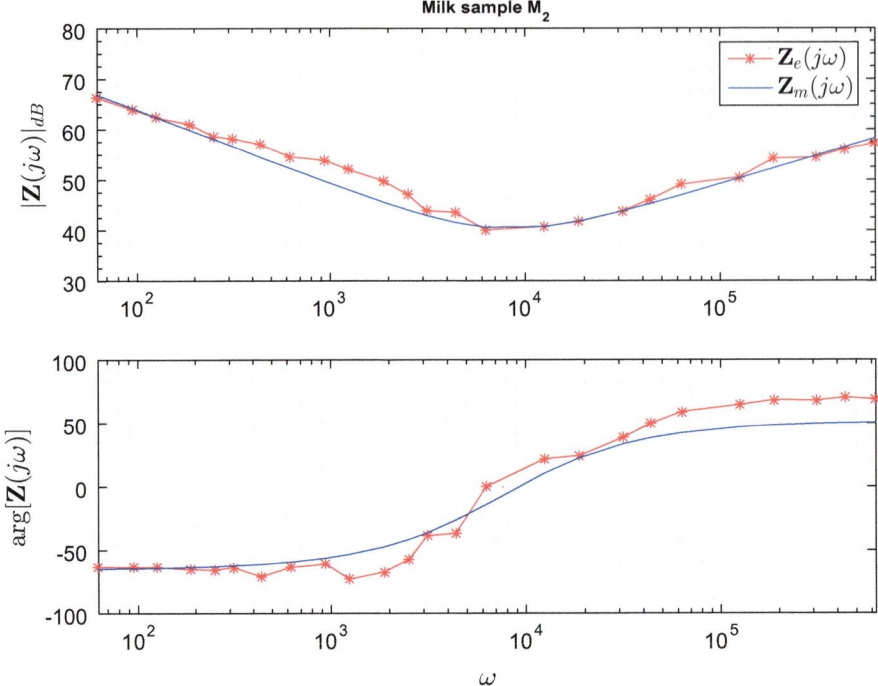

Fig. 5 The Bode diagram of the experimental, $\mathbf{Z}_e$, and model, $\mathbf{Z}_m$, impedances for the milk sample M_2

The results demonstrate that FO models yield a convincing description and reliable characterization of the samples, and that the EIS technique leads to a simple and straightforward procedure to characterize the specimen.

In conclusion, analyzing the results of Sects. 3.1 and 3.2 we verify the emergence of FO effects that are not captured by classical integer-order models.

4   Clustering and Visualization of Vegetable Tissues and Edible Drinks

4.1   HC of Vegetable Tissues and Edible Drinks

The HC processes a matrix $\mathbf{C} = [c_{ij}]$ based on the distance:

$$c_{ij} = \frac{1}{L} \sum_{k=1}^{L} \frac{\left| \Re[\mathbf{Z}_{e_i}(j\omega_k)] - \Re[\mathbf{Z}_{e_j}(j\omega_k)] \right|}{\left| \Re[\mathbf{Z}_{e_i}(j\omega_k)] \right| + \left| \Re[\mathbf{Z}_{e_j}(j\omega_k)] \right|} + \frac{\left| \Im[\mathbf{Z}_{e_i}(j\omega_k)] - \Im[\mathbf{Z}_{e_j}(j\omega_k)] \right|}{\left| \Im[\mathbf{Z}_{e_i}(j\omega_k)] \right| + \left| \Im[\mathbf{Z}_{e_j}(j\omega_k)] \right|},$$

$$(18)$$

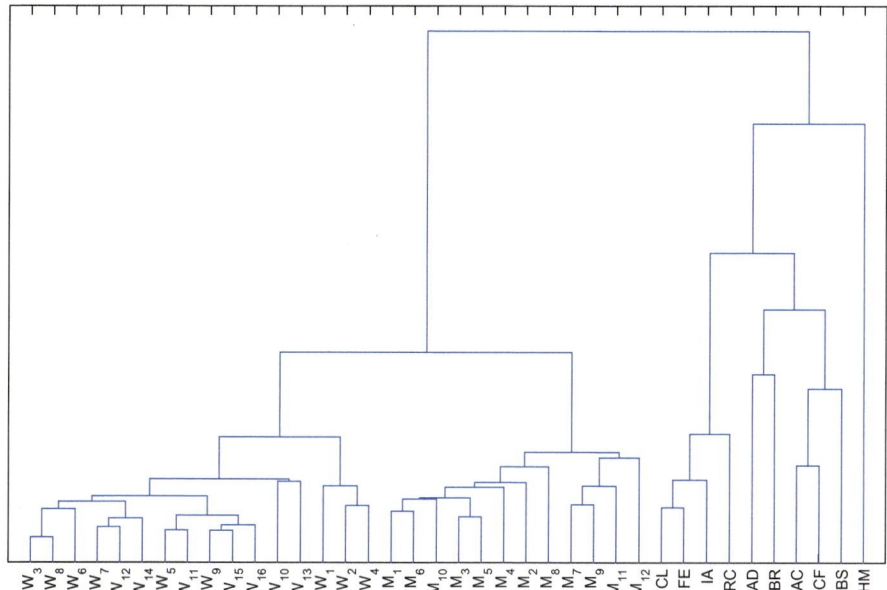

Fig. 6 Dendrogram generated by the HC for the $s = 38$ samples and matrix $\mathbf{C}$

where the indices $i, j = 1, \ldots, s$, so that s denotes the number of samples, and $\mathbf{Z}_e$ represents the impedances measured by means of EIS.

One must note that trying other alternative measures is a common procedure. In fact, one can choose distinct distances, and the corresponding charts, to obtain the best visualization. Nevertheless, several tests demonstrated that the Canberra distance leads to relevant results.

The successive (agglomerative) clustering and average-linkage method are used. Figure 6 depicts the dendrogram generated by the HC, with input $\mathbf{C} = [c_{ij}]$ and $s = N_l + N_v + N_w + N_m = 38$. One can note the emergence of patterns for vegetables, wine, and milk.

4.2 MDS of Vegetable Tissues and Edible Drinks

Figure 7 depicts the 2- and 3-dimensional maps of items obtained by the MDS, with input $\mathbf{C} = [c_{ij}]$ and $s = N_l + N_v + N_w + N_m = 38$, where three clusters composed of vegetables, wine, and milk emerge. Moreover, the $s = N_l + N_v = 10$ vegetable tissues and $s = N_w + N_m = 28$ edible drinks are compared apart from each other, and the corresponding 3-dimensional MDS maps are depicted in Fig. 8. The charts reveal blurred and clear clusters, respectively, confirming that leaves and vegetables are quite similar, while milk and wine have strong dissimilarities.

Fig. 7 The 2- and 3-dimensional MDS maps for the $s = 38$ samples and matrix **C**

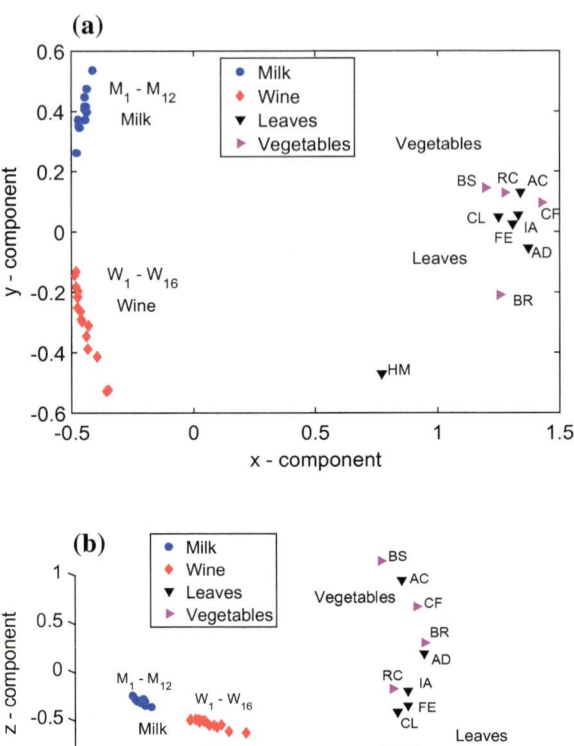

In conclusion, the dendrogram and MDS charts are alternatives with different characteristics, both leading to identical clusters, namely with a clear separation between the 3 groups, but some mixing between leaves and vegetables. Nevertheless, from the point of view of visualization, the 3-dimensional MDS is superior to the dendrogram technique.

5 Conclusions

In this chapter, the EIS technique was used to determine the electrical impedance spectra of different materials, and FO models were adopted to describe the experimental data. It was shown that FO transfer functions describe adequately the data. The potential use of simple, nonintrusive, and economical techniques in food pro-

Fig. 8 The 3-dimensional MDS maps for the $s = 10$ vegetable tissues and $s = 28$ edible drinks with matrix **C**

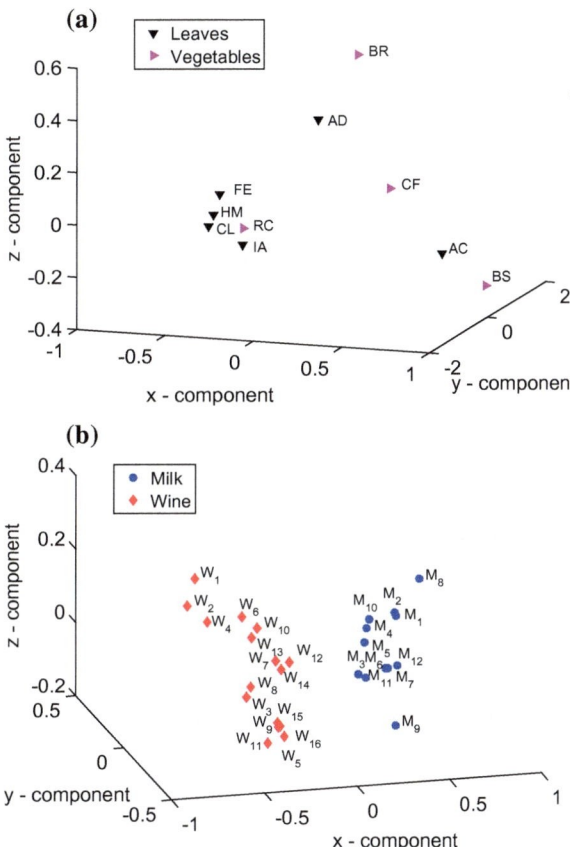

duction, biology, and medicine reveals possible directions to be further explored [27, 32, 38].

References

1. Åberg, P., Birgersson, U., Elsner, P., Mohr, P., Ollmar, S.: Electrical impedance spectroscopy and the diagnostic accuracy for malignant melanoma. Exp. Dermatol. **20**(8), 648–652 (2011)
2. Adachi, M., Sakamoto, M., Jiu, J., Ogata, Y., Isoda, S.: Determination of parameters of electron transport in dye-sensitized solar cells using electrochemical impedance spectroscopy. J. Phys. Chem. B **110**(28), 13872–13880 (2006)
3. Aggarwal, C.C., Hinneburg, A., Keim, D.A.: On the Surprising Behavior of Distance Metrics in High Dimensional Space. Springer (2001)
4. Ando, Y., Maeda, Y., Mizutani, K., Wakatsuki, N., Hagiwara, S., Nabetani, H.: Effect of air-dehydration pretreatment before freezing on the electrical impedance characteristics and texture of carrots. J. Food Eng. **169**, 114–121 (2016)

5. Biswas, K., Bohannan, G., Caponetto, R., Lopes, A., Machado, T.: Fractional Order Devices. Springer (2017)
6. Cha, S.H.: Comprehensive survey on distance/similarity measures between probability density functions. City $1(2)$, 1 (2007)
7. Cole, K.S., Cole, R.H.: Dispersion and absorption in dielectrics I. Alternating current characteristics. J. Chem. Phys. $9(4)$, 341–351 (1941)
8. Davidson, D., Cole, R.: Dielectric relaxation in glycerol, propylene glycol, and n-propanol. J. Chem. Phys. $19(12)$, 1484–1490 (1951)
9. Debye, P.J.W.: Polar Molecules. Chemical Catalog Company, Incorporated (1929)
10. Emmert, S., Wolf, M., Gulich, R., Krohns, S., Kastner, S., Lunkenheimer, P., Loidl, A.: Electrode polarization effects in broadband dielectric spectroscopy. Eur. Phys. J. B $83(2)$, 157–165 (2011)
11. Fraga, H., Malheiro, A., Moutinho-Pereira, J., Jones, G., Alves, F., Pinto, J.G., Santos, J.: Very high resolution bioclimatic zoning of Portuguese wine regions: present and future scenarios. Reg. Environ. Change $14(1)$, 295–306 (2014)
12. Freeborn, T.J.: A survey of fractional-order circuit models for biology and biomedicine. IEEE J. Emerg. Sel. Topics Circut. Syst. $3(3)$, 416–424 (2013)
13. Freeborn, T.J., Elwakil, A.S., Maundy, B.: Compact wide frequency range fractional-order models of human body impedance against contact currents. Math. Problems Eng. 2016 (2016)
14. Glatthaar, M., Riede, M., Keegan, N., Sylvester-Hvid, K., Zimmermann, B., Niggemann, M., Hinsch, A., Gombert, A.: Efficiency limiting factors of organic bulk heterojunction solar cellsidentified by electrical impedance spectroscopy. Solar Energy Mater. Solar Cells $91(5)$, 390–393 (2007)
15. Hartigan, J.A.: Clustering Algorithms. John Wiley & Sons, Inc. (1975)
16. Havriliak, S., Negami, S.: A complex plane analysis of α-dispersions in some polymer systems. J. Polymer Sci. Part C: Polymer Symp. 14, 99–117. Wiley Online Library (1966)
17. Ionescu, C., Lopes, A., Copot, D., Machado, J., Bates, J.: The role of fractional calculus in modelling biological phenomena: a review. Commun. Nonlinear Sci. Numer. Simul. (2017)
18. Irvine, J.T., Sinclair, D.C., West, A.R.: Electroceramics: characterization by impedance spectroscopy. Adv. Mater. $2(3)$, 132–138 (1990)
19. Lance, G.N., Williams, W.T.: Computer programs for hierarchical polythetic classification ("similarity analyses"). Comput. J. $9(1)$, 60–64 (1966)
20. Lance, G.N., Williams, W.T.: Mixed-data classificatory programs I—agglomerative systems. Aust. Comput. J. $1(1)$, 15–20 (1967)
21. Laufer, S., Ivorra, A., Reuter, V.E., Rubinsky, B., Solomon, S.B.: Electrical impedance characterization of normal and cancerous human hepatic tissue. Physiol. Measur. $31(7)$, 995 (2010)
22. Lopes, A.M., Machado, J.T.: Fractional order models of leaves. J. Vib. Control $20(7)$, 998–1008 (2014)
23. Lopes, A.M., Machado, J.T.: Modeling vegetable fractals by means of fractional-order equations. J. Vib. Control $22(8)$, 2100–2108 (2016)
24. Lopes, A.M., Machado, J.T., Ramalho, E.: On the fractional-order modeling of wine. Eur. Food Res. Technol. $6(243)$, 921–929 (2017)
25. Lopes, A.M., Machado, J.T., Ramalho, E.: Fractional-order model of wine. In: Chaotic, Fractional, and Complex Dynamics: New Insights and Perspectives, pp. 191–203. Springer (2018)
26. Lopes, A.M., Machado, J.T., Ramalho, E., Silva, V.: Milk characterization using electrical impedance spectroscopy and fractional models. Food Analyt. Methods $11(3)$, 901–912 (2018)
27. Martinsen, O.G., Grimnes, S., Schwan, H.P.: Interface phenomena and dielectric properties of biological tissue. Encycl. Surface Colloid Sci. 20, 2643–2653 (2002)
28. Maundy, B., Elwakil, A., Allagui, A.: Extracting the parameters of the single-dispersion Cole bioimpedance model using a magnitude-only method. Comput. Electron. Agric. 119, 153–157 (2015)
29. Nigmatullin, R., Nelson, S.: Recognition of the "fractional" kinetics in complex systems: dielectric properties of fresh fruits and vegetables from 0.01 to 1.8 GHz. Signal Process. $86(10)$, 2744–2759 (2006)

30. Riul, A., de Sousa, H.C., Malmegrim, R.R., dos Santos, D.S., Carvalho, A.C., Fonseca, F.J., Oliveira, O.N., Mattoso, L.H.: Wine classification by taste sensors made from ultra-thin films and using neural networks. Sens. Actuat. B: Chem. **98**(1), 77–82 (2004)
31. Saeed, N., Nam, H., Haq, M.I.U., Muhammad Saqib, D.B.: A survey on multidimensional scaling. ACM Comput. Surv. (CSUR) **51**(3), 47 (2018)
32. Sammer, M., Laarhoven, B., Mejias, E., Yntema, D., Fuchs, E.C., Holler, G., Brasseur, G., Lankmayr, E.: Biomass measurement of living lumbriculus variegatus with impedance spectroscopy. J. Electr. Bioimpedance **5**(1), 92–98 (2014)
33. Sokal, R.R., Rohlf, F.J.: The comparison of dendrograms by objective methods. Taxon, pp. 33–40 (1962)
34. Stanislavsky, A., Weron, K., Trzmiel, J.: Subordination model of anomalous diffusion leading to the two-power-law relaxation responses. EPL (Europhys. Lett.) **91**(4), 40003 (2010)
35. Vosika, Z., Lazarevic, M., Simic-Krstic, J., Koruga, D.: Modeling of bioimpedance for human skin based on fractional distributed-order modified Cole model. FME Trans. **42**(1), 74–81 (2014)
36. Watanabe, T., Orikasa, T., Shono, H., Koide, S., Ando, Y., Shiina, T., Tagawa, A.: The influence of inhibit avoid water defect responses by heat pretreatment on hot air drying rate of spinach. J. Food Eng. **168**, 113–118 (2016)
37. West, A.R., Sinclair, D.C., Hirose, N.: Characterization of electrical materials, especially ferroelectrics, by impedance spectroscopy. J. Electroceramics **1**(1), 65–71 (1997)
38. Wolf, M., Gulich, R., Lunkenheimer, P., Loidl, A.: Broadband dielectric spectroscopy on human blood. Biochimica et Biophysica Acta (BBA)-General Subjects **1810**(8), 727–740 (2011)
39. Zheng, S., Fang, Q., Cosic, I.: An investigation on dielectric properties of major constituents of grape must using electrochemical impedance spectroscopy. Eur. Food Res. Technol. **229**(6), 887–897 (2009)

Some Relations Between Bounded Below Elliptic Operators and Stochastic Analysis

Rémi Léandre

Abstract We apply Malliavin Calculus tools to the case of a bounded below elliptic right-invariant pseudo-differential operators on a Lie group. We give examples of bounded below pseudo-differential elliptic operators on $\mathbb{R}^d$ by using the theory of Poisson process and the Garding inequality. In the two cases, there are no stochastic processes because the considered semi-groups do not preserve positivity.

Keywords Malliavin calculus · Pseudo-differential operators · Generalized Poisson processes · Garding inequality

1 Introduction

Let G be a compact connected Lie group, with generic element g endowed with its biinvariant Riemannian structure and with its normalized Haar measure dg. e is the unit element of G.

Let f^i be a basis of $T_e G$. We can consider as right-invariant vector fields. This means that if we consider the action $R_{g_0} h \rightarrow (g \rightarrow h(gg_0))$ on smooth function h on G, we have

$$R_{g_0}(f^i h) = f^i(R_{g_0} h) \tag{1}$$

We consider a right-invariant elliptic pseudo-differential bounded below operator L of order larger than $2k$ on G. It generates by elliptic theory a semi-group P_t on $L^2(dg)$ and even on $C_b(G)$ the space of continuous functions on G endowed with the uniform norm.

R. Léandre (✉)
Laboratoire de Mathématiques, Université de Bourgogne-Franche-Comté, 25030 Besançon, France
e-mail: remi.leandre@univ-fcomte.fr

© Springer Nature Singapore Pte Ltd. 2019
P. Agarwal et al. (eds.), *Fractional Calculus*, Springer Proceedings
in Mathematics & Statistics 303, https://doi.org/10.1007/978-981-15-0430-3_3

Theorem 1 *If $t > 0$,*

$$P_t h(g_0) = \int_G p_t(g_0, g) h(g) dg \tag{2}$$

where $g \rightarrow p_t(g_0, g)$ is smooth if h is continuous.

This theorem is classical in analysis, but it enters in our general program to implement stochastic analysis tools in the theory of non-Markovian semi-group. See the review [7, 13] for that. See [10, 11] for another presentation where the Malliavin Matrix plays a key role. Here we don't use the Malliavin matrix. See [12] for the case of right-invariant differential operators.

Jump processes are generated by pseudo-differential operators, which satisfy the maximum principle. Unlike the Malliavin Calculus for jump processes [1, 5, 6], there is no limitation here on the size of jumps.

This theorem can be applied to any positive power of a right invariant strictly positive differential operator on G [14].

2 Pseudo-differential Operators

Let us recall what is a pseudo-differential operator on $\mathbb{R}^d$ [3, 5, 6, 15]. Let be a smooth function from $\mathbb{R}^d \times \mathbb{R}^d$ into $\mathbb{C}$ $a(x, \xi)$. We suppose that for all x

$$|D_x^r D_\xi^l a(x, \xi)| \leq C|\xi|^{m-l} + C \tag{3}$$

We suppose that for all x

$$|a(x, \xi)| \geq C|\xi|^{m'} \tag{4}$$

for $|\xi| > C$ for a suitable $m' > 0$. Let $\hat{h}$ be the Fourier transform of the continuous function h. We consider the operator L defines on smooth function h by

$$\hat{L}h(x) = \int a(x, \xi)\hat{h}(\xi)d\xi \tag{5}$$

L is said to be a pseudo-differential operator elliptic of order larger than m' with symbol a. This property is invariant if we do a diffeomorphism on $\mathbb{R}^d$ with bounded derivatives at each order. This remark allows to define by using charts a pseudo-differential operator elliptic of order larger than m' on a compact manifold M.

On a compact Riemannian manifold, we can consider the Riemannian measure. In local coordinates, the Riemannian metric is given by a smooth map

$$x \rightarrow g_{i,j}(x) \tag{6}$$

in the set of strictly positive matrix and the Riemannian measure is given by

$$dx = det(g_{.,.})^{-1/2} dx_1 \otimes .. \otimes dx_d \qquad (7)$$

We can normalize the Riemannian measure to be of total mass 1.
The fact that L is symmetric on $L^2(M)$ means that

$$\int_M < h_1(x), Lh_2(x) > dx = \int_M < Lh_1(x), h_2(x) > dx \qquad (8)$$

The fact that L is bounded below means that for some $C > 0$:

$$\int_M < h(x), Lh(x) > dx \geq -C \int_M < h(x), h(x) > dx \qquad (9)$$

In such a case L has a self-adjoint extension. This generates a semi-group of bounded operators P_t on $L^2(M)$ satisfying the heat equation

$$\frac{\partial}{\partial t} P_t h = -L P_t h \qquad (10)$$

for $h \in L^2(M)$ and $t > 0$. Moreover, we suppose that $P_0 h = h$. It generates moreover a semi-group on $C_b(M)$ by ellipticity.

An example can be given on $\mathbb{R}^d$ if we use the Garding inequality [15]. Suppose that we consider the Lebesgue measure on $\mathbb{R}^d$ and that for $|\xi| > C_0$ we have

$$Re(a(x, \xi)) > C|\xi|^{m'} \qquad (11)$$

for some $C > 0$. In such a case if we suppose L symmetric, it is bounded below.

3   Proof of the Theorem

3.1   Algebraic Scheme of the Proof: Malliavin Integration by Parts

We consider the family of operators on $C^\infty(G \times \mathbb{R}^n)$:

$$\tilde{L}_t^n = L + \sum_{i=1}^n f^{j_i} \frac{\partial}{\partial u_i} \alpha_t^i + \sum_{i=1}^n \frac{\partial^{2k}}{\partial u_i^{2k}} \qquad (12)$$

α_t^i are smooth function from $\mathbb{R}^+$ into $\mathbb{R}$. By elliptic theory, $\tilde{L}_t^n$ generates a semi-group $\tilde{P}_t^n$ on $C_b(G \times R^n)$. This semi-group is time inhomogeneous.

$$\tilde{P}_t^{n+1}[h(g)h^n(u)v](.,.,0) = \int_0^t \tilde{P}_{t,s}^n[f^{j+1}\alpha_s^{n+1}\tilde{P}_s^n[h(g)h^n(u)]](.,.) \qquad (13)$$

Moreover

$$\tilde{P}_t^{n+1}[uh(.)h^n(.)](.,.,u_{n+1}) = \tilde{P}_t^{n+1}[uh(.)h^n(.)](.,.,0) + \tilde{P}_t^n[h(.)h^n(.)](.,.)u_{n+1} \qquad (14)$$

h is a function of g, h^n a function of $u_1, ..., u_n$. This comes from the fact that $\frac{\partial}{\partial u_{n+1}}$ commutes with $\tilde{L}_t^{n+1}$.

Therefore, the two sides of (13) satisfy the same parabolic equation with second member. We deduce that

$$\tilde{P}_t^{n+1}[u_{n+1}\prod_{j=1}^n u_j h(.)](.,.,0) = \int_0^t ds\, \tilde{P}_{t,s}^n[f^{j_{n+1}}\alpha_s^{n+1}\tilde{P}_s^n[h\prod_{j=1}^n u_j]](.,.) \qquad (15)$$

This is an integration by parts formula. We would like to present this formula in a more appropriate way for our object.

We consider the operator

$$\overline{L}^n = L + \sum_{j=1}^n \frac{\partial^{2k}}{\partial u_j^{2k}} \qquad (16)$$

It generates a semi-group $\overline{P}_t^n$. In the sequel, we will skip the problem of sign coming if k is even or not. Since $\prod_{j=1}^n u_j$ is a polynomial, the Volterra expansion associated to $\tilde{P}_s[h\prod_{j=1}^n u_j]$ is finite and converge. We get

$$\tilde{P}_s[h\prod_{j=1}^n u_j](.,.) = \sum(-1)^l \int_{s>s_1>..>s_l>0} I_{s_1,..,s_l}^l ds_1..ds_l \qquad (17)$$

where

$$I_{s_1,..,s_l}^l = \overline{P}_{s-s_1}^n[\sum_{i=1}^n f^{j_i}\alpha_{s_1}^i \frac{\partial}{\partial u_i}[\overline{P}_{s_1-s_2}^n[\sum_{i=1}^n f^{j_i}\alpha_{s_2}^i$$
$$\frac{\partial}{\partial u_i}[\overline{P}_{s_3-s_2}^n[[\sum_{i=1}^n f^{j_i}\alpha_{s_2}^i \frac{\partial}{\partial u_i}[...[\overline{P}_{s_l}^n[h\prod_{j=1}^n u_j]..](.,.) \qquad (18)$$

Moreover

$$\overline{P}_s^n[h\prod_{j=1}^n u_j](g_0,.) = \overline{P}_s^n[h(.g_0)\prod_{j=1}^n u_j](e,.) \qquad (19)$$

such that

$$f^{i_j}\overline{P}^n_s[h\prod_{j=1}^n u_j](g_0,.) =$$

$$\overline{P}^n_s[f^{i_j}h(.g_0)\prod_{j=1}^n u_j](e,.) = \overline{P}^n_s[f^{i_j}h(.)\prod_{j=1}^n u_j](g_0,.) \quad (20)$$

We remark that in (17) the series is finite and stops at n because we consider a polynomial in v_i and because $\frac{\partial}{\partial u_i}$ commute with $\overline{P}_t$. If we consider $\overline{P}_t(h_1(g)h_2(v))$ it is a product of the $P_t(h_1)Q_t(h_2(v))$, where Q_t is generated by $\sum_{j=1}^n \frac{\partial^{2k}}{\partial u_j^{2k}}$. We deduce that in the term of the Volterra expansion of length l smaller than n, we get $(P_{t-s}(f^l h(g))Q_{t-s}(h_1(v))$ where $h_1(v)$ is an homogeneous polynomial with coefficient independent of g of degree $n-l$.

We do the following recursion hypothesis on l:

Hypothesis 1 There exists a positive real r_l such that if $(\alpha) = (i_{(\alpha)}, .., i_{(\alpha)})$. is a multi-index of length smaller than l constituted of $|(\alpha)|$ with the same element

$$|P_t[f^{(\alpha)}h\prod_{i=n}^n u_i](g,v.)| \leq Ct^{-r_l}\|h\|_\infty(1 + \prod_{i=n}^n |v_i|) \quad (21)$$

where $\|.\|_\infty$ is the uniform norm of h.

It is true for $l = 1$ by (13) and the next part.

If it is true for l, it is still true for $l + 1$, by using (15) and the Volterra expansion above for $f^{(\alpha)}h$ and taking $\alpha_s^{n+1} = s^{r_l}$.

By choosing suitable α_t^j, we have accordingly the framework of the Malliavin Calculus for any basis of the Lie algebra f^i, for any l

$$|P_t[\sum_i (f^i)^l h](g_0)| \leq C_{(\alpha)}\|h\|_\infty \quad (22)$$

in order to conclude, because the operator $\sum_i (f^i)^l$ is an elliptic operator whose degree tends to infinity when $l \to \infty$.

3.2   Estimates: the Davies Gauge Transform

We do as in [12] (16). The problem is that in $\tilde{P}^n_t[h\prod_{j=1}^n u_j](.,.)$ the test function u_j is not bounded and that $\tilde{P}^n_t$ acts only on $C_b(G \times \mathbb{R}^n)$. We do as in [12] the Davies gauge transform $\prod_l^n g(u_i)$ where

$$g(u) = (|u|) \tag{23}$$

if u is big and g is smooth strictly positive.

This gauge transform acts on the original operator by the simple formula $(\prod_{i=1}^{n} g(u_i))^{-1} \tilde{L}_1^n((\prod_{i=1}^{n} g(u_i).)$. On the semi-group it acts as

$$(\prod_{i=1}^{n} g(.))^{-1} \tilde{P}_t^n [(\prod_{i=1}^{n} g(u_i) h(.) h^n(.)](.,.) \tag{24}$$

But

$$(g(u_i))^{-1} \frac{\partial}{\partial u_i} (g(u_i).) = \frac{\partial}{\partial u_i} + C(u_i) \tag{25}$$

where the potential $C(u_i)$ is smooth with bounded derivatives at each order. Therefore, the transformed semi-group act on $C_b(G \times R^n)$. It remains to choose

$$h^n(u.) = \prod_{j=1}^{n} \frac{u_j}{g(u_j)} \tag{26}$$

in order to conclude. We deduce the bound:

$$|\tilde{P}_t^n|[h \prod_{j=1}^{n} |u_j|](.; v.) \leq C(\|h\|_\infty (1 + \prod_{i=n}^{n} |v_i|) \tag{27}$$

where $|\tilde{P}_t^n|$ is the absolute value of the semi-group $\tilde{P}_t^n$.

4   Study of an Example on the Linear Space

We give in this part a big category of examples on $\mathbb{R}^d$ of symmetric bounded below pseudo-differential operators which takes its origin in the theory of Poisson process [5, 6].

We consider the space $C^\infty(\mathbb{R}^d)$ of smooth functions h with bounded derivatives at each order.

We introduce a smooth function from $\mathbb{R}^d \times \mathbb{R}^d$ into $\mathbb{R}$ $(x, y) \rightarrow g(x, y)$ which equals to 0 for $|y| > C > 0$ for a small C and with bounded derivatives at each order. This allows us to introduce the integro-differential operator on $C^\infty(\mathbb{R}^d)$:

$$Lh(x) = (-1)^{l+1} \int_{\mathbb{R}^d} (h(x + y) - h(x)$$

$$- \sum_{i=1}^{2l} 1/i! < y^{\otimes i}, h^{(i)}(x)) g(x, y) |y|^{-(2l+d+\alpha)} dy \tag{28}$$

for $\alpha \in]-1, 0[$.

We do the following hypothesis: for all $x \in \mathbb{R}^d$, $h(x, 0) > C > 0$.

In such a case, we have shown [8, 9] that L is a pseudo-differential elliptic operator with symbol

$$a(x, \xi) = (-1)^{l+1} \int_{\mathbb{R}^d} (\exp[\sqrt{-1} < y, \xi >] -$$

$$\sum_{i=1}^{2l} 1/i!(\sqrt{-1} < y, \xi >)^i) g(x, y)|y|^{-(2l+d+\alpha)} dy \quad (29)$$

L is elliptic and satisfies Garding assumption (11) with $m' \to \infty$ when $l \to \infty$. We produce a large class of examples of such operators which are moreover symmetric in $L^2(dx)$.

Let be $X_j(x)$, $j = 1, .., d$ be some vector fields without divergence, with bounded derivatives of each order and which are uniformly in x in $\mathbb{R}^d$ a basis of $\mathbb{R}^d$.

Let $\phi_t(y)(x)$ be the dynamical system generated by the vector field $X(y, x) = \sum_{j=1}^{d} y_j X_j(x)$; $\phi_0(y)(x) = x$ and

$$d\phi_t(y)(x) = X(y, \phi_t(y))dt \quad (30)$$

We suppose $g(x, y) = g(y) = g(-y)$. We introduce the operator

$$L_1 h(x) = (-1)^{l+1} \int_{\mathbb{R}^d} (h(\phi_1(y)(x)) - h(x) -$$

$$\sum_{i=1}^{l} 1/(2i!)(X(y, x))^{(2i)} h(x)) g(y)|y|^{-(2l+d+\alpha)} dy \quad (31)$$

In the previous formula, the vector field $X(y, x)$ is considered as a one-order differential operator in x.

Lemma 2 *Under the symmetry condition on g, L_1 is symmetric and is defined on $C^\infty(\mathbb{R}^d)$.*

Proof The fact that L_1 is defined on $C^\infty(\mathbb{R}^d)$ comes from the fact that the asymptotic expansion of $y \to h(\phi_1(y)(x))$ near 0 is

$$h(x) + \sum_{i=1}^{2l} 1/i! X(y, x)^{(i)} h(x) \quad (32)$$

and from the fact that $g(y) = g(-y)$ such that only even integers remain in the sum (31).

The fact that L_1 is symmetric comes from two fact: the vector field $X(y, x)$ is divergence free such that

$$\int_{\mathbb{R}^d} h_1(x)X(y,x)^{(2i)}h_2(x)dx = \int_{\mathbb{R}^d} h_2(x)X(y,x)^{(2i)}h_1(x)dx \qquad (33)$$

by integrating by parts. Moreover, $x \to \phi_1(y)(x)$ preserves the Lebesgue measure such that

$$\int_{\mathbb{R}^d} h_1(x)h_2(\phi_1(y)(x))dx = \int_{\mathbb{R}^d} h_1(\phi_1(-y)(x))h_2(x)dx \qquad (34)$$

and the result arises from the equality $g(y) = g(-y)$. ◇

Theorem 3 L_1 *is an operator of the type (28) which is symmetric bounded below.*

Proof It remains only to show that L_1 is an operator of the type (28). For that we remark that the map

$$y \to \phi_1(y)(x) - x \qquad (35)$$

is a local diffeomorphism at every point y and a local diffeomorphism of a neighborhood of 0 in $\mathbb{R}^d$ onto a neighborhood of 0 in $\mathbb{R}^d$. ◇

Remark Let us give some heuristic explanation which explains this part. Let us consider a formal path measure dQ on a "space" of paths y_t with jumps starting from 0 which represents the semi-group P_t associated to the operator

$$Lh(x) = (-1)^{l+1} \int_{\mathbb{R}^d} (h(x+y) - h(x) -$$

$$\sum_{i=1}^{l} 1/(2i!) < y^{\otimes 2i}, h^{(2i)}(x) >)g(y)|y|^{-(2l+d+\alpha)}dy \qquad (36)$$

such that formally

$$P_t h(x) = \int h(y_t + x) \text{``}dQ(y_{\cdot})\text{''} \qquad (37)$$

We consider the "formal stochastic differential with jumps" whose solution (starting from x) $y_{1,t}(x)$ satisfies

$$\Delta y_{1,t}(x) = \phi_1((\Delta y_t))(y_{1,t-}(x)) - y_{1,t-}(x) \qquad (38)$$

where $y_{t-} = \lim_{s \to t-} y_s$ and $\Delta y_t = y_t - y_{t-}$. We should get

$$P_{1,t}h(x) = \int f(y_{1,t}(x) \text{``}dQ(y_{\cdot})\text{''} \qquad (39)$$

Moreover, a lot of compensation should appear in the formal equation giving $y_{1,t}$. We refer to [5] in the case where the path integrals are rigorously defined (In such a case only one compensation appears!).

References

1. Bismut, J.M.: Calcul des variations stochastiques et processus de sauts. Z. Wahr. Verw. Gebiete **63**, 147–235 (1983)
2. Chazarain, J., Piriou, A.: Introduction a la théorie des équations aux dérivées partielles linéaires. Gauthier-Villars, Paris, France (1981)
3. Hoermander, L.: The Analysis of Linear Partial Operators III. Springer, Berlin, Germany (1984)
4. Hoermander, L.: The Analysis of Linear Partial Operators IV. Springer, Berlin, Germany (1984)
5. Ishikawa, Y.: Stochastic Calculus of variations for jump processes, Basel. de Gruyter, Schweiz (2012)
6. Léandre, R.: Extension du théoreme de Hoermander a divers processus de sauts. Université de Besançon, France (1984). PHD Thesis
7. Léandre, R.: Stochastic analysis for a non-markovian generator: an introduction. Russ. J. Math. Phys. **22**, 39–52 (2015)
8. Léandre, R.: Large deviation estimates for a non-Markovian generator of Lévy type of big order. 4th Int. Conf. Math. Modern. Phys. Sciences, J. Phys.: conf. Ser. **633**, 012085, 2015 (E. Vagenas and al esds)
9. Léandre, R.: A Class of non-Markovian pseudo-differential operators of lévy type". Pseudo-differential operators: groups, geometry and applications. Birkhauser, 149–159 (M.W. Wong and al eds) (2017)
10. Léandre, R.: Perturbation of the Malliavin Calculus of Bismut type of large order. In: Gazeau, J.P. (ed. Physical and Mathematical Aspects of Symmetries, pp. 221–225. Springer (2017)
11. Léandre, R.: Malliavin Calculus of Bismut type for an operator of order four on a Lie group. J. Pseudo-differential Oper. Appl. **8**, 419–430 (2019)
12. Léandre, R.: Bismut's way of the Malliavin Calculus of large order generators on a Lie group. In: Tosun, M. (ed.) 6th International European Conference Mathematical Sciences and Applications 1926, 020026 A.I.P. Proceedings (2018)
13. Léandre, R.: Bismut's way of the Malliavin Calculus for non markovian semi-groups: an introduction. to appear in Analysis of pseudo-differential operators. M.W. Wong and al eds
14. Seeley, R.T.: Complex powers of an elliptic operator. In: Singular Integrals Proceedings Symposia in Pure Mathematics. Providence, U.S.A. A.M.S., pp. 288–307 (1966)
15. Taylor, M.: Partial Differential Equations II. Springer, Qualitative Studies of Linear Equations. Heidelberg, Germany (1997)

Discrete Geometrical Invariants: How to Differentiate the Pattern Sequences from the Tested Ones?

Raoul R. Nigmatullin and Artem S. Vorobev

Abstract Based on the new method (defined below as the discrete geometrical invariants—DGI(s)), one can show that it enables to find the statistical differences between random sequences that can be presented in the form of 2D curves. We generalized and considered the Weierstrass–Mandelbrot function and found the desired invariant of the fourth order that connects the WM-functions with different fractal dimensions. Besides, we consider an example based on real experimental data. A high correlation of the statistically significant parameters of the DGI obtained from the measured data (associated with reflection optical spectra of olive oil) with the sample temperature is shown. This new methodology opens wide practical applications in differentiation of the hidden interconnections between measured by the environment and external factors.

Keywords Weerstrass–Mandelbrot function · Discrete geometrical invariants · Equipment calibration · Nano-noise "reading"

2010 AMS Math. Subject Classification. Primary 40A05, 40A25; Secondary 45G05.

1 Introduction and Formulation of the Problem

If we follow for the modern tendencies in the applied sciences, one can notice that the efforts of many types of research were concentrated presumably on the analysis of complex systems. It implies the usage of methodology of many natural sciences as physics, chemistry, biology, economy, and improvement of the mathematical methods that should be more general and enables to describe the hierarchy of interactions,

R. R. Nigmatullin (✉) · A. S. Vorobev
Radioelectronics and Informative-Measurements Techniques Department, Kazan National Research Technical University (KNRTU-KAI), K. Marx str. 10, 420111 Kazan, Tatarstan, Russian Federation
e-mail: renigmat@gmail.com

A. S. Vorobev
e-mail: vartems14@gmail.com

© Springer Nature Singapore Pte Ltd. 2019
P. Agarwal et al. (eds.), *Fractional Calculus*, Springer Proceedings in Mathematics & Statistics 303, https://doi.org/10.1007/978-981-15-0430-3_4

intermittency between the structural organization levels of the considered complex systems. One of the main obstacles is the "invisible" boundary determination that divides the chaotic and deterministic behaviors of the complex systems. For better understanding of its behavior, a potential researcher needs to increase the deterministic part and decrease the part related to its chaotic and unpredictable behavior. If it is possible to solve this task then as a "reward" a researcher pulls apart the forecasting boundaries for prediction of the complex system behavior in the time evolution process. Many new features came from the fractal geometry with appearance of the B. Mandelbrot book [1] and its successful interpretation [2] that helps to consider and describe mathematically many new complex systems with self-similar/fractal geometry. This specific understanding helped to attract the mathematical tool as the fractional calculus [3, 4] and find for it its proper place in many practical applications. Now this powerful combination of the fractal geometry with the fractional calculus gives a new impact in the development of many natural sciences unifying them in one perfect instrument for knowledge and establishing new relationships that exist in the Mother Nature.

In this paper, we want to attract the attention of experts and many researches working in this "hot" spot as the fractal geometry and fractional calculus to "a discovery" made by Prof. Yu. I. Babenko in his books [5, 6]. Actually, he was able to generalize the well-known Pythagoras theorem and find new mathematical relationships between the lengths of many symmetrical sets/polyhedrons located in 2D and 3D spaces. After reading this instructive book, one of us (RRN) formulated the following problem: *is it possible to find some deterministic mathematical relationships between random sequences at least in 2D space and apply them for a more detailed comparison of the measured data?*

The obtained results showed that these DGI(s) really exist. Therefore, one can state that at least any two arbitrary random sets located in 2D space can relate with each other by means of their inter-correlations and integer moments. This generalization opens quite new possibilities in the reduced identification of different 2D curves (images) and comparison of various random curves with each other without the knowledge of a "true" fitting function, which, for many complex systems studied, it is absent. The preliminary results related to the application of the DGI in electrochemistry was published in paper [7]. In this paper, we present the complete invariant of the 4th order and show its possibilities for comparison of the WM-curves with different fractal dimensions and in finding of the hidden deterministic relationships between the measured data (reflectance optical olive oils spectra) and temperature changes during the experiment.

2   Basic Relationships and Description of the Algorithm

As it was reminded in the first section in the books [5, 6], it was shown that the well-known Pythagoras theorem can be generalized and propagated for a set of random points having coordinates (x_k, y_k) $(k = 1, 2, ..., n)$. Really, one can consider the square of the distance connecting an arbitrary point $M(x, y)$ with the kth point (x_k, y_k) belonging to the given set:

$$l_k^2 = (x - x_k)^2 - (y - y_k)^2, \tag{2.1}$$

one requires that

$$\frac{1}{n} \sum_{k=1}^{n} l_k^2 = I^2 \equiv const. \tag{2.2}$$

Inserting expression (2.1) into (2.2), one obtains

$$(x - \langle x \rangle)^2 + (y - \langle y \rangle)^2 = I^2 - R^2,$$

$$\langle x^p \rangle = \frac{1}{n} \sum_{k=1}^{n} x_p^k, \langle y^p \rangle = \frac{1}{n} \sum_{k=1}^{n} y_p^k, R^2 = \langle \Delta x^2 \rangle + \langle \Delta y^2 \rangle, \tag{2.3}$$

$$\langle \Delta V^2 \rangle \overset{def}{=} \langle V^2 \rangle - \langle V \rangle^2, V = x, y.$$

As one can notice from (2.3) that the set of circles can exist if the desired invariant $I^2 \geq R^2$, the equality sign corresponds to the circle with the zeroth radius. It is convenient to consider the invariant circle with radius $I^2 = 2R^2$. From another point of view, the requirement (2.2) corresponds to the reduction of the given set of points to the continuous circle with 4 statistical parameters ($\langle x^p \rangle$, $\langle y^p \rangle$, $p = 1, 2$). However, for practical purposes, this simplest requirement (2.2) is *not* sufficient and therefore, it has sense to consider other combinations.

2.1   The DGI of the Second Order (General Form)

In order to have reduction to the deterministic curve with sufficient number of statistical parameters we consider another combination that is a little complicated in comparison with the definition of the Euclidean distance (2.1):

$$L_k^2 = C^2(y - y_k)^2 - 2B(x - x_k) \cdot (y - y_k) + A^2(x - x_k)^2, k = 1, 2, ..., n. \tag{2.4}$$

The quadratic form (2.4) contains 5 statistical parameters ($\langle x^p \rangle$, $\langle y^p \rangle$, $p = 1, 2$), $\langle xy \rangle$ and 3 unknown parameters (A, B, C) figuring in (2.4). We subject this combination to the requirement:

$$\frac{1}{n} \sum_{k=1}^{n} L_k^2 = I^2 \equiv const. \tag{2.5}$$

Inserting (2.4) into (2.5) after simple algebraic manipulations one can obtain

$$C^2(y - \langle y \rangle)^2 - 2B(y - \langle y \rangle) \cdot (x - \langle x \rangle) + A^2(x - \langle x \rangle)^2 + E^2 \equiv I^2,$$
$$E^2 = C^2 \langle \Delta y^2 \rangle - 2B \langle \Delta x \Delta y \rangle + A^2 \langle \Delta x^2 \rangle. \tag{2.6}$$

As before, we put $I^2 = 2E^2$. In order to find three unknown parameters (A, B, C), it is convenient to use the obvious parameterization for the variables (x, y) relatively the angle φ:

$$y = \langle y \rangle + A \cos(\varphi - \alpha),$$
$$x = \langle x \rangle + C \cos(\varphi), 0 \leq \varphi \leq 2\pi. \tag{2.7}$$

Excluding the parameter φ from (2.7) and identifying expression (2.6) with relationship:

$$C^2 \langle (\Delta y)^2 \rangle - 2AC \cos \alpha \langle (\Delta x) \cdot (\Delta y) \rangle + A^2 \langle (\Delta x)^2 \rangle = A^2 C^2 - B^2,$$
$$E^2 = C^2 \langle \Delta y^2 \rangle - 2AC \cos \alpha \langle \Delta x \Delta y \rangle + A^2 \langle \Delta x^2 \rangle = A^2 C^2 \sin^2 \alpha, \tag{2.8}$$

one obtains

$$\cos \alpha = \frac{B}{AC}, \quad E^2 = A^2 C^2 - B^2. \tag{2.9}$$

In order to decrease the number of unknown parameters, we find from (2.7) the values A and C from the obvious conditions:

$$y_{max} = \langle y \rangle + A, y_{min} = \langle y \rangle - A, \rightarrow A = \frac{1}{2}(y_{max} - y_{min}),$$
$$x_{max} = \langle x \rangle + C, x_{min} = \langle x \rangle - C, \rightarrow C = \frac{1}{2}(x_{max} - x_{min}). \tag{2.10}$$

Parameter B is found from relationships (2.8) and (2.9) as a positive root of the quadratic equation written relatively B

$$B^2 - 2 \langle \Delta x \Delta y \rangle B - [A^2 C^2 - \langle \Delta x^2 \rangle A^2 - \langle \Delta y^2 \rangle C^2] = 0,$$
$$B = \langle \Delta x \Delta y \rangle + [(\langle \Delta x \Delta y \rangle)^2 + A^2 C^2 - \langle \Delta x^2 \rangle A^2 - \langle \Delta y^2 \rangle C^2]^{1/2}. \tag{2.11}$$

This single root is chosen from the comparison of two identity sequences $(x_k = y_k)$ that follows from the obvious requirement $B = A^2$, $(\alpha = 0)$. Therefore, one can say that with the help of the rotated counterclockwise ellipse (2.7) we reduced $2n$ random points figuring in (2.4) to 8 statistical parameters $(\langle x^p \rangle, \langle y^p \rangle, p = 1, 2), \langle xy \rangle, \alpha,$ $A, C)$. If it is necessary to include the higher moments $\langle x^p y^s \rangle, (p = 0, 1, 2, ...;$ $s = 0, 1, 2...)$ then other combinations of the type (2.4) should be considered.

Some generalizations for the invariant of the fourth order are considered below. It is easy to notice that the invariant (2.7) of the second order is *not* sufficient for a detailed comparison of two random sequences. Expression (2.9) is equivalent to the conventional Pearson Correlation Coefficient (PCC) that is related to the correlation of the second order $\langle \Delta x \Delta y \rangle$. Actually, we found the geometrical interpretation of the PCC and showed that the closed curve as an ellipse can be used for pictorial interpretation of the quadratic correlations between two random sequences. Therefore, it has a sense to consider the invariants of the higher orders for a more detailed and reliable comparison of a couple of random sequences with each other. Below, we want to consider the complete invariant of the fourth order. It is instructive also to give the basis of the proposed theory that will be useful for *quantitative* comparison of any two random sets located in 2D plane.

2.2 The General Theory of the Geometrical Invariants Based on the Higher Order Curves and the GDI of the Fourth Order

Unifying the ideas expressed in books [5, 6], one can consider the following combination:

$$L_k^{(m)} = \sum_{q=0, p=0}^{m} A_{q,p}(x - x_k)^q (y - y_k)^p. \tag{2.12}$$

This combination can be considered as the most *general* form that can be used for comparison of two random sets having coordinates (x_k, y_k) $(k = 1, 2, 3 \ldots, n)$. If one requires that

$$\frac{1}{n} \sum_{k=1}^{n} L_k^{(m)} = Inv, \tag{2.13}$$

then this form can be used for comparison of two random sequences of an arbitrary order in terms of different combinations of the integer moments. If we insert (2.12) into (2.13) and open the corresponding terms then one can obtain possible combinations of the integer moments of the type:

$$M_{q,p} = \langle (\Delta x)^q (\Delta y)^p \rangle \equiv \frac{1}{n} \sum_{k=1}^{n} (x - \langle x \rangle - \Delta x_k)^q (y - \langle y \rangle - \Delta y_k)^p,$$

$$\langle A \rangle = \frac{1}{n} \sum_{k=1}^{n} A_k, \ \Delta A_k = A_k - \langle A \rangle. \tag{2.14}$$

In this section, having in mind its practical application for comparison of the different experimental data with each other we consider the complete invariant of the fourth order that will be helpful for a more fine comparison of two sets.

2.3 The Complete Invariant of the Fourth-Order Admitting the Separation of Variables

A possible combination allowing to express the desired invariant in the analytical form can be written as

$$L_k^{(4)} = A_{40}(x - x_k)^4 + A_{31}(x - x_k)^3(y - y_k) - 2A_{22}(x - x_k)^2(y - y_k)^2 +$$
$$+ A_{13}(x - x_k)(y - y_k)^3 + A_{04}(y - y_k)^4.$$
$$(2.15)$$

Inserting expression (2.15) into (2.13) and equating the linear terms relatively the variables

$$X \equiv x - \langle x \rangle, \, Y \equiv y - \langle y \rangle, \tag{2.16}$$

to zero, we obtain the following combinations:

$$4A_{40}\langle (\Delta x)^3 \rangle + 3A_{31}\langle (\Delta x)^2 \Delta y \rangle + A_{13}\langle (\Delta y)^3 \rangle = 4A_{22}\langle (\Delta y)^2 \Delta x \rangle,$$
$$4A_{04}\langle (\Delta y)^3 \rangle + 3A_{13}\langle (\Delta y)^2 \Delta x \rangle + A_{31}\langle (\Delta x)^3 \rangle = 4A_{22}\langle (\Delta x)^2 \Delta y \rangle.$$
$$(2.17)$$

In order to decrease the number of the parameters entering in (2.15) we introduce the following ratios:

$$A_{31} = \sigma_x A_{22}, \, A_{13} = \sigma_y A_{22}, \, A_{40} = \theta_x A_{22}, \, A_{04} = \theta_y A_{22}. \tag{2.18}$$

These ratios help to cancel on an arbitrary constant A_{22} ($\neq 0$) and present system (2.17) in the form:

$$4\theta_x\langle (\Delta x)^3 \rangle + 3\sigma_x\langle (\Delta x)^2 \Delta y \rangle + \sigma_y\langle (\Delta y)^3 \rangle = 4\langle (\Delta y)^2 \Delta x \rangle,$$
$$4\theta_y\langle (\Delta y)^3 \rangle + \sigma_x\langle (\Delta x)^3 \rangle + 3\sigma_y\langle (\Delta y)^2 \Delta x \rangle = 4\langle (\Delta x)^2 \Delta y \rangle.$$
$$(2.19)$$

In order to find these four unknown parameters, it is necessary to find some additional relationships between them. One can notice that for identity relationships (x_k, y_k) $(k = 1, 2, 3 \ldots, n)$, the following relationships from (2.19) are valid:

$$4\theta_x + 3\sigma_x + \sigma_y = 4, \, 4\theta_y + \sigma_x + 3\sigma_y = 4$$

$$or$$
$$(2.20)$$
$$\theta_x = 1 - \frac{3}{4}\sigma_x - \frac{1}{4}\sigma_y, \, \theta_y = 1 - \frac{3}{4}\sigma_y - \frac{1}{4}\sigma_x.$$

The systems (2.19), (2.20) allow finding the unknown variables (ratios) and rewrite them by means of different correlations belonging of two compared sets.

$$\sigma_x = \frac{4}{\Delta}\left[3 \cdot \left(\left\langle \Delta x(\Delta y)^2 \right\rangle - \left\langle (\Delta y)^3 \right\rangle\right) \cdot \left(\left\langle \Delta x(\Delta y)^2 \right\rangle - \left\langle (\Delta x)^3 \right\rangle\right)+ \right.$$
$$\left. +\left(\left\langle (\Delta x)^3 \right\rangle - \left\langle (\Delta y)^3 \right\rangle\right) \cdot \left(\left\langle \Delta y(\Delta x)^2 \right\rangle - \left\langle (\Delta y)^3 \right\rangle\right)\right],$$

$$\sigma_y = \frac{4}{\Delta}\left[3 \cdot \left(\left\langle \Delta y(\Delta x)^2 \right\rangle - \left\langle (\Delta x)^3 \right\rangle\right) \cdot \left(\left\langle \Delta y(\Delta x)^2 \right\rangle - \left\langle (\Delta y)^3 \right\rangle\right)+ \right.$$

$$\left. +\left(\left\langle (\Delta y)^3 \right\rangle - \left\langle (\Delta x)^3 \right\rangle\right) \cdot \left(\left\langle \Delta x(\Delta y)^2 \right\rangle - \left\langle (\Delta x)^3 \right\rangle\right)\right],$$

$$\Delta = 9 \cdot \left(\left\langle \Delta y(\Delta x)^2 \right\rangle - \left\langle (\Delta x)^3 \right\rangle\right) \cdot \left(\left\langle \Delta x(\Delta y)^2 \right\rangle - \left\langle (\Delta y)^3 \right\rangle\right)+$$
$$+\left(\left\langle (\Delta x)^3 \right\rangle - \left\langle (\Delta y)^3 \right\rangle\right)^2.$$

(2.21)

Two other unknown parameters $\theta_{x,y}$ are found from (2.20).

Finally, we obtain the following invariant of the fourth order:

$$K(X, Y) = K_2(X, Y) + K_4(X, Y) = I_4,$$
$$K_2(X, Y) = A_x X^2 + B \cdot X \cdot Y + A_y Y^2,$$
$$K_4(X, Y) = \theta_x X^4 + \theta_y Y^4 - 2X^2 Y^2 + \sigma_x X^3 Y + \sigma_y X Y^3.$$

(2.22)

The following combinations shown below define the constants figuring in the DGI (2.22):

$$A_x = 6\theta_x \left\langle (\Delta x)^2 \right\rangle - 2\left\langle (\Delta y)^2 \right\rangle + 3\sigma_x \left\langle \Delta x \Delta y \right\rangle,$$
$$A_y = 6\theta_y \left\langle (\Delta y)^2 \right\rangle - 2\left\langle (\Delta x)^2 \right\rangle + 3\sigma_y \left\langle \Delta x \Delta y \right\rangle,$$
$$B = -8\left\langle \Delta x \Delta y \right\rangle + 3\sigma_x \left\langle (\Delta x)^2 \right\rangle + 3\sigma_y \left\langle (\Delta y)^2 \right\rangle.$$

(2.23)

The constant I_4 from (2.22) is defined by expression:

$$I_4 = \theta_x \left\langle (\Delta x)^4 \right\rangle + \theta_y \left\langle (\Delta y)^4 \right\rangle+$$
$$+\sigma_x \left\langle (\Delta x)^3 (\Delta y) \right\rangle + \sigma_y \left\langle (\Delta y)^3 (\Delta x) \right\rangle - 2\left\langle (\Delta x)^2 (\Delta y)^2 \right\rangle.$$

(2.24)

Finally, we obtain the eight parametric curve (2.22), which combines 6 correlations and 8 moments up to the fourth order inclusive:

$$\left\langle x \right\rangle, \left\langle y \right\rangle, \left\langle (\Delta x)^2 (\Delta y)^2 \right\rangle, \left\langle \Delta x(\Delta y)^2 \right\rangle, \left\langle \Delta y(\Delta x)^2 \right\rangle,$$
$$\left\langle \Delta x(\Delta y)^3 \right\rangle, \left\langle \Delta y(\Delta x)^3 \right\rangle, \left\langle (\Delta x)^4 (\Delta y)^4 \right\rangle, \left\langle (\Delta x)^{2,3,4} \right\rangle, \left\langle (\Delta y)^{2,3,4} \right\rangle.$$

(2.25)

The curve $K(X, Y)$ in (2.22) can be separated in the polar coordinate system. We present the desired curve in the form:

$$x(\varphi) = \langle x \rangle + r(\varphi) \cos \varphi,$$

$$y(\varphi) = \langle y \rangle + r(\varphi) \sin \varphi, \tag{2.26}$$

$$r(\varphi) = \left[\frac{\sqrt{q_2^2(\varphi) + 4I_4 q_4(\varphi)} - q_2(\varphi)}{2q_4(\varphi)} \right]^{1/2}.$$

The functions $q_{2,4}(\varphi)$ figuring in (2.26) are determined by expressions:

$$q_2(\varphi) = A_x \cos^2(\varphi) + B \sin \varphi \cos \varphi + A_y \sin^2(\varphi),$$

$$q_4(\varphi) = \theta_x \cos^4(\varphi) - 2\sin^2(\varphi) \cos^2(\varphi) + \theta_y \sin^4(\varphi) + \sigma_x \sin \varphi \cos^3(\varphi) + \sigma_y \sin^3(\varphi) \cos \varphi. \tag{2.27}$$

This curve determines statistical proximity/difference between 2D random curves/sets located in the plane. What happens if two random curves are identical to each other ($x_j = y_j$) for all numbers of the discrete points $j = 1, 2, ..., N$? In this case as it can be shown (the details are given in the *Mathematical Appendix*) that $\sigma_{x,y} = 4/3$, $\theta_{x,y} = -1/3$, $A_x = A_y$, $B = -2A_x$ and $I_4 = 0$. Hence, from (2.26) it follows that $r(\varphi) = 0$. In this case, expression (2.22) is degenerated into a *point* with coordinates $\langle x \rangle = \langle y \rangle$ located on the line $y = x$. In the *Mathematical Appendix*, it was found the form of the simplified curve (2.22) when two discrete sets x_k and y_k ($k = 1, 2, ..., n$) are becoming close to each other, i.e., $\Delta y_k = \Delta x_k \pm \Delta f_k (\Delta f_k = f_k - \langle f \rangle)$ - small factor distorting the set y_k).

Concluding this section, one can say that we propose the complete invariant of the fourth order (2.22), which enables to compare two random sets (sequences) located on 2D plane. In general, this result shows that two random sequences have at least the compact *deterministic* curve of the fourth order (2.22) combining 8 parameters I_4, $A_{x,y}$, B, $\sigma_{x,y}$, $\theta_{x,y}$. These parameters, in turn, depend on 14 statistical parameters (2.25) that help to compare one random set with another one. Therefore, we made a next step and generalized the conventional Pearson correlation coefficient (2.9) that is valid only for the correlations of the second order. A potential researcher receives a new statistical tool for more "fine" analysis and comparison of a couple of random sets with each other. In subsequent chapters, we want to show *how* to apply this new tool for comparison of random sequences of different nature.

Finishing this section, it is necessary to remind about another important possibility of a new approach that makes this DGI-tool more significant and general. The previous results (expression (2.15) and below) were obtained for vectors (x_k, y_k) ($k = 1, 2, 3 ..., n$). Are the previous results conserved if one replaces the vectors for matrices $(M_{i,j}, L_{i,j})$ ($i = 1, 2, ..., I; j = 1, 2, ..., J$)? Attentive analysis of the results obtained above shows that one can obtain a positive answer. Really, for this case the mean values for $\langle x \rangle$, $\langle y \rangle$ are rewritten in the following form:

$$\langle x \rangle = \frac{1}{I \cdot J} \sum_{i,j}^{I,J} M_{i,j}, \quad \langle y \rangle = \frac{1}{I \cdot J} \sum_{i,j}^{I,J} L_{i,j}, \tag{2.28}$$

and different moments and inter-correlations keep formally their forms:

$$Q_{q,p} = \left\langle (\Delta x)^q (\Delta y)^p \right\rangle \equiv \frac{1}{I \cdot J} \sum_{i,j=1}^{I,J} \left(x - \langle x \rangle - \Delta x_{i,j} \right)^q \left(y - \langle y \rangle - \Delta y_{i,j} \right)^p,$$

$$\langle A \rangle = \frac{1}{I \cdot J} \sum_{i,j=1}^{I,J} A_{i,j}, \quad \Delta A_{i,j} = A_{i,j} - \langle A \rangle.$$

(2.29)

As it follows from last expression, the subsequent algebraic transformations remained the same and, therefore the final result (2.26) and its simplified expression (6.6) keep their structures, as well. This important generalization allows applying the DGI-tool for analysis of different $2D$-images and $3D$-projections, especially in cases when it is necessary to compare the random trajectories generated by unpredictable movements of molecules, viruses, and other "small" objects. This new possibility, undoubtedly, merits the separate research.

3   New "Reading" of the Weierstrass–Mandelbrot Function

As it is known [2], the WM-function is defined by the following relationship:

$$S(z) = \sum_{n=-N}^{N} b^n \cdot F(z \cdot \xi^n), \quad F(z) = 1 - \cos(z), \, b = \frac{1}{\xi^v},$$

(3.1)

$$v = 2 - D, \, N \gg 1.$$

The dependencies of these parameters with respect to the chosen parameter D are shown in Figs. 6, 7, and 8. We omit the dependence $x_c(D)$ which keeps its constant value equal to -0.02444 for all values of D from [0.5, 2.0]. As one can notice from (3.1), one can generalize the conventional definition of the WM-function and propagate it for more wide class of the functions varying the function $F(z)$. The sum (3.1) has the obvious property:

$$S(z\xi) = \frac{1}{b} + Up(z) - Dn(z),$$

(3.2)

$$Up(z) = b^N f(z\xi^{N+1}), \quad Dn(z) = b^{-N-1} f(z\xi^{-N}).$$

Expression (3.2) satisfies approximately the functional equation (3.3):

$$S(z\xi) \cong \frac{1}{b} S(z), \quad S(z) = z^v Pr(ln(z)), \quad v = \frac{ln(1/b)}{ln(\xi)} = 2 - D,$$

(3.3)

$$Pr(ln(z) \pm ln(\xi)) = Pr(ln(z)),$$

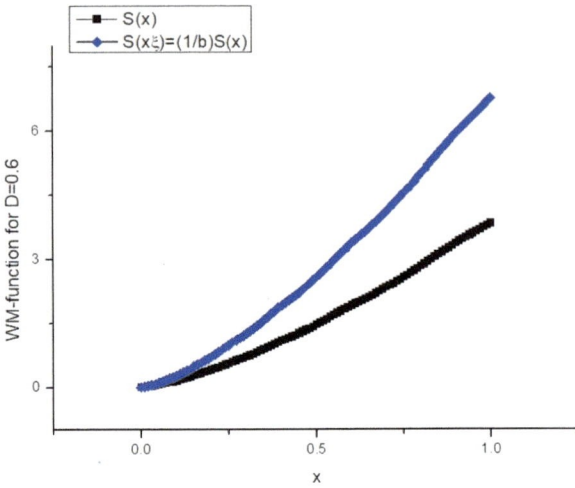

Fig. 1 Verification of the relationships (3.2) for the WM-function for $D = 0.6$. The contribution of the functions $Up(x)$ and $Dn(x)$ are negligible and therefore they are not shown

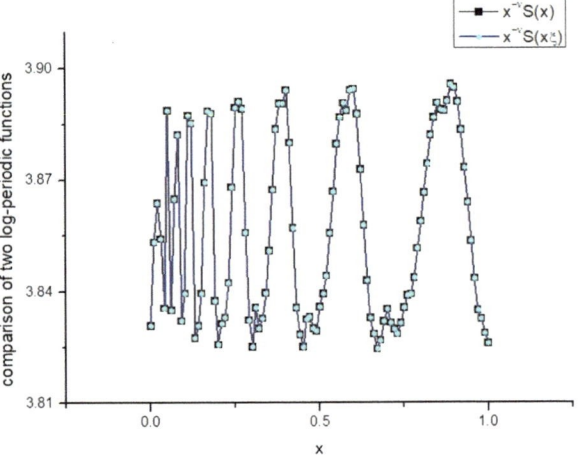

Fig. 2 Verification of two log-periodic functions $Pr(x) = Pr(z\xi)$ entering in expression (3.3) for $D = 0.6$

if contributions of the functions $Up(z)$ and $Dn(z)$ on the ends of the corresponding intervals are negligible. We want to stress here that expressions (3.3) are more correct in comparison with expression (2.16) given in the book [2] that was found in the results of numerical calculations [8]. Now it has a sense to formulate a problem that can be solved with the help of the DGI approach. Is it possible to relate the parameter D from (3.3) with parameters (20), (21), (23), and (24) forming the desired curve (26) and test the relationships (3.2) and (3.3) numerically? For the aim we chose $N = 60$ in (3.1) and select the interval for D as [0.5–2.0]. Then we compare successively the curve corresponding to the $D = 0.5$ with other curves from the interval [0.6, 2.0] with step $h = 0.1$. The verifications of relationships (3.1) and (3.2) for the limiting cases $D = 0.5$ and $D = 2.0$ are given in Figs. 1, 2, 3 and 4, correspondingly.

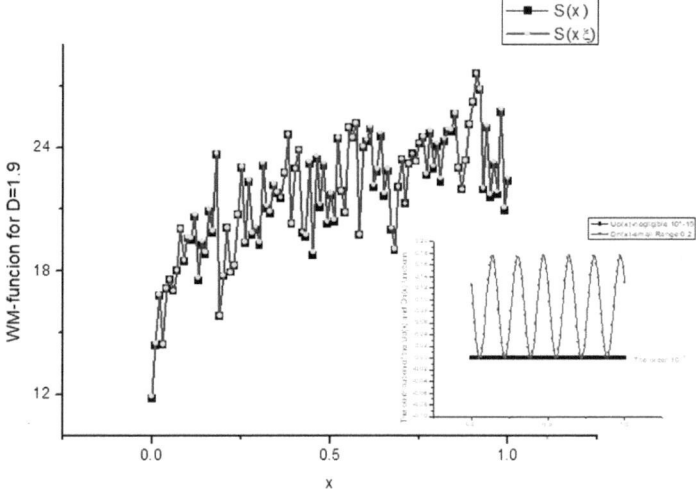

Fig. 3 Verification of the relationships (3.2) for the WM-function for $D = 1.9$. The contribution of the functions $Up(x)$ and $Dn(x)$ are shown inside the small figure on the right. Their contributions are small

Fig. 4 Verification of two log-periodic functions $Pr(x) = Pr(z\xi)$ entering in expression (3.3) for $D = 1.9$

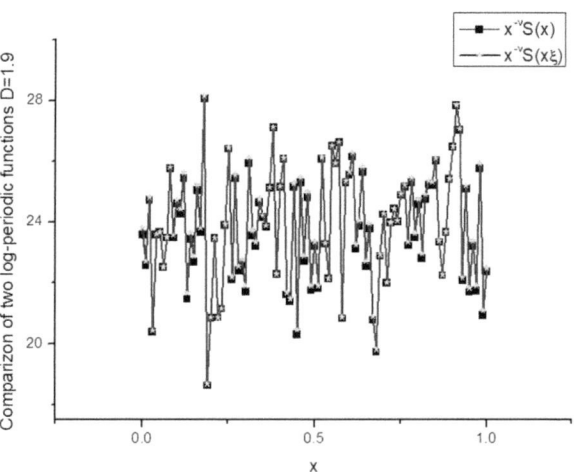

Figure 5 shows that the deviation factor (6.6) for all selected ranges of D that do not exceed the unit value. It allows to apply the simplified version of the DGI (6.5) that contains only 4 parameters (x_c, y_c, $\varepsilon/3Q_0$, B, I_4).

Finishing this section, we want to stress one important point. As it has been mentioned at the end of the section two, the DGI-tool can be applied successfully for comparison of different images having random fractal dimensions or their distributions. It helps to "read" quantitatively two sequences or $2D$ projections: one of them can be considered as the pattern one and another image can be defined as tested. The

Fig. 5 This dependence demonstrates the criterion of applicability of the simplified version of the DGI shown in the *Mathematical Appendix*. It is calculated in accordance with expression (6.6) and signifies that for all values of D from the interval [0.6–2.0] the simplified version of the DGI from (6.5) is applicable

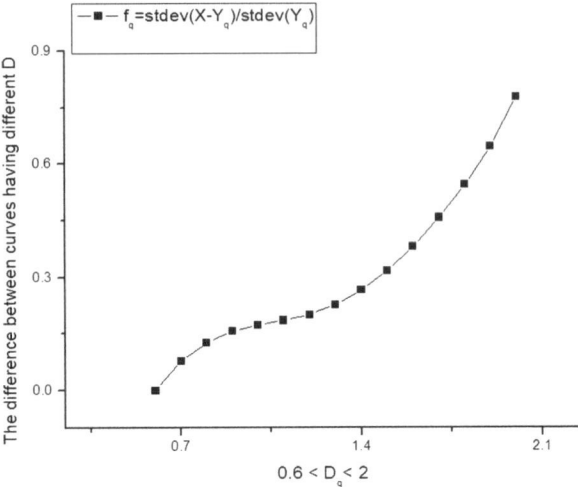

Fig. 6 The behavior of the mean value of the compared curve relative to the WM-function having $D = 0.5$

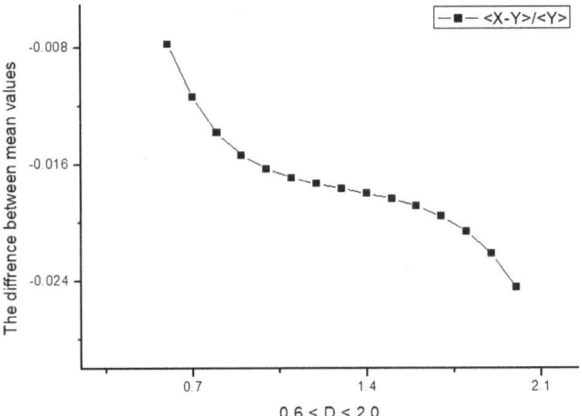

new methodology helps to analyze two images in terms of the reduced number of parameters (8 or 4) and express this comparison in the form of the DGI curves (2.26) or (6.5), correspondingly.

This instructive example shows that the DGI can serve an *additional source of information* that connects the fractal parameters of the WM function with the deterministic curve as the DGI of the fourth order. It helps also to reduce the initial set of data points (equaled 100) to some small number (4!) of significant parameters that helps to compare the initial curves in terms of the integer moments and their mutual cross-correlations defined by expression (2.25). Figure 9 demonstrates the form of the DGI for two limiting cases ($D = 0.5, 2.0$), including also the intermediate case $D = 1.5$. The next section demonstrates the results of the DGI application to real data.

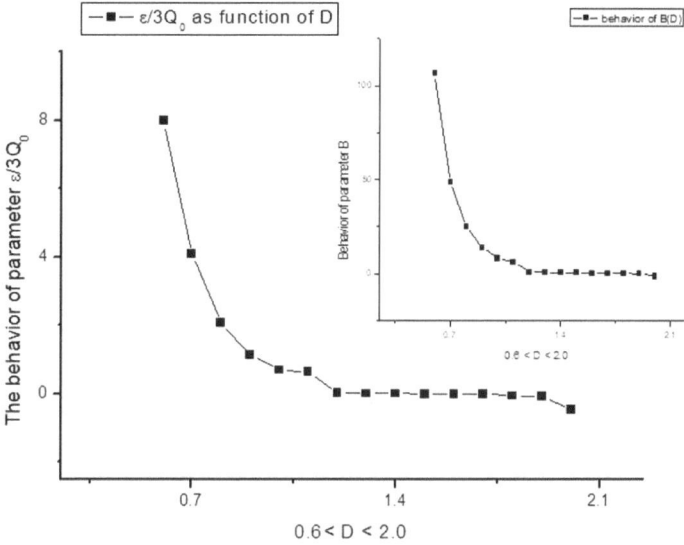

Fig. 7 In this plot, we show the dependence of two other parameters as $\varepsilon/3Q_0$ (central plot) and $B(D)$ (small plot above) with respect to the parameter D from the interval [0.6–2.0]. One can notice that these parameters keep their monotone behavior with respect to D

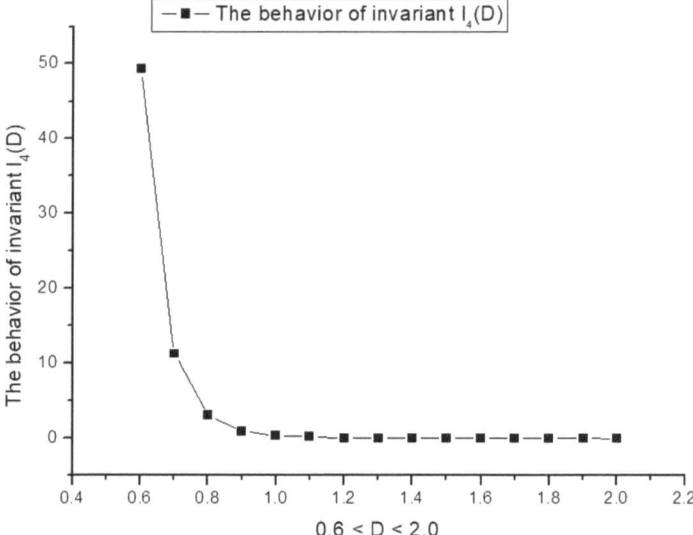

Fig. 8 Finally, this plot shows the behavior of the invariant $I_4(D)$ defined by expression (6.4). This is the biggest parameter keeping its monotone behavior

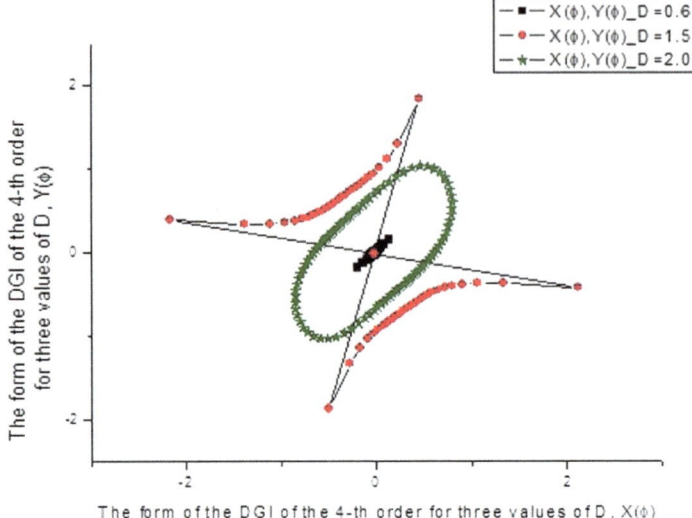

Fig. 9 In this plot, we show the parametric dependence of the function $X(\varphi)$ and $Y(\varphi)$ or three values of $D = 0.6$ (small segment in the middle), $D = 1.5$ (two "hyperbolic" red branches) and $D = 2.0$ (green ellipse-like curve). It is hard to imagine that these different curves are generated by a monotone set of parameters shown on the previous figures

4 Some Examples Based on Real Data

The question of the finding of additional relationships between input predominant factor and the measured response/output was raised in many papers [9–11]. We faced the same problem while studying the optical reflectance experiments associated with changes in chemical properties of extra virgin olive oil with respect to some external factor. One of the main input factors that can change the quality of the olive oil studied is the influence of the surrounding temperature. The finding of the desired olive oil parameter with the temperature is not new and was considered in some papers [12–14]. The description of the experiment and the data processing algorithm with the results obtained are described below.

4.1 Details of Experiment

The testing of the DGI method was carried out on the measured data, which we obtained in the process of olive oil temperature variations by nonchemical method. The results of the experiment were obtained in accordance with optical characteristics of the used equipment. The task was in measuring the reflectance optical spectra intensity with respect to the measurement of the internal olive oil temperature and

(a) **(b)**

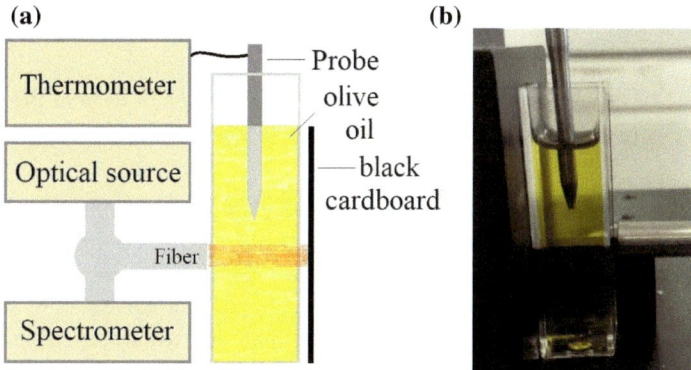

Fig. 10 **a** Schematic ensemble of the experimental setup and **b** photo of the cuvette with the optical fiber (located on the right-hand side) and the temperature probe (located above and placed inside the cuvette)

in finding their expected correlations in the frame of new approach. Figure 10a, b show the image and sketch of the experimental setup, respectively. The food olive oil sample was chosen as a complex fluid. It has the Trade mark—"Desantis" (class: "olio extra virgine di olive", Italy).

Optical reflectance spectra in the wavelength range from 195 nm to 1117 nm were obtained with the usage the commercial software "SpectraSuite" and the "MAT-LAB" script that, in turn, were controlled by the optical spectrometer (Ocean Optics HR4000CG-UV-NIR). The volume of 4 ml of the plastic (polystyrene) cuvette was filled with 3.5 ml of the olive oil and then was illuminated with a light source (Edmund MI150). The optical fiber (QR400-7-UV/BX) connected the optical source with cuvette and then the reflected signal was transmitted to the spectrometer. Black cardboard was placed on the back of the cuvette to avoid undesirable optical reflection. The temperature was monitored using a digital thermometer (Probe thermometer TFA LT-101), its sensitive part was placed inside the olive oil (see Fig. 10b). All optical absorptions were registered at the fixed temperature (T) in the range from 9.3 to 21.8 °C in dark conditions. Each measurement took a time of 3 s, and the time duration for the whole experiment occupied $3 \cdot 301 = 903$ s, i.e. 15 min. The measurement time for the thermometer consists of one second; therefore, the averaging over 3 points was carried out. This temporal requirement creates a possibility to "feel" possible changes in the structure of the olive oil in accordance with possible temperature variations.

The "zone of interest" for our experiments was in the range up to 20 °C. This zone determines some chemical processes that take place in extra virgin olive oil. One of the temperature ranges where these chemical reactions take place corresponds to the interval 5–15 °C [15].

4.2   The Data Treatment Procedure. Algorithm

In the frame of the new algorithm, we use the DGI approach to treat the measured data. Without loss of generality, the proposed algorithm can be divided into the following six steps:

(1) The intensity reflectance optical spectra were measured in the full wavelength range (195.98–1117.05 nm) and for each measuring cycle containing 3647 wavelength points were obtained. As a final result, we had 301 measurement cycles, corresponding to the measured T (see Fig. 11). Taking into account the averaging procedure over 3 points, i.e., 903 (duration of the whole experiment) $\rightarrow$ 301 (cycles), we obtain the desired plot T(number of cycle) shown in Fig. 12.

(2) Reduction to 12 measured data points. One of the aims of our measurements was to increase the speed of measurements (number of repetitions/per time). Therefore, we tried to avoid the measurement repetitions for each data point several times. However, keeping the desired speed it is possible to improve the quality of data through the reduction procedure. In this case, all data were reduced to 12 points, i.e., every 12 points from each measurement were reduced to one averaged point (this was done for all 301 measurement cycles). We can do this due to the fact that we used a "food" thermometer with an error of $\pm 0.5\,^{\circ}\mathrm{C}$ ($Err.$). The admissible temperature interval for our measurements was $12.5\,^{\circ}\mathrm{C}$ (from 9.3 to $21.8\,^{\circ}\mathrm{C}$), the number of measurement cycles was 301 (N). Therefore, the number of reduced points ($R.p.$) can be calculated as $R.p = N/(Range/Err) \approx 12$. Therefore, we obtained finally the number of the reduced cycles equal to $301/12 \approx 25$. In accordance with this requirement, we obtained the same number of temperature and optical measurement cycles. This reduction procedure helps to average the unwanted data fluctuations also.

(3) The DGI (x, y_m) tool was used for data obtained by the optical method, where x is the first reduced cycle ($x = y_1 = const$), y_m is the subsequent cycle ($m = 2, 3, \ldots, M$). Therefore, 24 calculations of the desired DGI were obtained from the total $M = 25$ y_m reduced cycles. Each compared curve has four quantitative parameters ($f_q, I_4, \varepsilon/3Q, B$) depending on T.

(4) In order to decrease the number of plots, it is necessary to establish the maximal correlations between the calculated parameters ($f_q, I_4, \varepsilon/3Q, B$) with T_m. We want to notice that all these parameters and temperature had the same number of data points equal to 24. In order to obtain the reliable correlations of these parameters with temperature, we used the value of the Complete Correlation Factor ($CCF(Pr, T)$; $Pr = (f_q, I_4, \varepsilon/3Q, B)$) that is calculated with the usage of all admissible set of the fractional moments [16]. In the result of this evaluation, the following values were obtained: $CCF_{f_q} = 0.8623$, $CCF_{I_4} = 0.8946$, $CCF_{\varepsilon/3Q} = 0.8413$, $CCF_B = 0.8563$. The highest value of the CCF (in terms of the correlation degree with temperature change) belongs to the parameter I_4 (below we use for it the simplified abbreviation I). Its variations with temperature is shown the Fig. 13 (black points).

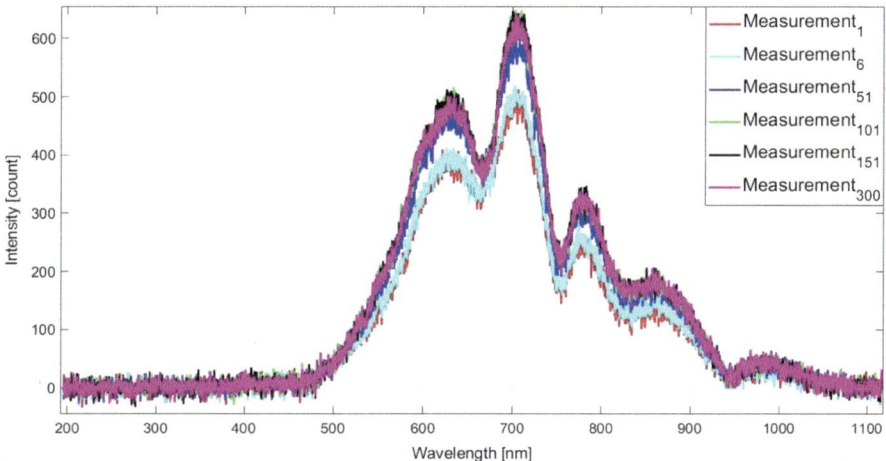

Fig. 11 The visually selected 6 curves (chosen from a set of 301 measurement cycles) of the reflectance optical spectra covering the range [196, 1117] nm

(5) Using the Procedure of the Optimal Linear Smoothing (POLS), it becomes possible to obtain a monotone and smoothed curve $I(T)$ [17]. It is depicted in Fig. 13 by green points (CCF for $I_{POLS} = 0.8964$.)

(6) In order to test the reduction procedure, we realized the same calculation, however, for non-reduced data. In this case, we have 301 measurements cycles including the same number of the temperature points with possible temperature deviations ($\Delta T = \pm 0.5\%C$). We realized the similar steps (4 and 5) and after the application of the POLS we obtained two other curves that take into account the limiting values of temperature fluctuations. These limiting curves ($max_I(+0.5\%C)$, $min_I(-0.5\%C)$) are shown by red and blues points, accordingly.

The obtained results demonstrate undoubtedly the effectiveness of the DGI approach that can be applied as a new working tool for the quantitative finding of the "hidden" relationships between the correlation parameters of the DGI with the predominant input factor (T). In addition, these relationships can be ordered with the help of the CCF and smoothed by the POLS. Besides, one can confirm a similar temperature trend found by researchers from Bulgaria [15] that discovered a quasi-linear temperature dependence of the reflectance spectrum intensity with respect to temperature in the range [9, 12–17] °C.

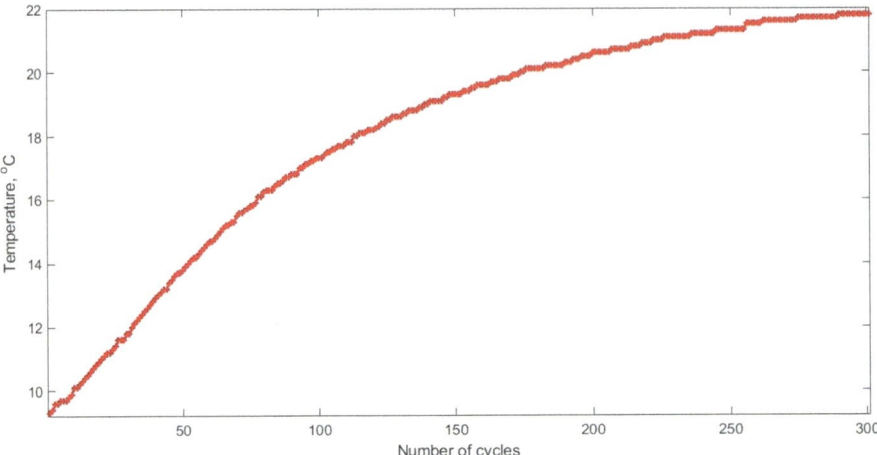

Fig. 12 The temperature plot covering the given optical range and located in the interval [9.3, 21.8] °C

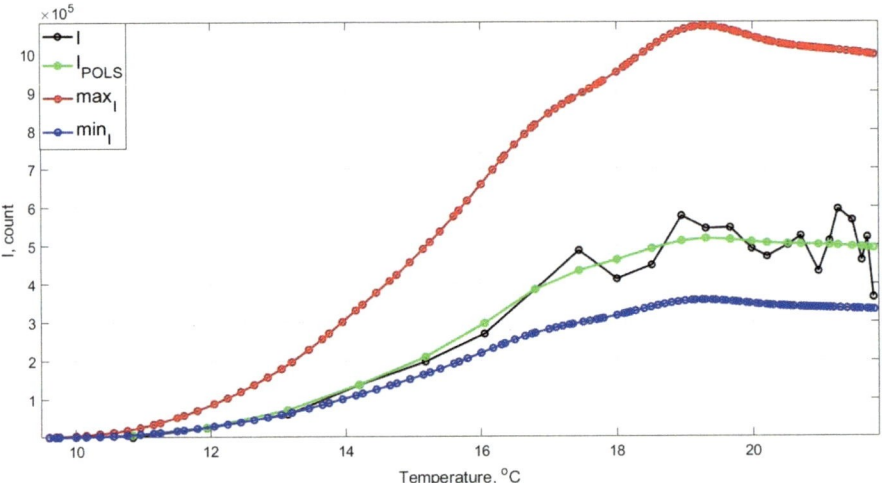

Fig. 13 This figure demonstrates the variations of the curves $I(T)$. Black points characterize a "true" curve obtained with the help of reduction procedure. The green, red, and blue curves show the temperature variations of the curve $I(T)$ obtained with the help of the POLS procedure. The red, blue curves (max_I, min_I) correspond to the limiting cases (obtained without reduction procedure)

5 Results and Discussion

In this paper, we showed possible applications of the DGI of the fourth order that admits its separation and presentation in the parametric form (2.26), (2.27). Definitely, the frame of this paper does not allow demonstrating all possibilities of this new instrument. These new possibilities will be a subject of the further research. However, based on the obtained results one can say the following:

1. The DGIs will help to compare two curves (including 2D random sets) with each other. This comparison is universal and it will be useful especially in cases when analytical expression for description of model/real data is absent.

2. The DGI helps to realize "automatically" the reduction procedure of $2N$ data points to 8 statistical parameters: $(I_4, A_{x,y}, B, \sigma_{x,y}, \theta_{x,y})$. These parameters are tightly related with 14 integer moments and inter-correlations (2.25) that signify about the statistical proximity and differences of two compared random sets.

3. The simplified DGIs help to find the relationships between fractal dimension D (3.2) and the reduced set of the statistical parameters $(x_c, y_c, I_4, \varepsilon/3Q, B)$ in the case of statistical proximity of the compared initial curves. It will give an additional possibility for better understanding the random fractal sets and relate their basic parameters with the inter-correlations of two random sets compared.

4. The simplified DGIs can help in finding additional relationships between three basic parameters $(I_4, \varepsilon/3Q, B)$ with respect to temperature T, which are in optical measurements described above, it was used as the predominant input factor. The DGI approach can take into account the temperature fluctuations and confirm independently some specific peculiarities found by other researches [15].

6 Mathematical Appendix

Expression for the invariant of the fourth order in the case when two sets are close to each other.

In this Appendix, we want to obtain an approximate expression for the general invariant (2.22) when the first set x_k is distorted by the function f_k and the second set y_k is expressed in the form $\Delta y_k = \Delta x_k \pm \Delta f_k$ $(\Delta f_k = f_k - \langle f \rangle)$. The evaluation of this curve is not trivial because we should open the limit 0/0 that appears in calculations of the parameters $\sigma_{x,y}$ in expressions (2.21). We take into account that deviations of the factor Δf_k in the both sides relates its mean value equal to zero and therefore the average value $\pm \langle A \Delta f_k \rangle \approx 0$ (where A represents some value). However, the combination $(\pm \langle A \Delta f_k \rangle) \cdot (\pm \langle B \Delta f_k \rangle) = (\langle A \Delta f_k \rangle) \cdot (\langle B \Delta f_k \rangle) \neq 0$ because the positive and negative compensations in this product do not take place. Taking into account this remark, one can evaluate the expressions for Δ and $\sigma_{x,y}$ in expressions (2.21). After some cumbersome and long calculations, one can obtain

$$\sigma_x = \frac{4}{\Delta} \left[3\left\langle \Delta x (\Delta f)^2 \right\rangle^2 - \left\langle (\Delta x)^2 \Delta f \right\rangle \cdot \left\langle (\Delta f)^3 \right\rangle + \left\langle (\Delta f)^3 \right\rangle^2 \right],$$

$$\sigma_y = \frac{4}{\Delta} \left[3\left\langle \Delta x (\Delta f)^2 \right\rangle^2 - \left\langle (\Delta x)^2 \Delta f \right\rangle \cdot \left\langle (\Delta f)^3 \right\rangle \right], \qquad (6.1)$$

$$\Delta \cong 9\left\langle \Delta x (\Delta f)^2 \right\rangle^2 - 3\left\langle (\Delta x)^2 \Delta f \right\rangle \cdot \left\langle (\Delta f)^2 \right\rangle + \left\langle (\Delta f)^3 \right\rangle^2.$$

As one can notice from (6.1) we kept the values proportional to $O((\Delta f)^4, (\Delta f)^6)$, inclusively. If we introduce the values:

$$Q_0 = 3\left\langle \Delta x (\Delta f)^2 \right\rangle^2 - \left\langle (\Delta x)^2 \Delta f \right\rangle \left\langle (\Delta f)^3 \right\rangle,$$

$$\varepsilon = \left\langle (\Delta f)^3 \right\rangle^2,$$

(6.2)

then the parameters $\sigma_{x,y}$, $\theta_{x,y}$ are expressed in the compact form:

$$\sigma_x \cong \frac{4}{3} + \frac{8\varepsilon}{9Q_0}, \quad \sigma_y \cong \frac{4}{3} - \frac{4\varepsilon}{9Q_0}$$

$$\theta_x \cong -\frac{1}{3} - \frac{5\varepsilon}{9Q_0}, \quad \theta_y \cong -\frac{1}{3} + \frac{\varepsilon}{9Q_0}.$$

(6.3)

The second terms in (6.3) are considered as corrections and have the order $O(\Delta f)^2$. Based on (6.3), one can evaluate other parameters keeping the same accuracy:

$$A_x = A_y \cong -2\left\langle (\Delta f)^2 \right\rangle - \frac{2}{3}\frac{\varepsilon}{Q_0}\left\langle (\Delta x)^2 \right\rangle, \quad B = -2A_x$$

$$I_4 = \frac{2}{3}\frac{\varepsilon}{Q_0}\left\langle (\Delta x)^2 (\Delta f)^2 \right\rangle.$$

(6.4)

As one can notice from expressions (6.3) and (6.4), the simplified invariant curve contains four parameters: $\langle (\Delta x)^2 \rangle$, $\langle (\Delta f)^2 \rangle$, $\langle (\Delta x)^2 (\Delta f)^2 \rangle$, ε / Q_0.

Finally, the invariant curve of the fourth order takes the following form:

$$x = \left\langle x \right\rangle + X(\varphi), \quad X(\varphi) = r(\varphi)\cos(\varphi),$$

$$y = \left\langle y \right\rangle + \left\langle f \right\rangle + Y(\varphi), \quad Y(\varphi) = r(\varphi)\sin(\varphi),$$

$$r(\varphi) = \left[\frac{\sqrt{P_2^2(\varphi) + 4I_4 P_4(\varphi)} - P_2(\varphi)}{2P_4(\varphi)} \right]^{1/2},$$

$$P_2(\varphi) = 3B\left(\cos(\varphi) - \sin(\varphi) \right)^2,$$

$$P_4(\varphi) = \left(\cos(\varphi) - \sin(\varphi) \right)^4 + \frac{\varepsilon}{3Q_0}\left[5\left(\cos(\varphi) \right)^4 + 8\left(\cos(\varphi) \right)^3 \sin(\varphi) \right] +$$

$$+ \frac{\varepsilon}{3Q_0}\left[4\left(\sin(\varphi) \right)^3 \cos(\varphi) - \left(\sin(\varphi) \right)^4 \right].$$

(6.5)

As it follows from these expressions, when two sets coincide with each other they are reduced to the point $x = y = \langle x \rangle$. Finally, it is necessary to demonstrate the criterion for application of the simplified invariant:

$$R = \left[\frac{stdev(X - Y)}{stdev(Y)} \right] < 1. \tag{6.6}$$

The functions X and Y are defined by expressions (2.16) and (6.5) and the operation "$stdev$" is defined by the conventional expression:

$$stdev(F) = \left[\frac{1}{2} \sum_{j=1}^{N} (F_j - \langle F \rangle)^2 \right]^{1/2}. \tag{6.7}$$

The model examples considered in section (3.2) satisfy to criterion (6.6).

References

1. Mandelbrot, B.B.: The Fractal Geometry of Nature. Freeman and Company, San Francisco (1983)
2. Feder, J.: Fractals. Plenum Press, Ny and London (1988)
3. Samko, S.B., Kilbas, A.A., Marichev, I.: Fractional Integrals and Derivatives—Theory and Applications. Gordon and Breach, New York (1993)
4. Uchaikin, V.V. : Method of the Fractional Derivatives. Artishock Publishing House, Ulyanovsk (2008)
5. Babenko, Y.I.: Power Relations in a Circumference and a Sphere. Norell Press Inc., USA (1997)
6. Babenko, Yu.I.: The power law invariants od the point sets, Professional, S-Petersburg, ISBN 978-5-91259-095-5, www. naukaspb.ru, (Russian Federation) (2014)
7. Nigmatullin, R.R., Budnikov, H.C., Sidelnikov, A.V., Maksyutova, E.I.: Application of the discrete geometrical invariants to the quantitative monitoring of the electrochemical background, Res. J. Math. Comput. Sci. (RJMCS). eSciPub LLC, Houston, TX USA. Website: http://escipub.com/
8. Berry, M.V., Lewis, Z.V.: On the weierstrass mandelbrot fractal function. Proc. R. Soc. London A **370**, 459–484 (1980)
9. Butler, J.M., Johnson, J.E., Boone, W.R.: The heat is on: room temperature affects laboratory equipment–an observational study. J. Assis.t Reprod Genet., 1389–1393. Published online 2013 Aug 7. https://doi.org/10.1007/s10815-013-0064-4
10. Hoffman, G.R., Birtwistle, J.K.: Factors affecting the performance of a thin film magnetoresistive vector magnetometer. J. Appl. Phys. **53**, 8266 (1982). https://doi.org/10.1063/1.330303
11. Mostafa Mohamed Abd El-Raheem, Hoda Hamid Al-Ofi, Abdullah Alhuthali, Ateyyah Moshrif AL-Baradi, Effect of preparation condition on the optical properties of transparent conducting oxide based on zinc oxide. Optics. **4**(3), 17–24 (2015). https://doi.org/10.11648/j.optics.20150403.11
12. Saleem, M., Ahmad, N., Ali, H., Bilal, M., Khan, S., Ullah, R., Ahmed, M., Mahmood, S.: Investigating temperature effects on extra virgin olive oil using fluorescence spectroscopy. IOP Publishing: Laser Phys. 27 (2017), pp. 1–10. https://doi.org/10.1088/1555-6611/aa8cd7
13. Giuffrè, A.M., Zappia, C., Capocasale, M.: Effects of High Temperatures and Duration Of Heating on Olive Oil Properties for Food use and Biodiesel Production. Springer: Journal of the American Oil Chemists' Society, June 2017, Volume 94, Issue 6, pp. 819–830 (2017)
14. Clodoveo, M.L., Delcuratolo, D., Gomes, T., Colelli, G.: Effect of different temperatures and storage atmospheres on Coratina olive oil quality, Elsevier. Food Chemistry **102**(3), 571–576 (2007)

15. Bodurov, I., Vlaeva, I., Marudova, M., Yovcheva, T., Nikolova, K., Eftimov, T., Plachkova, V.: Detection of adulteration in olive oils using optical and thermal methods. Bulg. Chem. Commun. **45**(Special Issue B), 81–85 (2013)
16. Nigmatullin, R.R., Ceglie, C., Maione, G., Striccoli, D.: Reduced fractional modeling of 3D video streams: the FERMA approach. Nonlinear Dyn. (2014). https://doi.org/10.1007/s11071-014-1792-4
17. Nigmatullin, R.R.: New Noninvasive Methods for "Reading" of Random Sequences and Their Applications in Nanotechnology. Springer: New Trends in Nanotechnology and Fractional Calculus Applications, pp. 43–56

Nonlocal Conditions for Semi-linear Fractional Differential Equations with Hilfer Derivative

Benaouda Hedia

Abstract This paper studies the existence of solutions for nonlocal semi-linear fractional differential equations of Hilfer type in Banach space by using the non-compact measure method in the weighted space of continuous functions. The main result is illustrated with the aid of an example.

Keywords Semi-linear differential equations · Nonlocal initial value problems · Hilfer fractional derivative · Fixed point theorems · Measure of non-compactness · Condensing map

AMS (MOS) Subject Classifications: 26A33 · 34K37 · 37L05 · 34B10.

1 Introduction

Differential equations of fractional order have recently proved to b e valuable tools in the modeling of many physical phenomena [8]. There has been a significant theoretical development in fractional differential equations in recent years; see the monographs of Kilbas et al. [16], Zhou [19, 20]. In [13], Hilfer proposed a generalized Riemann–Liouville fractional derivative, for short, Hilfer fractional derivative, which is an interpolator between Riemann–Liouville and Caputo fractional derivatives. This operator appeared in the theoretical simulation of dielectric relaxation in glass-forming materials [14].

Recently, considerable attention has been given to the existence of solution of initial and boundary value problems for fractional and semi-linear-fractional differential equations and inclusions involving Hilfer fractional derivative, see [12]. On the other hand, when an existence result is proved for the fractional Cauchy problem where the solutions are not unique, it is natural to discuss the topological structure of the solution set [9, 10].

B. Hedia (✉)
Loboratory of Mathematics and Informatics, University Ibn Khaldoun of Tiaret, PO BOX 78, 14000 Tiaret, Algeria
e-mail: nilpot_hedia@yahoo.fr

© Springer Nature Singapore Pte Ltd. 2019
P. Agarwal et al. (eds.), *Fractional Calculus*, Springer Proceedings in Mathematics & Statistics 303, https://doi.org/10.1007/978-981-15-0430-3_5

Motivated by the papers cited above, in this paper, we consider a class of nonlocal initial semi-linear fractional differential equation of Hilfer type described by the form

$$D_{0^+}^{\alpha,\beta} x(t) = Ax(t) + f(t, x(t)), \quad t \in (0, b], \tag{1.1}$$

$$I_{0^+}^{1-\gamma} x(t) = \sum_{i=1}^{m} \lambda_i x(\tau_i), \ \alpha \le \gamma = \alpha + \beta - \alpha\beta, \ \tau_i \in (0, b], \tag{1.2}$$

where the two-parameter family of fractional derivative $D^{\alpha,\beta}$ denote the left-sided Hilfer fractional derivative introduced in [13, 14], $0 < \alpha \le 1, 0 \le \beta \le 1$. The state $x(.)$ takes value in a Banach space E with norm $\|.\|$, A is the infinitesimal generator of semigroup of bounded linear operators (i.e., C_0 semigroup) $T(t)_{t \ge 0}$ that will be specified later in Banach space E. The operator $I_{0^+}^{1-\gamma}$ denotes the left-sided Riemann–Liouville fractional integral, $f : (0, b] \times E \to E$ will be specified in later sections. $\tau_i, i = 1, 2, \ldots, m$ are prefixed points satisfying $0 < \tau_1 \le \cdots \le \tau_m < b$ and $\Gamma(\gamma) \ne \sum_{i=1}^{m} \lambda_i \tau_i$ where $\Gamma(\gamma) = \int_0^{+\infty} x^{1-\gamma} e^{-x} dx$.

Physically, condition (1.2) says that some initial measurements were made at the times 0 and $\tau_i, i = 1, \ldots, m$, and the observer uses this previous information in their model. This type of situation can lead us to a better description of the phenomenon. For example, [6], Deng considers the phenomenon of diffusion of a small amount of gas in a tube and assumes that the diffusion is observed via the surface of the tube. The nonlocal condition allows additional measurement which is more precise than the measurement just at $t = 0$.

Our main aim in this work is to extend the result given in [18], by using a fixed point principle for condensing maps combined with Browder–Gupta approach [4] in a general setting, namely when the function right-hand side has values in infinite-dimensional Banach space.

This paper is organized in the following way. In Sect. 2, we give some general results and preliminaries and in Sect. 3 we present our main results.

I wish you the best of success.

Tiaret

February 14, 2018

2 Preliminary Results

In this section, we introduce some notation and technical results which are used throughout this paper [5].

Let $J := [0, b]$, $b > 0$ and $(E, \| \cdot \|)$ be a Banach space. $C(J, E)$ be the space of E-valued continuous functions on J endowed with the uniform norm topology

$$\|x\|_\infty = \sup\{\|x(t)\|, \ t \in J\}.$$

$L^1(J, E)$ the space of E-valued Bochner integrable functions on J with the norm

$$\|f\|_{L^1} = \int_0^b \|f(t)\| dt.$$

We consider the Banach space of continuous functions

$$C_{1-\gamma}([0, b], E) = \{x \in C((0, b], E) : \lim_{t \to 0^+} t^{1-\gamma} x(t) < +\infty\}.$$

A norm in this space is given by

$$\|x\|_\gamma = \sup_{t \in [0,b]} t^{1-\gamma} \|x(t)\|.$$

Obviously, $C_{1-\gamma}([0, b], E)$ is a Banach space. For Ω a subset of the space $C_{1-\gamma}$ $([0, b], E)$, define Ω_γ by

$$\Omega_\gamma = \{x_\gamma : x \in \Omega\}, \tag{2.1}$$

where

$$x_\gamma = \begin{cases} t^{1-\gamma} x(t), & \text{if } t \in (0, b]; \\ \lim_{t \to 0^+} t^{1-\gamma} x(t), & \text{if } t = 0. \end{cases} \tag{2.2}$$

It is clear that $x_\gamma \in C(J, E)$. We note the following Ascoli-Arzelà-type criteria.

Lemma 1 *A set $\Omega \subset C_{1-\gamma}([0, b], E)$ is relatively compact if and only if Ω_γ is relatively compact in $C([0, b], E)$.*

Proof See, for instance, [1].

Definition 1 Let X and Y be two topological vector spaces. We denote by $\mathcal{P}(Y)$ the family of all nonempty subsets of Y and by

$$\mathcal{P}_k(Y) = \{C \in \mathcal{P}(Y) : \textbf{compact}\},$$

$$\mathcal{P}_b(Y) = \{C \in \mathcal{P}(Y) : \textbf{bounded}\}.$$

Let $G : [0, b] \to \mathcal{P}(E)$ be a multifunction. It is called

(i) integrable, if it admits a Bochner integrable selection $g : [0, b] \to E$, $g(t) \in G(t)$ for a.e. $t \in [0, b]$;

(ii) integrably bounded, if there exists a function $\zeta \in L^1([0, b]; R_+)$ such that

$$\|G(t)\| := \sup\{\|g\| : g \in G(t)\} \le \zeta(t) \quad \text{a.e. } t \in [0, b].$$

We give some concepts of fractional calculus. Let $0 < \alpha < 1$. A function $x : J \to E$ has a fractional integral if the following integral:

$$I^\alpha x(t) = \frac{1}{\Gamma(\alpha)} \int_0^t (t-s)^{\alpha-1} x(s)ds$$

is defined for $t \geq 0$. The Riemann–Liouville fractional derivative of x of order α is defined as

$$D^\alpha x(t) = \frac{1}{\Gamma(1-\alpha)} \frac{d}{dt} \left(\int_0^t (t-s)^{-\alpha} x(s)ds \right) = \frac{d}{dt} I^{1-\alpha} x(t),$$

where $\Gamma(\cdot)$ is the Gamma function, provided it is well defined for $t \geq 0$. The previous integral is taken in Bochner sense.

The left-sided Hilfer fractional derivative of order $0 < \alpha \leq 1$ and $0 \leq \beta \leq 1$ is defined by

$$D_{0^+}^{\alpha,\beta} x(t) = \left(I^{\beta(1-\alpha)} \frac{d}{dt} \left(I^{(1-\beta)(1-\alpha)} x \right) \right)(t).$$

for functions such that the expression on the right-hand side exists.

Lemma 2 ([7]) *Let $\alpha, \beta \in R_+$. Then*

$$\int_0^1 t^{\alpha-1}(1-t)^{\beta-1}dt = \frac{\Gamma(\alpha)\Gamma(\beta)}{\Gamma(\alpha+\beta)},$$

and hence

$$\int_0^x t^{\alpha-1}(x-t)^{\beta-1}dt = x^{\alpha+\beta-1} \frac{\Gamma(\alpha)\Gamma(\beta)}{\Gamma(\alpha+\beta)}.$$

The integral in the first equation of Lemma 2 is known as Beta function $B(\alpha, \beta)$. Let us recall the following definitions and results that will be used in the sequel.

Definition 2 Let E be a real Banach space and $(Y, \leq)$ a partially ordered set. A function $\beta : \mathcal{P}(E) \to Y$ is called a measure of non-compactness in E if

$$\beta(\Omega) = \beta(\overline{co}\Omega)$$

for every $\Omega \subset \mathcal{P}(E)$, where $\overline{co}\Omega$ denotes the closed convex hull of Ω.

Definition 3 ([15, 17]) A measure of non-compactness β is called

(i) monotone if $\Omega_0, \Omega_1 \in \mathcal{P}(E)$, $\Omega_0 \subset \Omega_1$ implies $\beta(\Omega_0) \leq \beta(\Omega_1)$;
(ii) nonsingular if $\beta(\{a\} \cup \Omega) = \beta(\Omega)$ for every $a \in E$, $\Omega \in \mathcal{P}(E)$;
(iii) invariant with respect to union with compact sets, if $\beta(\{K\} \cup \Omega) = \beta(\Omega)$ for every $K \in \mathcal{P}_k(E)$ and $\Omega \in \mathcal{P}(E)$;

If Y is a cone in a normed space, we say that the MNC is

(iv) regular if $\beta(\Omega) = 0$ is equivalent to the relative compactness of Ω;

(v) algebraically semi-additive, if $\beta(\Omega_0 + \Omega_1) \leq \beta(\Omega_0) + \beta(\Omega_1)$ for each $\Omega_0, \Omega_1 \in \mathcal{P}(E)$.

One of most important examples of a measure of non-compactness possessing all these properties is the Hausdorff measure of non-compactness defined by

$$\chi(\Omega) = \inf\{\varepsilon > 0 : \Omega \text{ has a finite } \varepsilon - \text{net}\}.$$

Definition 4 A continuous map $F : X \subset E \to E$ is said to be condensing with respect to a MNC β (β-condensing) if for every bounded set $\Omega \subset X$, that is,

$$\beta(F(\Omega)) < \beta(\Omega),$$

we have Ω is relatively compact.

Lemma 3 ([3, 15]) *If* $\{u_n\}_{n=1}^{+\infty} \subset L^1(J, E)$ *satisfies* $\|u_n(t)\| \leq \kappa(t)$ *a.e. on* J *for all* $n \geq 1$ *with some* $\kappa \in L^1(J, R_+)$. *Then the function* $\chi(\{u_n(t)\}_{n=1}^{+\infty})$ *belongs to* $L^1(J, R_+)$ *and*

$$\chi\left(\left\{\int_0^t u_n(s)ds : n \geq 1\right\}\right) \leq 2 \int_0^t \chi(u_n(s)ds : n \geq 1)ds. \tag{2.3}$$

The application of the topological degree theory for condensing maps implies the following fixed point principle.

Theorem 1 ([2, 15]) *Let* $V \subset E$ *be a bounded open neighborhood of zero and* $\Gamma : \overline{V} \to E$ *a* β-condensing map with respect to a monotone nonsingular MNC β in E. *If* Γ *satisfies the boundary condition*

$$x \neq \lambda\Gamma(x)$$

for all $x \in \partial V$ *and* $0 < \lambda \leq 1$, *then the fixed point set* $\mathfrak{F}ix\Gamma = \{x : x = \Gamma(x)\}$ *is nonempty and compact.*

3 Main Result

Definition 5 The Wright function $M_q(\theta)$ defined by

$$M_q(\theta) = \sum_{n=1}^{\infty} \frac{(-\theta)^{n-1}}{(n-1)!\Gamma(1-qn)}$$

is such that

$$\int_0^\infty \theta^\delta M_q(\theta)d\theta = \frac{\Gamma(1+\delta)}{\Gamma(1+q\delta)}, \quad \text{for } \delta \geq 0.$$

Define the operators $\mathcal{K}_\alpha$, $\mathcal{S}_{\alpha,\beta}$

$$\mathcal{K}_\alpha(t) = t^{\alpha-1} P_\alpha(t), \quad P_\alpha(t) = \int_0^\infty \alpha\theta M_\beta(\theta) T(t^\alpha\theta) d\theta,$$

$$\mathcal{S}_{\alpha,\beta} = I_{0^+}^{\beta(1-\alpha)} \mathcal{K}_\alpha(t).$$

The properties of these operators were explored by Zhou [19, 20]. Suppose that there exists the bounded operator $B : E \to E$ given by

$$B = \left[I - \sum_{i=1}^m \lambda_i \mathcal{S}_{\alpha,\beta}(\tau_i) \right]^{-1}. \tag{3.1}$$

Lemma 4 *The operator B defined in (3.1) exists and is bounded if one of the following two conditions holds:*

(i) The reals numbers λ_i satisfies $M \sum_{i=1}^m |\lambda_i| < 1$.

(ii) $T(t)$ is compact for each $t > 0$ and the homogeneous linear nonlocal problem

$$D_{0^+}^{\alpha,\beta} x(t) = Ax(t), \quad t \in (0, b], \ 1 < \alpha \le 1, \ 0 < \beta \le 1$$

$$I_{0^+}^{1-\gamma} x(t) = \sum_{i=1}^m \lambda_i x(\tau_i), \ \alpha \le \gamma = \alpha + \beta - \alpha\beta, \ \tau_i \in (0, b]$$

has no nontrivial mild solutions.

Definition 6 A function $x \in C_\gamma([0, b], E)$ is called mild solution of the problem (1.1)–(1.2), if it satisfies the following equation $D_{0^+}^{\alpha,\beta} x(t) = Ax(t) + f(t, x(t))$, $t \in (0, b]$ and the condition (1.2).

Lemma 5 (See (5) Theorem 2.3) *Let $f(., u(.)) \in C_{1-\gamma}([a, b])$ for any $u \in C_{1-\gamma}$ $[a, b]$. A function $u \in C_{1-\gamma}[a, b]$ is solution of the fractional initial value problem*

$$\begin{cases} D^{\alpha,\beta} x(t) = f(t, u(t)), \ 0 < \alpha \le 1, \ 0 < \beta \le 1, \\ I_{a^+}^{1-\gamma} = u_a, \qquad\qquad \gamma = \alpha + \beta - \alpha\beta. \end{cases}$$

If and only if u satisfies the the following Volterra integral equation.

$$x(t) = \frac{t^{\gamma-1} u_a}{\Gamma(\gamma)} + \frac{1}{\Gamma(\alpha)} \int_0^t (t-s)^{\alpha-1} (Ax(x) + f(s, x(s))) ds. \tag{3.2}$$

According to the Lemma (4) and (5) we have the following lemma which will be useful in the sequel:

Lemma 6 *Let h be a continuous function, x is solution for the fractional integral equation*

$$x(t) = \mathcal{S}_{\alpha,\beta} |T| \sum_{i=1}^{m} \lambda_i B(g(\tau_i)) + g(t) \text{ if and only if}$$

$$D_{0+}^{\alpha,\beta} x(t) = Ax(t) + h(t), \quad t \in (0, b],$$

$$I_{0+}^{1-\gamma} x(t) = \sum_{i=1}^{m} \lambda_i x(\tau_i), \ \alpha \leq \gamma = \alpha + \beta - \alpha\beta, \ \tau_i \in (0, b],$$

where

$$g(\tau_i) = \int_0^{\tau_i} \mathcal{K}_\beta(\tau_i - s) h(s) ds,$$

$$g(t) = \int_0^t \mathcal{K}_\beta(t - s) h(s) ds$$

$$T := \frac{1}{\Gamma(\gamma) - \sum_{i=1}^{m} \lambda_i \tau_i}.$$

We prove an Aronszajn-type result for this problem. We need to make the following assumptions:

(H1) $T(t)$ is continuous in the uniform operator topology for $t > 0$, and $\{T(t)\}_{t \geq 0}$ of is uniformly bounded, i.e., there exists $M > 1$ such that $\sup\limits_{t \in [0,+\infty)} |T(t)| < M$, $M \sum_{i=1}^{m} < 1$.

(H2) The map $f : [0, b] \times E \to E$ is continuous.

(H3) There exists a function $p \in C([0, b], \mathbb{R}_+)$ such that

$$\|f(t, x)\| \leq p(t)(1 + t^{1-\gamma} \|x\|), \quad \text{for all } t \in [0, b] \text{ and } x \in E.$$

(H4) There exists a constant $c > 0$ such that for each nonempty, bounded set $\Omega \subset C_{1-\gamma}([0, b], E)$

$$\chi(f(t, \Omega)) \leq c\chi(\Omega(t)), \quad \text{for all } t \in [0, b],$$

where χ is the Hausdorff measure of non-compactness in E.

(H5) There is a constant $\overline{M} > 0$ such that

$$\frac{\overline{M}}{\left(\dfrac{M\|p\|_\infty b^{1-\gamma+\beta}}{\Gamma(\beta+1)}\right)\left(\dfrac{M\|B\|T\displaystyle\sum_{i=1}^{m}\lambda_i}{\Gamma(\gamma)}+1\right)(1+\overline{M})} > 1. \qquad (3.3)$$

To prove the existence of solutions to (1.1)–(1.2), we need the following auxiliary lemmas.

Lemma 7 ([11]) *Under assumption $(H1)$, $P_\beta(t)$ is continuous in the uniform operator topology for $t > 0$.*

Lemma 8 ([11]) *Under assumption (H_1), for any fixed $t > 0$, $\{\mathcal{K}_\beta(t)\}_{t>0}$ and $\{\mathcal{S}_{\alpha,\beta}(t)\}_{t>0}$, are linear operators, and for any $x \in X$*

$$\|\mathcal{K}_\beta(t)x\| \le \frac{Mt^{\beta-1}}{\Gamma(\beta)}\|x\|, \quad \|\mathcal{S}_{\alpha,\beta}(t)x\| \le \frac{Mt^{(\beta-1)(\alpha-1)}}{\Gamma(\alpha(1-\beta)+\beta)}\|x\|$$

Lemma 9 ([11]) *Under assumption (H_1), $\{\mathcal{K}_\beta(t)\}_{t>0}$, and $\{\mathcal{S}_{\alpha,\beta}(t)\}_{t>0}$ are strongly continuous, which means that for any $x \in X$ and $0 < t' < t'' \le b$ we have*

$$\|\mathcal{K}_\beta(t')x - \mathcal{K}_\beta(t'')x\| \to 0, \quad \|\mathcal{S}_{\alpha,\beta}(t')x - \mathcal{S}_{\alpha,\beta}(t'')x\| \to 0,$$

as $t', t'' \to 0$,

Theorem 2 *Assume that (H1)–(H5) are satisfied. Then the set $S(f, \{\tau_i\}_{i=1}^n)$ is nonempty and compact.*

We transform the problem (1.1)–(1.2) into a fixed point problem. Consider the operator $N : C_{1-\gamma}([0, b], E) \to C_{1-\gamma}([0, b], E)$ defined by

$$N(x)(t) = \mathcal{S}_{\alpha,\beta}(t)\left(T\sum_{i=1}^{n}\lambda_i B\left[\int_0^{\tau_i}\mathcal{K}_\beta(\tau_i - s)f(s, y(s))ds\right]\right)$$
$$+ \int_0^t \mathcal{K}_\beta(t - s)f(s, y(s))ds.$$

Clearly, from Lemma 1.12 [18], the operator N is well defined and the fixed points of N are solutions to 1.1–1.2. Thus $\mathcal{F}ixN = S(f, \{\tau_i\}_{i=1}^n)$. Next, we subdivide the operator N into two operators P and Q as follows:

$$(Px)(t) = \mathcal{S}_{\alpha,\beta}(t)\left(T\sum_{i=1}^{n}\lambda_i B\left[\int_0^{\tau_i}\mathcal{K}_\beta(\tau_i - s)f(s, y(s))ds\right]\right).$$

and

$$(Qx)(t) = \int_0^t \mathcal{K}_\beta(t-s) f(s, y(s)) ds.$$

Now, we show that $S(f, \{\tau_i\}_{i=1}^n) \neq \emptyset$, the proof is devised into several steps.
Step 1. P is continuous.

Let $\{x_n\}$ be a sequence such that $x_n \to x$ in $C_{1-\alpha}([0, b], E)$. Then

$$t^{1-\gamma} \| P(x_n)(t) - P(x)(t) \|$$

$$\leq \frac{M|T|}{\Gamma(\gamma)}$$

$$\sum_{i=1}^m \lambda_i B \left[\int_0^{\tau_i} \| \mathcal{K}_\beta(\tau_i - s) \| \| f(s, x_n(s)) - f(s, x(s)) \| ds \right]$$

$$\leq \frac{M|T| \|B\|}{\Gamma(\beta)\Gamma(\gamma)}$$

$$\sum_{i=1}^m \lambda_i \int_0^{\tau_i} |(\tau_i - s)^{\beta-1} s^{\gamma-1} \| f(\cdot, x_n(\cdot)) - f(\cdot, x(\cdot)) \|_\gamma ds$$

$$\leq \frac{M|T| \|B\|}{\Gamma(\beta)\Gamma(\gamma)}$$

$$\sum_{i=1}^m \lambda_i \tau_i^{\gamma+\beta-1} B(\gamma, \beta) \| f(\cdot, x_n(\cdot)) - f(\cdot, x(\cdot)) \|_\gamma$$

$$\leq \left[\frac{M|T| \|B\| \sum_{i=1}^m \lambda_i \tau_i^{\gamma+\beta-1}}{\Gamma(\gamma)\Gamma(\beta)} \right]$$

$$B(\gamma, \beta) \| f(\cdot, x_n(\cdot)) - f(\cdot, x(\cdot)) \|_\gamma$$

and

$$t^{1-\gamma} \| Q(x_n)(t) - Q(x)(t) \| \tag{3.4}$$

$$\leq t^{1-\gamma} \int_0^t \| \mathcal{K}_\beta(t-s) \| \| f(s, x_n(s)) - f(s, x(s)) \| ds \tag{3.5}$$

$$\leq \frac{M t^{1-\gamma}}{\Gamma(\beta)} \int_0^t (t-s)^{\beta-1} s^{\gamma-1} s^{1-\gamma} \| f(s, x_n(s)) - f(s, x(s)) \| ds \tag{3.6}$$

$$\leq \frac{M t^{1-\gamma}}{\Gamma(\beta)} \int_0^t (t-s)^{\beta-1} s^{\gamma-1} \| f(\cdot, x_n(\cdot)) - f(\cdot, x(\cdot)) \|_\gamma ds \tag{3.7}$$

$$\leq \frac{M b^\beta}{\Gamma(\beta)} B(\gamma, \beta) \| f(\cdot, x_n(\cdot)) - f(\cdot, x(\cdot)) \|_\gamma ds. \tag{3.8}$$

Hence

$$\|N(x_n) - N(x)\|_\alpha \leq \left[\frac{\|B\|\|T\|}{\Gamma(\gamma)} \sum_{i=1}^m \lambda_i b^\beta + b^\beta \right]$$

$$\times \frac{M \ B(\gamma, \beta)}{\Gamma(\beta)} \|f(\cdot, x_n(\cdot)) - f(\cdot, x(\cdot))\|_\gamma.$$

Using the hypothesis (H_2), we have

$$\|N(x_n) - N(x)\|_\alpha \to 0, \quad \text{as } n \to +\infty.$$

Step 2. N maps bounded sets into bounded sets in $C_{1-\gamma}([0, b], E)$.
Indeed, it is enough to show that there exists a positive constant ℓ such that for each $x \in B_\eta = \{x \in C_{1-\alpha}([0, b], E) \ : \ \|x\|_\gamma \leq \eta\}$ one has $\|N(x)\|_\gamma \leq \ell$.
Let $x \in B_\eta$. Then for each $t \in (0, b]$, by (H_3) we have

$$t^{1-\gamma} \|Nx(t)\| \leq$$

$$\left(\frac{M \ T}{\Gamma(\gamma)} \sum_{i=1}^m \lambda_i B \left[\int_0^{\tau_i} \|\mathcal{K}_\beta(\tau_i - s)\| \|f(s, x(s))\| ds \right) \right]$$

$$+ t^{1-\gamma} \int_0^t |\mathcal{K}_\beta(t - s)| \|f(s, x(s))\| ds$$

$$\leq \frac{\|B\|T \ M^2 \|p\|_\infty \sum_{i=1}^m \lambda_i}{\Gamma(\beta)\Gamma(\gamma)} \int_0^{\tau_i} (\tau_i - s)^{\beta-1}(1 + s^{1-\gamma}\|x(s)\|)ds$$

$$+ \frac{\|p\|_\infty M \ t^{1-\gamma}}{\Gamma(\beta)} \int_0^t (t - s)^{\beta-1}(1 + s^{1-\gamma}\|x(s)\|)ds$$

$$\leq \frac{\|B\|\|p\|_\infty T \ M^2(1 + \|x\|_\gamma) \sum_{i=1}^m \lambda_i \tau_i^\beta}{\Gamma(\beta + 1)\Gamma(\gamma)}$$

$$+ \frac{\|p\|_\infty M b^{1+\beta-\gamma}(1 + \|x\|_\gamma)}{\Gamma(\beta + 1)}$$

$$\leq \frac{(1 + \eta)\|p\|_\infty}{\Gamma(\beta + 1)} \left[\frac{\|B\|T \ M^2 \sum_{i=1}^m \lambda_i \tau_i^\beta}{\Gamma(\gamma)} + M b^{1-\gamma+\beta} \right] := \ell.$$

Step 3. N maps bounded sets into equicontinuous sets.
First, we prove $\{Px, \ x \in B_\eta\}$ is equicontinuous. Let $t_1, t_2 \in (0, b]$, $t_1 \leq t_2$, let B_η be a bounded set in $C_{1-\gamma}([0, b], E)$ as in Step 2, and let $x \in B_\eta$, we have

$$\|t_2^{1-\gamma} Px(t_2) - t_1^{1-\gamma} Px(t_1))\|$$

$$\leq \left\| t_2^{1-\gamma} \mathcal{S}_{\alpha,\beta}(t_2) \left(T \sum_{i=1}^{n} \lambda_i B \left[\int_0^{\tau_i} \mathcal{K}_\beta(\tau_i - s) f(s, y(s)) ds \right] \right) \right.$$

$$\left. - t_1^{1-\gamma} \mathcal{S}_{\alpha,\beta}(t_1) \left(T \sum_{i=1}^{n} \lambda_i B \left[\int_0^{\tau_i} K_\alpha(\tau_i - s) f(s, y(s)) ds \right] \right) \right\|$$

$$+ \left\| t_2^{1-\gamma} \mathcal{S}_{\alpha,\beta}(t_2) \left(T \sum_{i=1}^{n} \lambda_i B \left[\int_0^{\tau_i} K_\alpha(\tau_i - s) f(s, y(s)) ds \right] \right) \right.$$

$$\left. - t_1^{1-\gamma} \mathcal{S}_{\alpha,\beta}(t_1) \left(T \sum_{i=1}^{n} \lambda_i B \left[\int_0^{\tau_i} K_\alpha(\tau_i - s) f(s, y(s)) ds \right] \right) \right\|.$$

from the fact that $t_1^{1-\gamma} \mathcal{S}_{\alpha,\beta}(t)$ is uniformly continuous on J, we deduce then $\{Px, \ x \in B_\eta\}$ is equicontinuous. Using condition (H_2), one has

$$\|t_2^{1-\gamma} Qx(t_2) - t_1^{1-\gamma} Qx(t_1)|$$

$$\leq \left\| \int_{t_2}^{t_1} t_2^{1-\gamma}(t_2 - s)^{\beta-1} P_\beta(t_2 - s) f(s, x(s)) ds \right\|$$

$$+ \left\| \int_0^{t_1} t_2^{1-\gamma}(t_2 - s)^{\beta-1} P_\beta(t_2 - s) f(s, x(s)) ds \right.$$

$$- \int_0^{t_1} t_1^{1-\gamma}(t_1 - s)^{\beta-1} P_\beta(t_2 - s) f(s, x(s)) ds \Big\|$$

$$+ \Big| \int_0^{t_1} t_1^{1-\gamma}(t_1 - s)^{\beta-1} P_\beta(t_2 - s) f(s, x(s)) ds$$

$$- \int_0^{t_1} t_1^{1-\gamma}(t_1 - s)^{\beta-1} P_\beta(t_1 - s) f(s, x(s)) ds \Big\|$$

$$\leq \frac{M\|p\|_\infty (1 + \|x\|_\gamma)}{\Gamma(\beta)} \Big| \int_{t_2}^{t_1} t_2^{1-\gamma}(t_2 - s)^{\beta-1} ds \Big\| +$$

$$\frac{M\|p\|_\infty (1 + \|x\|_\gamma)}{\Gamma(\beta)}$$

$$\int_0^{t_1} \left[t_1^{1-\gamma}(t_1 - s)^{\beta-1} - t_2^{1-\gamma}(t_2 - s)^{\beta-1} \right] ds$$

$$+ \|p\|_\infty (1 + \|x\|_\gamma) |$$

$$\int_0^{t_1} t_1^{1-\gamma}(t_1 - s)^{\beta-1} \left[P_\beta(t_2 - s) - P_\beta(t_1 - s) \right] ds \|$$

$$I_1 + I_2 + I_3,$$

where

$$I_1 = \frac{M\|p\|_\infty(1 + \|x\|_\gamma)}{\Gamma(\beta)} \| \int_{t_2}^{t_1} t_2^{1-\gamma}(t_2 - s)^{\beta-1} ds \|$$

$$I_2 = \frac{M\|p\|_\infty(1 + \|x\|_\gamma)}{\Gamma(\beta)}$$

$$\int_0^{t_1} \left[t_1^{1-\gamma}(t_1 - s)^{\beta-1} - t_2^{1-\gamma}(t_2 - s)^{\beta-1} \right] ds$$

$$I_3 = (1 + \|x\|_\gamma)\|p\|_\infty|$$

$$\int_0^{t_1} t_1^{1-\gamma}(t_1 - s)^{\beta-1} \left[P_\beta(t_2 - s) - P_\beta(t_1 - s) \right] ds \|,$$

which yields $\lim_{t_2 \leftrightarrow t_1} I_1 = 0$. Similarly, we can prove that $\lim_{t_2 \to t_1} I_2 = \lim_{t_2 \to t_1} I_3 = 0$

Thus $\{Qx, x \in B_\eta\}$ is equicontinuous.

Step 4. N is ν-condensing.

We consider the measure of non-compactness defined in the following way. For every bounded subset $\Omega \subset C_{1-\gamma}([0, b], E)$

$$\nu(\Omega) = \max_{\Omega \in \Delta(\Omega)} (\widetilde{\gamma}(\Omega), \mathrm{mod}_{C_{1-\gamma}}(\Omega)). \tag{3.9}$$

$\Delta(\Omega)$ is the collection of all countable subsets of Ω and the maximum is taken in the sense of the partial order in the cone R_+^2. $\widetilde{\gamma}$ is the damped modulus of fiber non-compactness

$$\widetilde{\gamma}(\Omega) = \sup_{t \in [0,b]} e^{-Lt} \chi(\Omega_\gamma(t)), \tag{3.10}$$

where $\Omega_\gamma(t) = \{x_\gamma(t) : x \in \Omega\}$. $\mathrm{mod}_{C_{1-\gamma}}(\Omega)$ is the modulus of equicontinuity of the set of functions Ω given by the formula

$$\mathrm{mod}_{C_{1-\gamma}}(\Omega) = \lim_{\delta \to 0} \sup_{x \in \Omega} \max_{|t_1 - t_2| \leq \delta} \|x_\gamma(t_1) - x_\gamma(t_2)\|. \tag{3.11}$$

Let

$$q(L) := \sup_{t \in [0,b]} \int_0^t (t - s)^{\alpha-1} s^{\alpha-1} e^{-L(t-s)} ds. \tag{3.12}$$

It is clear that

$$\sup_{t \in [0,b]} \int_0^t (t - s)^{\alpha-1} s^{\alpha-1} e^{-L(t-s)} ds \xrightarrow[L \to +\infty]{} 0.$$

We can choose L such that

$$\bar{q}_1 := \frac{2cTM^2 \sum_{i=1}^m \lambda_i \|B\| q(L)}{\Gamma(\beta)\Gamma(\gamma)} < \frac{1}{2} \tag{3.13}$$

and

$$\bar{q}_2 := \frac{2cMb^{1-\gamma}q(L)}{\Gamma(\beta)} < \frac{1}{2}. \tag{3.14}$$

From Lemma 1, the measure ν is well defined and gives a monotone, nonsingular, semi-additive, and regular measure of non-compactness in $C_{1-\gamma}([0, b], E)$.

Let $\Omega \subset C_{1-\gamma}([0, b], E)$ be a bounded subset such that

$$\nu(N(\Omega)) \geq \nu(\Omega). \tag{3.15}$$

We will show that (3.15) implies that Ω is relatively compact. Let the maximum on the left-hand side of the inequality (3.15) be achieved for the countable set $\{y^n\}_{n=1}^{+\infty}$ with

$$y^n(t) = S_1 f_n(t) + S_2 f_n(t), \quad \{x^n\}_{n=1}^{+\infty} \subset \Omega \tag{3.16}$$

with

$$S_1 f_n(t)$$

$$= S_{\alpha,\beta}(t) \left(T \sum_{i=1}^{n} \lambda_i B \left[\int_0^{\tau_i} K_\alpha(\tau_i - s) f(s, y_n(s)) ds \right] \right),$$

$$S_2 f_n(t) = \int_0^t K_\alpha(t - s) f(s, y_n(s)) ds$$

and $f_n(t) = f(t, x^n(t))$. So that

$$\tilde{\gamma}(\{y^n\}_{n=1}^{+\infty}) \leq \tilde{\gamma}(\{S_1 f_n\}_{n=1}^{+\infty}) + \tilde{\gamma}(\{S_2 f_n\}_{n=1}^{+\infty}). \tag{3.17}$$

We give now an upper estimate for $\tilde{\gamma}(\{y^n\}_{n=1}^{+\infty})$. By using (H_4) we have

$$\chi(\{\mathcal{K}_\beta(t - s) f_n(s)\}_{n=1}^{+\infty})$$

$$\leq \frac{cM}{\Gamma(\beta)}(t - s)^{\alpha-1}\chi(\{x^n(s)\}_{n=1}^{+\infty})$$

$$= \frac{cM}{\Gamma(\beta)}(t - s)^{\alpha-1}s^{\gamma-1}\chi(\{x_\gamma^n(s)\}_{n=1}^{+\infty}) \tag{3.18}$$

$$\leq \frac{cM}{\Gamma(\beta)}(t - s)^{\alpha-1}s^{\gamma-1}e^{Ls} \sup_{0 \leq s \leq t} e^{-Ls}\chi(\{x_\gamma^n(s)\}_{n=1}^{+\infty})$$

$$= \frac{cM}{\Gamma(\beta)}(t - s)^{\alpha-1}s^{\gamma-1}e^{Ls}\tilde{\gamma}(\{x^n\}_{n=1}^{+\infty})$$

for all $t \in [0, b]$, $s \leq t$. Then applying Lemma 3, we obtain

$$t^{1-\gamma}\chi(\{S_1 f_n(t)\}_{n=1}^{+\infty}) \leq$$

$$\frac{2cM^2 T \|B\|}{\Gamma(\beta)\Gamma(\gamma)} \sum_{i=1}^{m} \lambda_i \int_0^{t_i} (t_i - s)^{\alpha-1}s^{\gamma-1}e^{Ls}\tilde{\gamma}(\{x^n\}_{n=1}^{+\infty}) ds$$

and

$$t^{1-\gamma}\chi(\{\mathcal{S}_2 f_n(t)\}_{n=1}^{+\infty}) \le$$
$$\frac{2cMb^{1-\gamma}}{\Gamma(\beta)} \int_0^t (t-s)^{\alpha-1} s^{\gamma-1} e^{Ls} \widetilde{\gamma}(\{x^n\}_{n=1}^{+\infty}) ds.$$

Taking (3.13) and (3.16) into account, we derive

$$\widetilde{\gamma}(\{y^n\}_{n=1}^{+\infty}) \le (\bar{q}_1 + \bar{q}_2) \widetilde{\gamma}(\{x^n\}_{n=1}^{+\infty}). \qquad (3.19)$$

Combining the last inequality with (3.15), we have

$$\widetilde{\gamma}(\{x^n\}_{n=1}^{+\infty}) \le (\bar{q}_1 + \bar{q}_2) \widetilde{\gamma}(\{x^n\}_{n=1}^{+\infty}).$$

Therefore

$$\widetilde{\gamma}(\{x^n\}_{n=1}^{+\infty}) = 0.$$

Hence by (3.19), we get

$$\widetilde{\gamma}(\{y^n\}_{n=1}^{+\infty}) = 0.$$

Furthermore, from Step 3, we know that $\mathrm{mod}_{C_{1-\gamma}}(N(\Omega)) = 0$ and (3.15) yields $\mathrm{mod}_{C_{1-\gamma}}(\Omega) = 0$. Finally,

$$\nu(\Omega) = (0, 0),$$

which proves the relative compactness of the set Ω.

Step 5. A priori bounds.
Let $x = \lambda N(x)$ for some $0 < \lambda < 1$. This is implied by (H_3), (H_5) follows

$$\|x\|_\gamma \ne \overline{M}.$$

Set

$$U = \{x \in: \|x\|_\gamma < \overline{M}\}.$$

From the choice of U, there is no $x \in \partial U$ such that $x = \lambda N(x)$ for some $\lambda \in [0, 1]$ yielding the desired a priori boundedness.

By Theorem (1), $\mathcal{F}ixN = S(f, \{\tau_i\}_{i=1}^n)$ is nonempty compact subset of $C_{1-\alpha}([0, b], E)$.

4 Conclusion

In this paper, I have given a new result concerning the existence of solution of a class of semi-linear Hilfer fractional differential equation with nonlocal conditions using a measure of non-compactness combined with condensing map in Banach space.

Acknowledgements The author would like to express his warmest thanks to all members of $ICFDA18$ International Conference on Fractional Differentiation and its Applications 2018 for his/her valuable comments and suggestions.

References

1. Agarwal, R.P., Hedia, B., Beddani, M.: Structure of solutions sets for imlpulsive fractional differential equation. J. Fract. Calc. Appl **9**(1), 15–34 (2018)
2. Andres, J., Górniewicz, L.: Topological Fixed Point Principles for Boundary Value Problems. Kluwer, Dordrecht (2003)
3. Bothe, D.: Multivalued perturbations of m-accretive differential inclusions. Israel J. Math. **108**, 109–138 (1998)
4. Browder, F.E., Gupta, G.P.: Topological degree and nonlinear mappings of analytic type in Banach spaces. J. Math. Anal. Appl. **26**, 390–402 (1969)
5. Deimling, K.: Nonlinear Functional Analysis. Springer (1985)
6. Deng, K.: Exponential decay of solutions of semilinear parabolic equations with nonlocal initial conditions. J. Math. Anal. Appl. **179**, 630–637 (1993)
7. Diethelm, K.: Analysis of Fractional Differential Equations. Springer, Berlin (2010)
8. Diethelm, K., Freed, A.D.: On the solution of nonlinear fractional order differential equations used in the modeling of viscoplasticity. In: Keil, F., Mackens, W., Voss, H., Werther, J. (eds.) Scientific Computing in Chemical Engineering II-Computational Fluid Dynamics, Reaction Engineering and Molecular Properties, pp. 217–224. Springer, Heidelberg (1999)
9. Djebali, S., Górniewicz, L., Ouahab, A.: Solutions Sets for Differential Equations and Inclusions. De Gruyter, Berlin (2013)
10. Dragoni, R., Macki, J.W., Nistri, P., Zecca, P.: Solution Sets of Differential Equations in Abstract Spaces, Pitman Research Notes in Mathematics Series 342. Longman, Harlow (1996)
11. Gu, H., Trujillo, J.J.: Existence of mild solution for evolution equation with Hilfer fractional derivative. Appl. Math. Comput. **257**, 344–354 (2015)
12. Furati, K.M., Kassim, M.D., Tatar, N.E.: Existence and uniqueness for a problem involving Hilfer fractional derivative. Comput. Math. Appl. **64**, 1616–1626 (2012)
13. Hilfer, R.: Applications of Fractional Calculus in Physics. World Scientific, Singapore, pp. 87–429 (2000)
14. Hilfer, R.: Experimental evidence for fractional time evolution in glass materials. Chem. Phys. **284**, 399–408 (2002)
15. Kamenskii, M., Obukhovskii, V., Zecca, P.: Condensing Multivalued Maps and Semilinear Differential Inclusions in Banach Spaces. De Gruyter, Berlin (2001)
16. Kilbas, A.A., Srivastava, H.M., Trujillo, J.J.: Theory and Applications of Fractional Differential Equations. North-Holland Mathematics Studies, 204. Elsevier Science B.V., Amsterdam (2006)
17. Toledano, J.M.A., Benavides, T.D., Azedo, G.L.: Measures of Noncompactness in Metric Fixed Point Theory. Birkhauser, Basel (1997)
18. Wang, J.R., Zhang, Y.: Nonlocal initial value problems for differential equations with Hilfer fractional derivative. Appl. Math. Comput. **266**, 850–859 (2015)
19. Zhou, Y.: Basic Theory of Fractional Differential Equations. World Scientific, Singapore (2014)
20. Zhou, Y.: Fractional Evolution Equations and Inclusions: Analysis and Control. Academic Press (2016)

Offshore Wind System in the Way of Energy 4.0: Ride Through Fault Aided by Fractional PI Control and VRFB

Rui Melicio, Duarte Valério and V. M. F. Mendes

Abstract This chapter presents a simulation of a study to improve the ability of an offshore wind system to recover from a fault due to a rectifier converter malfunction. The system comprises: a semi-submersible platform; a variable-speed wind turbine; a PMSG; a 5LC-MPC; a fractional PI controller using the Carlson approximation. Recovery is improved by shielding the DC link of the converter during the fault using as further equipment a redox vanadium flow battery, aiding the system operation as desired in the scope of Energy 4.0. Contributions are given for: (i) the fault influence on the behavior of voltages and currents in the capacitor bank of the DC link; (ii) the drivetrain modeling of the floating platform by a three-mass modeling; (iii) the vanadium flow battery integration in the system.

This work is funded by: European Union through the European Regional Development Fund, included in the COMPETE 2020 (Operational Program Competitiveness and Internationalization) through the ICT project (UID/GEO/04683/2019) with the reference POCI010145FEDER007690; FCT, through IDMEC, under LAETA, Project UID/EMS/50022/2019; Portuguese Foundation for Science and Technology (FCT) under Project UID/EEA/04131/2019, and grant SFRH/BSAB/142920/2018.

R. Melicio (✉)
ICT, Instituto de Ciências da Terra, Universidade de Évora, Evora, Portugal
e-mail: ruimelicio@gmail.com

R. Melicio · D. Valério
IDMEC, Instituto Superior Técnico, Universidade de Lisboa, Lisbon, Portugal

R. Melicio · V. M. F. Mendes
Departmento de Física, Escola de Ciências e Tecnologia, Universidade de Évora, Évora, Portugal

V. M. F. Mendes
Department of Electrical Engineering and Automation, Instituto Superior de Engenharia de Lisboa, Lisbon, Portugal

CISE, Electromechatronic Systems Research Centre, Universidade da Beira Interior, Covilha, Portugal

© Springer Nature Singapore Pte Ltd. 2019
P. Agarwal et al. (eds.), *Fractional Calculus*, Springer Proceedings
in Mathematics & Statistics 303, https://doi.org/10.1007/978-981-15-0430-3_6

Keywords Offshore wind system · VRFB · MPC five-level converter · Ride through capability · Simulation · Energy 4.0 · Fractional control

1 Introduction

Wind energy conversion into electric energy through wind power systems either onshore or offshore plays a significant role in a future shaped by the need for sustainable development concerning energy usage [1–40] and smart energy in the context of Energy 4.0. The wind system quota has increased in capacity into the mixed generation of electric grids, improving the diversification of resources and contributing to matching the needs in the usage of electric energy [6]. Fluctuation on the side of the conversion into electric energy is expected to increase due to the uncertainty inherent to the intermittency of exploitation of wind or solar energy sources [16]. Fluctuation on the side of the usage of electric energy is also expected to increase in the future, for instance, due to the use of electric vehicles. Both fluctuations are prone to lead to new challenges and threats to be faced in the scope of smart energy in the context of Energy 4.0. Particularly, if not properly conduced, the exploitation of wind energy sources for conversion into electric energy is a power system interconnection menacing the quality of energy and the transient stability of electric grid [37] (Table 1).

Grid codes establish interconnecting guidelines, i.e., instructions specifying technical and operative requirements to conduce power production and other parties involved in the production, transportation, and usage of electric energy. Indeed, in the context of sustainable power production in the scope of Energy 4.0, more restrictive grid codes are expected to be in force to conduce the operation of wind systems to avoid abnormal behavior leading to menace, for instance, as loss of power quality or of stability- appropriated margin into the electric grid. Wind systems must cope with acceptable performance regulated in grid codes to integrate the electric grid. So, after a failure, the recovery of normal operation in due time is of great importance to avoid eventual coming off from the electric grid. Rethinking how a wind system can satisfy grid codes and capture more value from the participation into the mixed production of an electric grid while mitigating faults are challenges of research in the way of Energy 4.0. Furthermore, abnormal behavior in the operation which is not avoided in due time can lead to a fault going into a failure, needing human intervention on the wind system. This is hampered for offshore wind systems by the access of the place of exploitation, often impossible in days with severe weather in fall or winter seasons [17, 20]. So, improving the recovery of normal operation, particularly Fault Ride Through capability [25], is of vital importance for offshore wind systems. Research contributing to the operation continuity of wind systems in the occurrence of an eventual failure is a promising line of research. A way of ensuring operation continuity and capture more value often stated and justified as an advantageous option is the use of energy storage. Energy storage through the technology of VRFB has the advantage of a full depth of discharge without affecting performance

Table 1 Nomenclature

VRFB	Vanadium redox flow battery
BMS	Battery management system
IGBT	Insulated gate bipolar transistor
MPC	Multiple point clamped
PMSG	Permanent magnet synchronous generator
5LC	Five-level power converter
u	Wind speed value with disturbance
u_0	Average wind speed
n	Index of eigenswing excited
A_n	Magnitude of the eigenswing n
ω_n	Eigenfrequency of the eigenswing n
η	Wave elevation at the point x, y
η_a	Vector of harmonic wave amplitudes
ϑ	Vector of harmonic wave frequencies
ζ	Vector of harmonic wave phases (random)
ϕ	Vector of harmonic wavenumbers
ψ	Vector of harmonic wave directions
P_t	Mechanical power with perturbation
P_{tt}	Mechanical power without perturbation
m	Order of the harmonic in an eigenswing
a_{nm}	Normalized magnitude of g_{nm}
g_{nm}	Distribution of m-order harmonic in eigenswing n
h_n	Modulation of eigenswing n
φ_{nm}	Phase of m-order harmonic in eigenswing n
u_{sk}	Rectifier input or inverter output voltages
U_{cj}	Voltage in the capacitor bank j
i_{cj}	Current in the capacitor bank j
U_{dc}	Total DC voltage
C_j	Capacitance of the capacitor bank j
i_{fk}	Currents injected into the electric grid
L_n	Inductance of the electric grid
R_n	Resistance of the electric grid
u_{fk}	Voltage at the filter
i_k	Input or output current in MPC five-level converter
u_k	Voltage at the electric grid, $k \in \{4, 5, 6\}$
μ	Fractional order of the derivative or of the integral

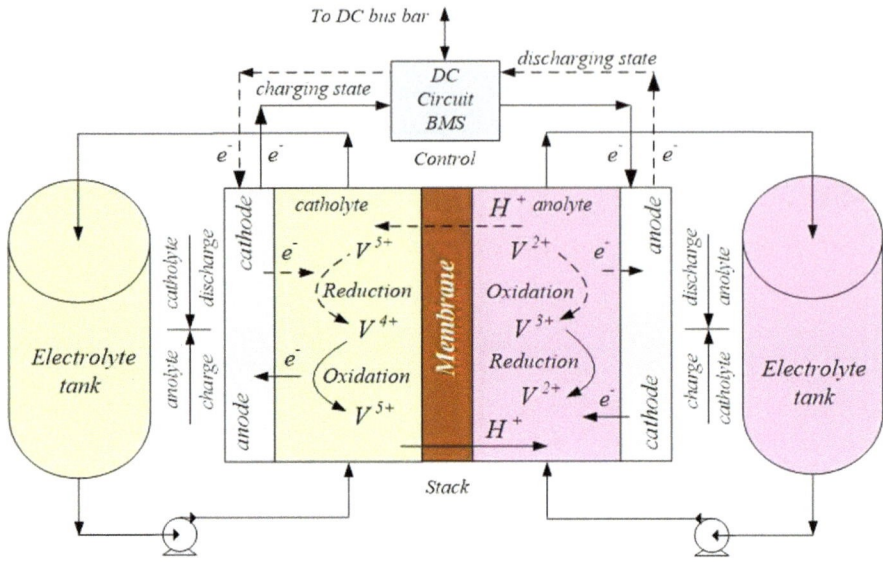

Fig. 1 VRFB reaction, charge/discharge process

and the useful life of the battery, customizable scalability high power, long duration of about 15,000 cycles of charge/discharge, fast response, large capacity, and a fair balance of energy efficiency and costs [12]. In comparison with other technologies, VRFB technology is pointed out as an appropriate storage for electric grids, [2, 3, 14], contributing to smooth the impact of uncertainty on the availability of renewable energy and allowing satisfaction of conditions imposed by grid codes [15]. A schematic diagram of the electrochemical reactions charge/discharge processes of the VRFB [2] is shown in Fig. 1.

In Fig. 1 are shown the two tanks to store the electrolytes to be gradually pumped into the stack of the electrochemical cells, where by chemical reactions the electrolytes are charged or discharged. So, augmenting the volume of the tanks allows scaling up the storage capacity. The membrane carries a selective transmission of protons from the two sides of the VRFB. Each vanadium ion by a process of oxidation drops at the positive terminal an electron during charge. The flow of electrons is controlled by the BMS and directed to be collected by a vanadium ion at the cathode by a process of reduction during charge, during discharge the process is reversed. A VRFB is recognized by having not only capability to respond nearly instantaneously to demand, standing in for the traditional means of meeting peak demand, but also the ability to convey energy when required over significant periods of time, for instance, a time of twelve hours as pointed out in [10]. So, the technology of VRFB at utility scale is expected to be the way of the future for energy storage in the scope of Energy 4.0. The technology of VRFB is already in the way of application at utility scale. In the Isle of Gigha, Scotland, a VRFB is in use due to the ability to balance variable generation from renewable sources and due to the cost-effective time shifting. In Dalian, China, a 200 MW/800 MWh VRFB utility scale is to be implemented for peak-shaving and grid stabilization. Advantages pointed out are reliability, full recyclability of the electrolyte, and more than 20-year useful life-time [9].

Many of the faults in wind systems can occur in power control and sensor electronic devices [39]. Faults are susceptible to disturb the operation and cause an inability to perform within specified requirements, i.e., grid codes, or going into failure. Hence, not only fault avoidance, but also tolerance to fault, i.e. the ability to continue to perform within specified performance in the presence of a fault is regarded as crucial. In this regard, a VRFB can be used during a short time span to replace the delivery of energy coming from the generator interrupted, for instance, by a fault in the rectifier converter. This replacement is intended to preserve the wind system connection to the electric grid during a time span that ends with the recovery of the full operation, avoiding a failure, implying the disconnection of the wind system [35]. The VRFB can be incorporated into a wind system by a direct current circuit a BMS required for: monitoring the state; processing and reporting secondary data; protecting and controlling the environment of the VRFB [2]. However, a convenient control strategy for the selection of voltage vectors to maintain the equilibrium of the voltages in the capacitor bank must be carried out to further aid in avoiding failure of the wind system. One part of the scope of the research in this chapter is the above convenient control strategy for the selection of voltage vectors, aiding in avoiding a failure of the wind system subject to a rectifier converter fault. This control strategy is implemented by a convenient exploration of the use of the redundant vectors and the dominant currents in the capacitor bank.

The chapter is concerned with a wind system equipped with a MPC-5LC and the main contributions are: (i) the study of the fault influence on the behavior of voltages and currents in the capacitor bank; (ii) the consideration of the dynamics of the floating platform by a three-mass modeling drivetrain; (iii) the study of the incorporation a VRFB assisting the recovery of operation. The rest of the chapter is organized as follows: Sect. 2 is concerned with the integration of wind systems in the scope of Energy 4.0. Section 3 presents the model. Section 4 presents the control method. Section 5 presents a case study and discusses the consequence of the results. Section 6 presents the concluding remarks.

2   Energy in a Sustainable Way

The industry of energy is at the brink of a new industrial revolution, not only shifting the present, but also the way of the future, in what regards information processing, control and action of intervenient agents. Availability of energy has been of paramount importance and a key influence on all Industrial Revolutions and is expected to remain so in the future. But another important influence is the paradigm for the organization of the energy business, which is expected to play a substantial role.

The mix of sources of energy used in the power systems has changed, because of the integration in an unprecedented scale of intermittent renewable sources of energy, of nuclear phase-out, and of the appearance of utility-scale technology for storage of electric energy. Also, new concepts for grids at the level of transmission and distribution of electric energy are expected to happen together with the operation of power systems in a sustainable way under Energy 4.0.

The increasing use of intermittent renewable energy, and in particular of wind systems, requires control acting in due time, and needs to be balanced with flexible generation, demand management, energy storage, or interconnection devices. Wind systems must embrace Energy 4.0 concepts to cope with the future, implementing convenient monitoring, transferring, and analyzing data in a smart grid and IoT way. Systems interconnection and energy sustainability is the ambition of the future energy business, using information not only from monitoring systems of the energy industry, but also from other industrial systems [4, 28]. Advanced smart technology and better control systems for wind systems operation connected with electric grid allow some flexibility on the requirements for the performance of electric generators. But one of the most severe requirements in the case of offshore wind system could be the black start capability, which requires the ability to recover from a total or partial shutdown within a set timeframe, without any external supply. Under the Energy 4.0 framework, the integration of offshore wind systems in a smart grid is expected to be aided by real-time monitoring and safety actions to mitigate the impact of faults [34], improving efficiency and sustainability.

3 Modeling

The wind system under consideration is equipped with the following main components: a semi-submersible platform of the category used in the WindFloat project [29] anchored to the seabed by suspended cables; a platform where is placed the variable-speed wind turbine and the equipment for power control by blade pitch angle; a PMSG; the MPC-5LC; and an energy storage system assumed to be, but not necessary a VRFB. What is necessary is that the energy storage system device has enough energy to aid in avoiding failure due to the fault.

3.1 Wind and Marine Wave

The wind speed is modeled by a sum of harmonics ranging from 0.1 Hz to 10.0 Hz as in [35] given by

$$u = u_0 \left[1 + \sum_n A_n \sin (\omega_n t) \right] \tag{1}$$

The marine wave elevation is described as a phase/amplitude model as in [32] considered by a convenient sum of harmonic waves given by

$$\eta(x, y, z) = \sum_i \eta_a(i) \cos\left[\vartheta_i t + \zeta_i - \upsilon_i(x, y)\right] \tag{2}$$

where

$$\upsilon_i(x, y) = \phi_i \left(x \cos \psi_i + y \sin \psi_i\right) \tag{3}$$

In (3) is computed the inner product of the position vector in the horizontal plane (x, y) by the wave vector, having the information concerned with the displacement of the wave, pointing in the normal direction to the wave front.

3.2 Wind Turbine

The turbine mechanical power is modeled as a sum of a function of the magnitude of three eigenswings as in [31] given by

$$P_t = P_{tt}\left[1 + \sum_{n=1}^{3} A_n \left(\sum_{m=1}^{2} a_{nm} g_{nm}(t)\right) h_n(t)\right] \tag{4}$$

where

$$g_{nm} = \sin\left(\int_0^t m\omega_n(t') \, dt' + \varphi_{nm}\right) \tag{5}$$

The data considered for (4) and (5) are reported in [1].

3.3 Drivetrain Model

The aerodynamic loads to which wind turbines are subject have an important influence in the design of their structural components, due to the need to resist fatigue. Fatigue-oriented design [13] is needed for the tower, the blades, and also the drivetrain. The model followed in this chapter (consisting of three masses coupled by elastic elements) can be found to be reported in technical literature (e.g., [23, 36]).

The reason why the drivetrain is modeled with several masses is its flexibility, which is needed to increase reliability in the presence of fatigue [5, 13]. In fact, the vibrations of the structure, revealed by noise in the aerodynamic blades, can be reduced by a better aerodynamic design, and so the efforts are now shifted to the drivetrain itself.

3.4 Generator

The generator is a PMSG, and its model is that usually employed for a normal synchronous electric machine. This has been reported in technical literature such as [22]. An additional constraint is needed: the direct component of the electric current in the stator is imposed to be zero. This constraint is intended to prevent the demagnetization of the permanent magnet in a PMSG [33].

3.5 Five-Level Power Converter

Figure 2 shows the details of the MPC 5LC rectifier. It shows that the MPC-5LC rectifier and the inverter have 24 unidirectional commanded IGBTs. The IGBTs are modeled as ideal components. Branch k of the converter consists of a group of eight IGBTs connected to the same phase. IGBTs are identified by S_{ik} with $i \in \{1, \ldots, 8\}$, corresponding to a branch $k \in \{1, 2, 3\}$ (in the case of the rectifier) or $k \in \{4, 5, 6\}$ (in the case of the inverter) [32]. Figure 3 shows the MPC-5LC, equipped with the VRFB. For details, see [18].

The voltages in the rectifier (input, $k \in \{1, 2, 3\}$) and the inverter (output, $k \in \{4, 5, 6\}$) are given by

$$u_{sk} = \frac{1}{3} \sum_{j=1}^{p-1} \left(2\delta_{jn_k} \sum_{a=1, a\neq k}^{3} \delta_{jn_a} \right) U_{cj} \tag{6}$$

On each capacitor bank C_j, the current i_{cj} can be found as

$$i_{cj} = \sum_{k=1}^{3} \delta_{nk}i_k - \sum_{k=4}^{6} \delta_{nk}i_k, \quad k \in \{1, \ldots, 6\} \tag{7}$$

The currents in each capacitor banks are, in (7), the input and output currents of the MPC-5LC. The state equation the DC bar voltage U_{dc} is

$$\frac{dU_{dc}}{dt} = \sum_{j=1}^{p-1} \frac{1}{C_j} i_{cj}, \quad j \in \{1, \ldots, p-1\} \tag{8}$$

Note that (8) is valid even if the BMS has called the VRFB.

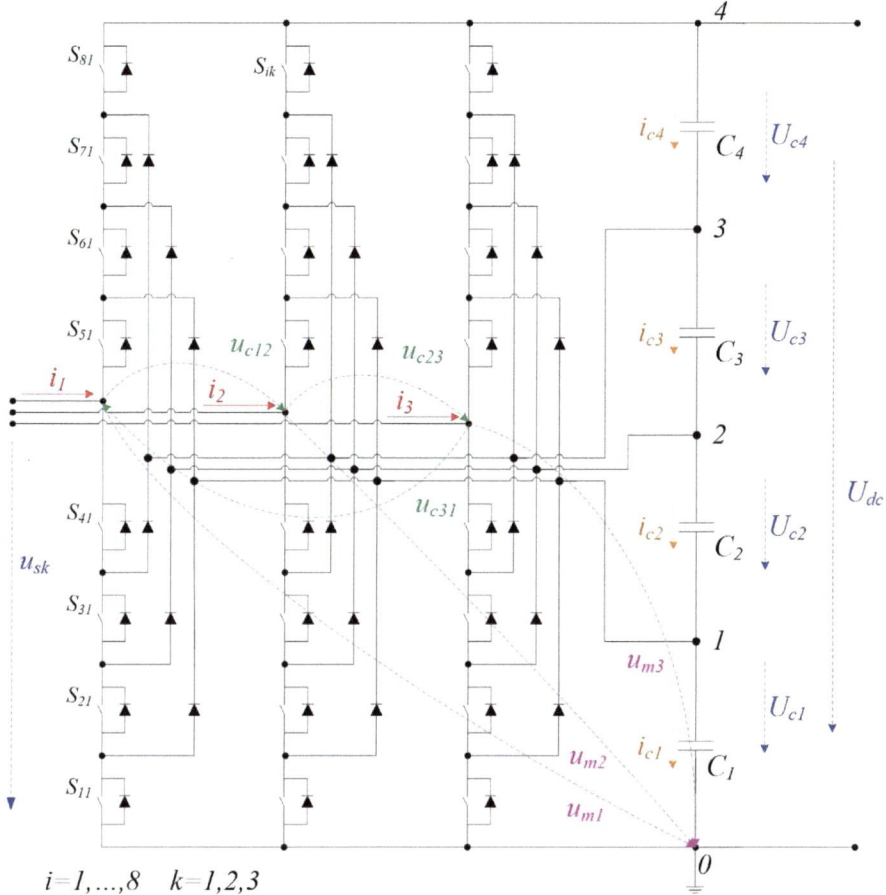

Fig. 2 MPC 5LC rectifier

3.6 Electric Grid

The current injected into the electric grid is modeled by the following state equation:

$$\frac{di_{fk}}{dt} = \frac{1}{L_n}\left(u_{fk} - R_n i_{fk} - u_k\right), \quad k \in \{4, 5, 6\} \tag{9}$$

This model consists of an ideal voltage source, in series with the short-circuit impedance.

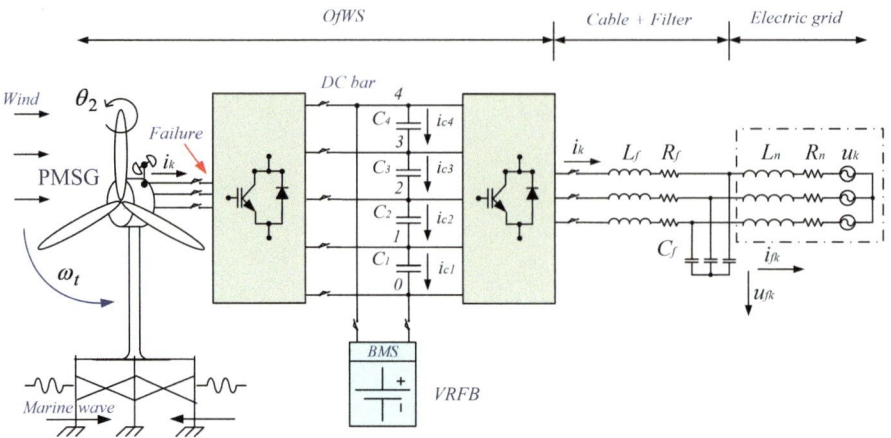

Fig. 3 Offshore wind system equipped with VRFB

3.7 VRFB

The VRFB is a DC source with a value of 7.14 kV, 2% plus of the DC link voltage. Although this type of source can have significant discharge times [15], the capability to respond nearly instantaneously is the most important for the purpose of the chapter.

4 Control Method

4.1 Fractional-Order Controllers

Fractional-order controllers are those designed by the application of fractional-order derivatives, having advantageous robustness as the main reason for being useful in the applications. Usual fractional-order controllers include fractional PIDs, introduced in [27] and so-called because of their similitude with PID controllers, and CRONE (Commande Robuste d'Ordre Non Entier) controllers [19, 26]. The CRONE controllers are designed according to a methodology conceived with control robustness in mind. In the system under study, the controller for the variable-speed operation is fractional PI controller [35], from the PID family, combining a fractional derivative with a constant gain at low frequencies. The transfer function for the controller for the variable-speed operation is given by

$$C(s) = 2.6 + 0.6\frac{1}{s^{0.5}} \tag{10}$$

Several tuning methods have been proposed for fractional PIDs, including tuning rules such as in [11], or numerical methods as in [24]. Controller (16) was designed according to the rules in [21]. The term $s^{0.5}$ is the Laplace transform of a half-derivative. More about fractional derivatives is presented in [30, 38]. An implementation of a term such as s^α is usually carried out by means of an approximation. The most usual approximations [38] are the CRONE approximation, due to the work of Alain Oustaloup, the Matsuda approximation, and the Carlson approximation, the one used for the system under study. It was introduced in [7], and is based on the Newton–Raphson method for finding numerical solutions of equations, which can be used to obtain numerical values for $a^{1/n}$ finding the roots of $f(x) = x^n - a$. The method can also be used if a is not a positive real number, but rather the Laplace transform variable s instead. In this way, iterative approximations of $s^{1/n}, n \in \mathbb{N}$ can be found. Some calculations show that, beginning with the trivial (and far from accurate) approximation

$$s^{1/n} \approx G_1(s) = 1, \tag{11}$$

further approximations given by

$$s^{1/n} \approx G_{k+1}(s) = G_k(s)\frac{(n-1)G_k^n(s) + (n+1)s}{(n+1)G_k^n(s) + (n-1)s}, \tag{12}$$

can be found, which are rather accurate in a frequency range that increases with n, centered on frequency 1 rad/s. The approximation used to implement (10) was obtained with two iterations:

$$s^{\frac{1}{-2}} \approx G_3(s) = \frac{s^4 + 36s^3 + 126s^2 + 84s + 9}{9s^4 + 84s^3 + 126s^2 + 36s + 1} \tag{13}$$

Transfer function (13) provides a good approximation a limited frequency range $[\omega_l, \omega_h] = [10^{-1}, 10]$ rad/s. Finally, (10) was implemented as $C(s) \approx 2.6 + 0.6G_3(s)$.

4.2 Power Converter Control

The modeling to be considered for the MPC-5LC control is of fractional order complemented with a sliding mode control associated with PWM by space vector modulation. The output voltage vectors in the (α, β) space for the 5LC are shown in Fig. 4.

In Fig. 4, the required selection for the output voltage vector in the (α, β) space is carried out in function of the discrepancy between the current of the stator and the reference current. A power converter is a time variable structure due to the IGBTs switching blockage/conduction states [32]. The operation of time-variable structures subject to uncertainties and external disturbances must be suitably complemented by sliding mode control as is reported in the literature. This operation of the power con-

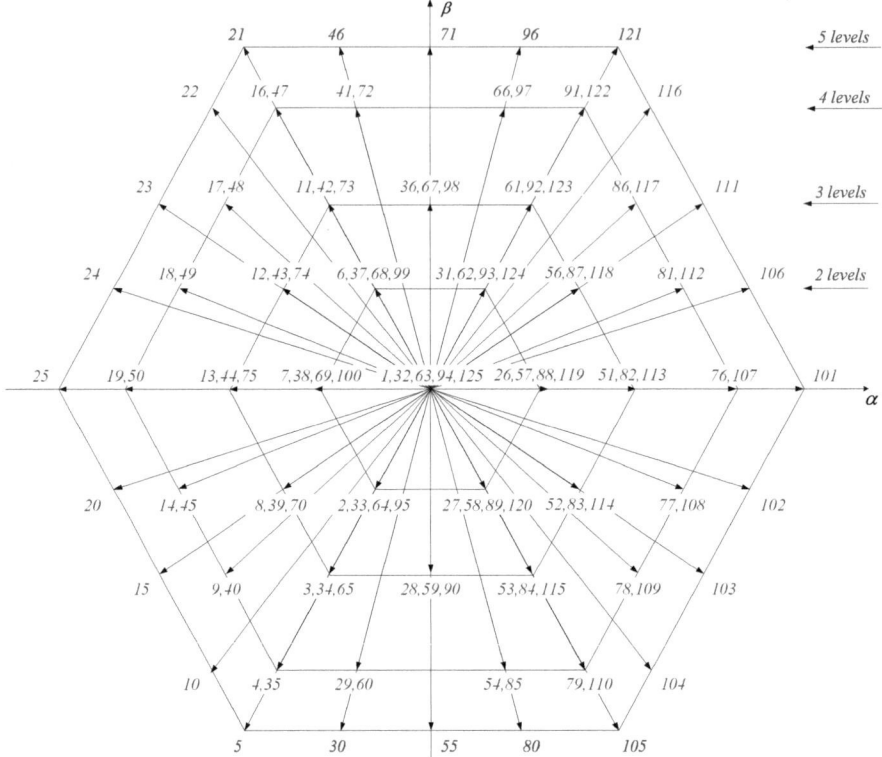

Fig. 4 Output voltage vectors for the MPC-5LC

verter as a time variable structure subject to uncertainties and external disturbances
is the one presented by the authors in [35].

5 Case Study

The case study is carried out using the computer application MATLAB/Simulink and
has a time horizon of 10 s. The electric grid voltage is of 5 kV at 50 Hz. A 10 kHz
switching frequency is assumed for the IGBTs. The capacitor bank reference voltage
U_{dc}^* is of 7 kV. The main data concerned with the wind system [36] is summarized
in Table 2.

The wind speed is shown in Fig. 5. The marine wave elevation is shown in Fig. 6
that shows a significant perturbation due to marine wave with a period of about 10 s
subjecting the wind system to a significant perturbation. The data for (1) and (4) are
$u_0 = 14.5$ m/s; $A_1 = 0.01$; $A_2 = 0.08$; $A_3 = 0.15$; $\omega_1(t) = \omega_t(t)$; $\omega_2(t) = 3\omega_t(t)$;
and $\omega_3(t) = [g_{11}(t) + g_{21}(t)]/2$, $g_{11}(t)$, $g_{21}(t)$ given by (5).

Table 2 Wind system data

Turbine moment of inertia	5.5×10^6 kg m^2
Turbine rotor diameter	90 m
Hub height	45 m
Tip speed	17.64–81.04 m/s
Rotor angular velocity	6.9–31.6 rpm
PMSG rated power	2 MW
PMSG inertia moment	400×10^3 kg m^2

Fig. 5 Wind speed

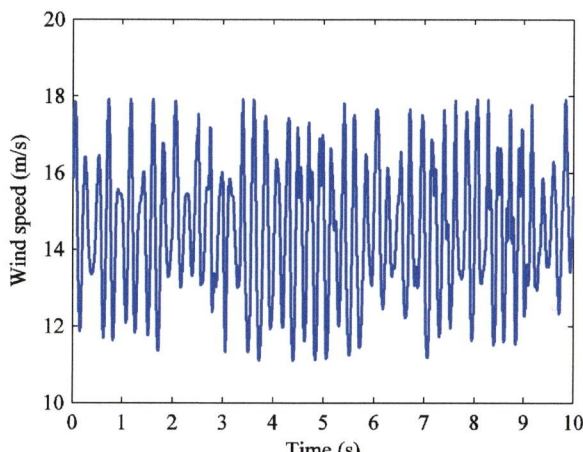

Fig. 6 Marine wave elevation

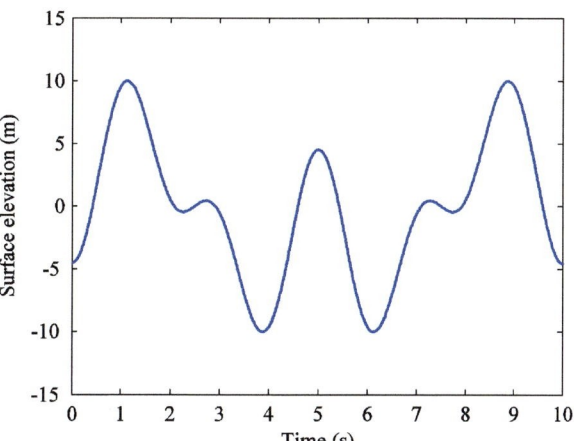

Fig. 7 Power coefficient

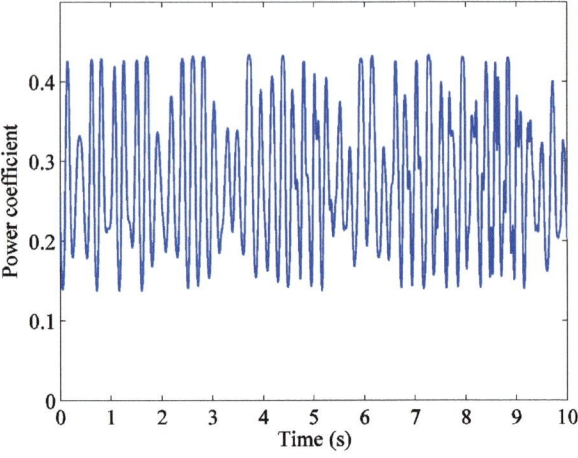

Fig. 8 Voltage in the capacitor bank

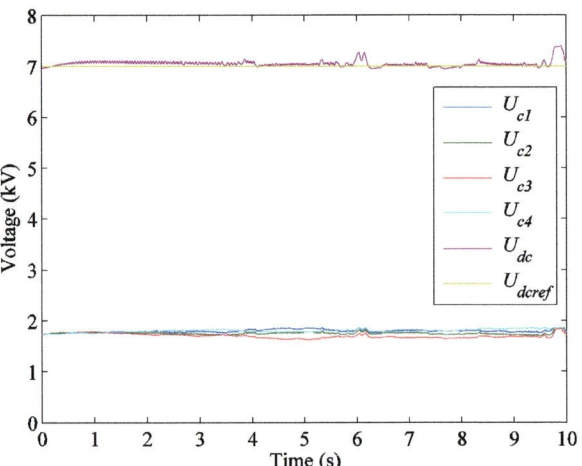

5.1   No Fault

This simulation is carried out with no-fault consideration for purpose of comparison, i.e., is a normal operation with intermittent availability of wind energy, having the wind speed shown in Fig. 5. The power range satisfaction due to the action of the maximum power point tracking imposes the power coefficient shown in Fig. 7. The voltages in the capacitor bank are shown in Fig. 8. The current in the capacitor bank is shown in Fig. 9. These last two figures show that the control of the converter surmounts the imbalance voltages in the capacitor bank, having appropriated currents in the capacitors and almost a steady behavior for voltage in the DC link. So, normal operation i s pursued.

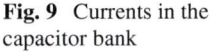

Fig. 9 Currents in the capacitor bank

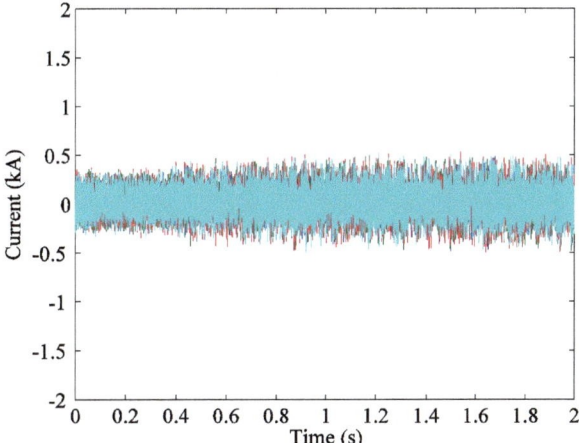

5.2 Failure

In what concerns the data, this simulation has the same wind speed and the marine wave as the first one. It also has the same equipment for the wind system, but with a fault. This fault imposes that a circuit breaker at the input of the rectifier is open between 1.15 s and 1.65 s. So, during this period no energy flows to the capacitors, i.e., the capacitors are not charged by the rectifier, implying a hovering menace of fault going into failure on the wind system. This menace is in accordance with the simulation results given for the voltages in the capacitors as shown in Fig. 10. In this figure, it is shown that the voltage drop across capacitors is significant and the level of voltage is not recovered in due time to avoid the disconnection. So, this simulation is in accordance with the wind system having an inevitable fault going into failure, i.e., the system goes into the necessary disconnection to avoid further worst consequences.

The input voltages in the rectifier are shown in Fig. 11. Here it is once more shown that the wind system is unable to recover voltage for feasible operation in the rectifier. Again, this simulation results are in accordance with the wind system not being able to avoid disconnection.

The currents in the capacitor bank are shown in Fig. 12. This figure shows that after 1.15 s the currents in the capacitor bank have a smaller positive oscillation than the negative one, meaning that the capacitors are discharging. The control of the converter is trying to surmount the imbalance on voltages in the capacitor bank, but there is not enough electric charge to sustain the drop on the voltages dropping significantly near 1.65 s.

The main conclusion of this simulation is in accordance with the fact that the system is unable to maintain the interconnection with the electric grid. So, ride through fault is feasible for the wind system. The ability to perform within specified

Fig. 10 Voltage in the capacitors of the capacitor bank

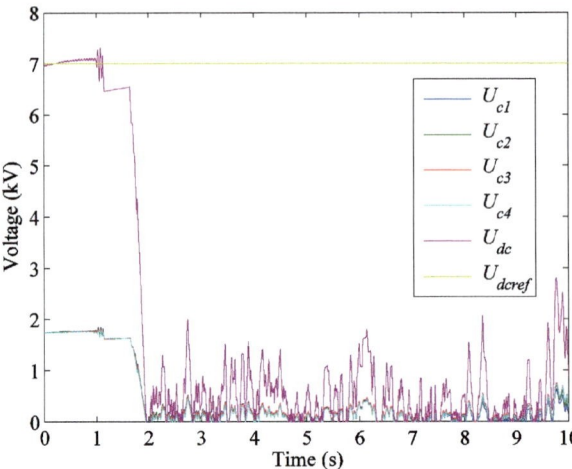

Fig. 11 Input voltages in the rectifier

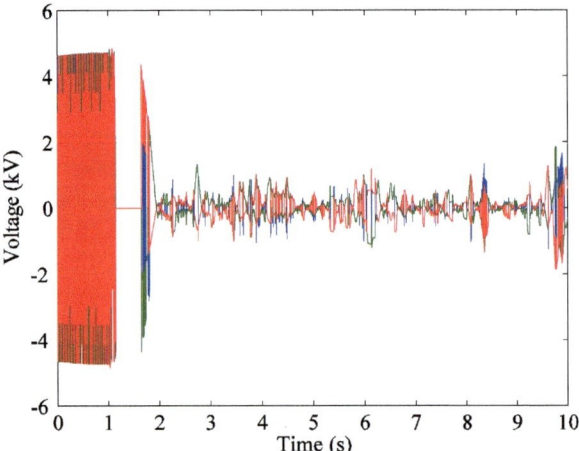

performance requirements, i.e., in the way of fault tolerance in the way of the future at Energy 4.0, is not feasible and disconnection is to happen to avoid further worst consequences of equipment damage.

5.3 Ride Through Fault

In what concerns the data, this simulation has the same wind speed and the marine wave as the first one. It also has the same equipment for the wind system, and a fault considered as imposing that a circuit breaker at the input of the rectifier is open between 1.15 s and 1.65 s. That is to say, the simulation has the same data as that of

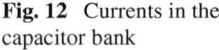
Fig. 12 Currents in the capacitor bank

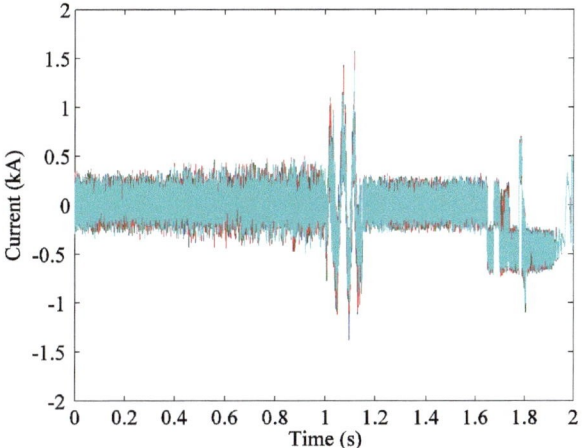

Sect. 5.2, but with the addition of an energy storage system device assumed—but not necessarily—to be given by the technology of VRFB. What is important to assume is that the energy storage system device is designed to have enough energy to aid in avoiding the fault going into failure during the time without energy flowing from the rectifier converter. So, the simulation is concerned with the wind system calling due to the fault the aid of the energy storage system device by the BMS to conveniently charge the capacitor bank during the period of transient operation between 1.15 s and 1.65 s. The behavior of the system during this period tends to be described by the arrangement shown in Fig. 13.

In Fig. 13 is shown the VRFB call during the period of transient operation to contribute with a convenient charge of the capacitors. So, recovering of normal operation is feasible after fault clearing. The voltages in the capacitor bank are shown in Fig. 14.

In Fig. 14 is shown that when the VRFB is called a transient occurs while the returning of the VRFB to standby is done more smoothly. The voltage drop across capacitors is not significant and the level of voltage is recovered in due time to avoid the disconnection. So, this simulation is in accordance with the wind system having a fault but not going into failure. The system can ride through the fault due to the halt dropping of the level of voltages in the capacitor bank. The control surmounts imbalance voltages on the capacitors, giving a recovering of the appropriated behavior of the voltage in the DC link in a way of fault tolerance feasibility. The input voltages in the rectifier, the currents in the rectifier, and the current in the capacitor bank are, respectively, shown in Figs. 15, 16 and 17.

In Fig. 15, it is shown that the wind system can recover the normal operation of the rectifier converter, the system went into a state of the fault and goes into normal operation after fault clearing. In Fig. 16, it is shown as expected that the input currents in the rectifier are null during the period 1.15 s to 1.65 s. After this period, the currents accommodate values given by the wind system control to achieve maximization of

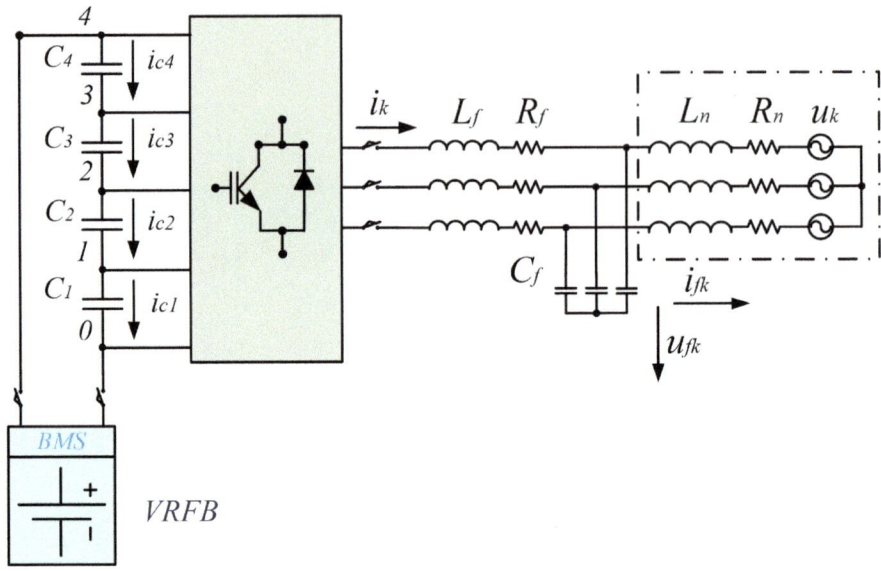

Fig. 13 System during the transient

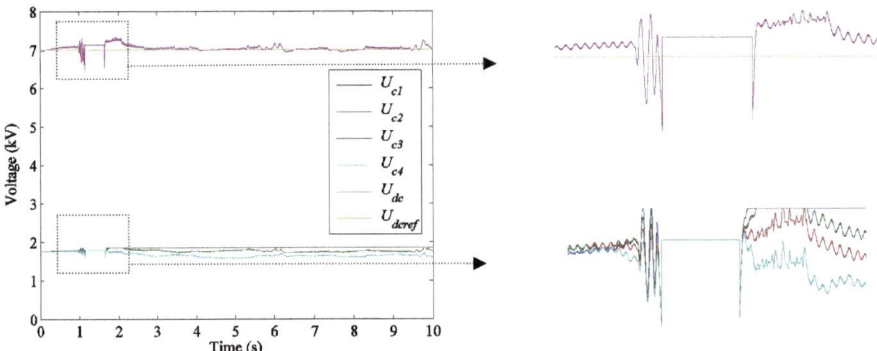

Fig. 14 Voltage in the capacitor bank

energy conversion in the power range of the wind system. In Fig. 17 is shown that when the VRFB is called a transient occurs and the returning of the VRFB to a standby state has smooth picks on currents in the capacitor bank. After the disconnection of the VRFB, the rectifier can provide an electric charge to the capacitor bank to avoid the imbalance voltages. Hence, the wind system can recover the feasible operation after the fault clearing due to the assistance of the VRFB, conveniently charged. The total harmonic distortion (THD) of the electric current injected into the electric grid for the simulations with No fault and with Ride through fault are shown in Table 3. These figures show, as expected, that the THD of the electric current into the electric grid is not significantly greater and the difference is faded after fault clearing.

Fig. 15 Input voltages in the rectifier

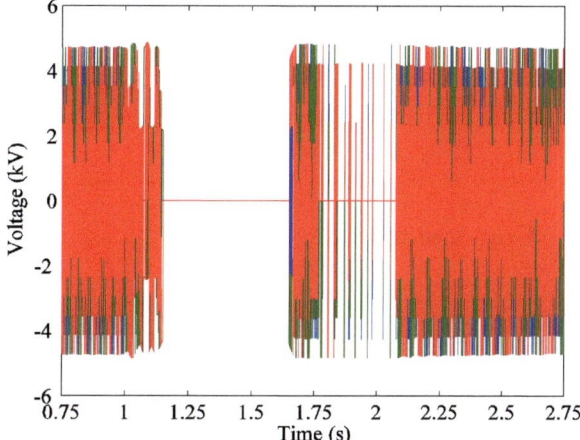

Fig. 16 Input currents in the rectifier

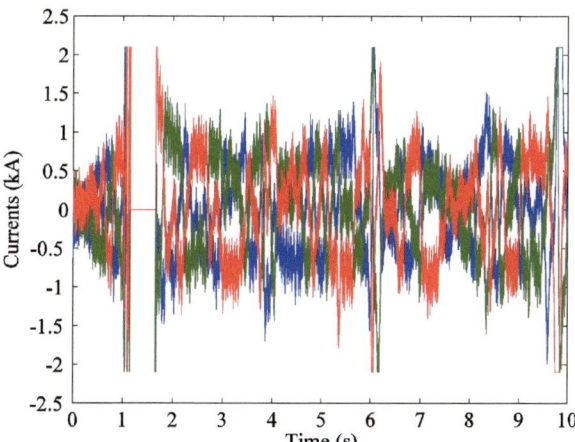

Fig. 17 Current in the capacitor bank

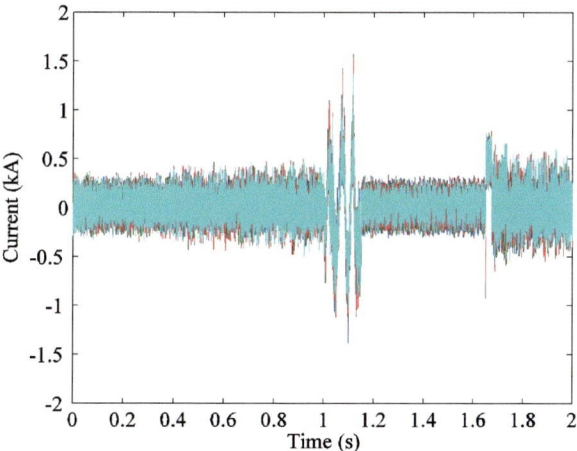

Table 3 Average THD of the current injected in the electric grid

Case study	THD
No fault	0.51
Fault with VRFB	0.54

6 Conclusions

Grid codes are important tools to mitigate threats coming into the electric grid. But faults in wind systems are likely to disable the ability to perform within grid codes. Worst still, due to their location, offshore wind systems may require a considerable time of disconnection due to the maintenance and repairs needed because of a fault going into failure. If a wind system in fault does not have the *Ride through fault* capability and is not disconnected in due time, then equipment damage and negative impacts are to be expected in the electric grid. Consequently, not only fault avoidance, but also tolerance to faults are of crucial importance for offshore wind systems, to avoid the need of disconnecting from the grid, to reduce the economic consequence of being not able of injecting energy into the grid, and to remain within the requirements of grid codes.

A model is proposed for the simulation of a wind system having a threatening on the continuity of the operation due to a fault in the rectifier converter. The fault creates an interruption in the energy delivery from the PMSG to the electric grid. The simulation of the fault is in accordance with an inability to maintain energy injection into the grid, tapping the system in a state of no recovery, i.e., the system goes from fault into failure and disconnection is expected. Then a VRFB controlled by a BMS is suggested as an aid to introduce Ride through fault in a strategic hardware solution with the MPC-5LC to satisfy grid codes in what regards continuity of the operation. The strategic hardware solution of the VRFB with the MPC 5LC must be reinforced with a convenient selection of voltage vectors to maintain the equilibrium of the voltages to further aid in avoiding failure of the wind system when subject to the rectifier fault. This strategic hardware solution and the reinforcement is in the line of the objectives of Energy 4.0. The simulation carried out shows that normal operation is recovered after rectifier fault clearing and imbalance voltages on the capacitor bank are circumvented due to the action of the convenient selection of voltage vectors. Also, the quality of energy injected in what regards the total harmonic distortion is not significantly affected.

References

1. Akhmatov, V., Knudsen, H., Nielsen, A.H.: Advanced simulation of windmills in the electric power supply. Int. J. Electr. Power Energy Syst. **22**, 421–434 (2000)
2. Arribas, B.N., Melicio, R., Teixeira, J.G., Mendes, V.M.F.: Vanadium redox flow battery storage system linked to the electric grid. Renew. Energy Power Qual. J. **1**(14), 1025–1030 (2016)
3. Banham-Hall, D.D., Taylor, G.A., Smith, C.A.: Flow batteries for enhancing wind power integration. IEEE Trans. Power Syst. **27**(3), 1690–1697 (2012)
4. Batista, N.C., Melicio, R., Mendes, V.M.F.: Services enabler architecture for smart grid and smart living services providers under industry 4.0. Energy Build. **141**, 16–27 (2017)
5. Brooke, J.: Wave Energy Conversion Systems, 1st edn. Elsevier Science, UK (2003)
6. Campanile, A., Scamardella, V.P.A.: Mooring design and selection for floating offshore wind turbines on intermediate and deep water depths. Ocean Eng. **148**, 349–360 (2018)
7. Carlson, G.E., Hajlijak, C.A.: Approximation of fractional capacitors $(1/s)^{1/n}$ by a regular Newton process. IEEE Trans. Circuit Theory **11**, 210–213 (1964)
8. Castro-Santos, L., Diaz-Casas, V.: Economic influence of location in floating offshore wind farms. Ocean Eng. **107**, 13–22 (2015)
9. CNESA China Energy Storage Alliance: China's top 10 storage headlines of 2016 (2017). URL http://en.cnesa.org/featured-stories/2017/1/6/chinas-top-10-storage-headlines-of-2016/. Accessed 1 June 2017
10. Community Energy Scotland: Gigha battery project (2017). URL http://www.communityenergyscotland.org.uk/gigha-battery-project.asp/. Accessed 22 May 2017
11. D., D.V., da Costa, J.S.: Tuning of fractional PID controllers with ziegler-nichols type rules. Signal Process. **86**, 2771–2784 (2006)
12. Divya, K.C., Østergaard, J.: Battery energy storage technology for power systems—an overview. Electr. Power Syst. Res. **79**, 511–520 (2009)
13. El-Kafafy, M., Devriendt, C., Guillaume, P., Helsen, J.: Automatic tracking of the modal parameters of an offshore wind turbine drivetrain system. Energies **10**(4), 1–15 (2017)
14. Gomes, L.L.R., Pousinho, H.M.I., Melcio, R., Mendes, V.M.F.: Stochastic coordination of joint wind and photovoltaic systems with energy storage in day-ahead market. Energy **124**, 310–320 (2017)
15. Guarnieri, M., Mattavelli, P., Petrone, G., Spagnuolo, G.: Vanadium redox flow batteries: potentials and challenges of an emerging storage technology. IEEE Indu. Electron. Mag. **10**(4), 20–31 (2016)
16. Haas, J., Olivares, M.A., Palma-Behnke, R.: Grid-wide subdaily hydrologic alteration under massive wind power penetration in Chile. J. Environ. Manag. **154**, 183–189 (2015)
17. Jichuan, K., Liping, S., Hai, S., Chulin, W.: Risk assessment of floating offshore wind turbine based on correlation-FMEA. Ocean Eng. **129**, 382–388 (2017)
18. Khomfoi, S., Tolbert, L.M.: Multilevel power converters. In: Rashid, M.H. (ed.) Power Electronics Handbook, 2nd edn, pp. 451–482. Academic Press, USA (2007)
19. Lanusse, P., Malti, R., Melchior, P.: CRONE control system design toolbox for the control engineering community: tutorial and case study. Philos. Trans. Royal Soc. A: Math. Phys. Eng. Sci. **371**, 1–14 (2013)
20. Lin, L., Kai, W., Vassalos, D.: Detecting wake performance of floating offshore wind turbine. Ocean Eng. **156**, 263–276 (2018)
21. Maione, G., Lino, P.: New tuning rules for fractional PI-alfa controllers. Nonlinear Dyn. **49**, 251–257 (2007)
22. Melicio, R., Mendes, V.M.F., ao, J.P.S.C.: A pitch control malfunction analysis for wind turbines with permanent magnet synchronous generator and full-power converters: proportional integral versus fractional-order controllers. Electric Power Components Syst. **38**, 387–406 (2010)
23. Melicio, R., Valério, D., Mendes, V.: Fractional control of an offshore wind system. In: Proceedings of The International Conference on Fractional Differentiation and its Applications, pp. 1–6. Amman (2018)

24. Monje, C.A., Vinagre, B.M., Feliu, V., Chen, Y.Q.: Tuning and auto-tuning of fractional order controllers for industry applications. Control Eng. Pract. **16**, 798–812 (2008)
25. Nanou, I.S., Papathanassiou, G.N.P.S.A.: Assessment of communication-independent grid code compatibilitysolutions for vschvdc connected offshore wind. Electric Power Syst. Res. **121**, 28–51 (2015)
26. Oustaloup, A., Levron, F., Matthieu, B., Nanot, F.M.: Frequency-band complex noninteger differentiatot: characterization and synthesis. IEEE Trans. Circuit. Syst. I: Funda. Theory Appl. **47**, 25–39 (2000)
27. Podlubny, I.: An Introduction to Fractional Derivatives, Fractional Differential Equations, to Methods of Their Solution and Some of Their Applications. Academic Press, San Diego (1999)
28. Robison, R.P., Sengupta, M., Rauch, D.: Intelligent energy industrial systems 4.0. IT Professional **17**, 17–24 (2015)
29. Roddier, D., Cermelli, C., Aubault, A., Weinstein, A.: Windfloat: a floating foundation for offshore wind turbines. J. Renew. Sustain. Energy **2**(3), 033,104 (2010)
30. Samko, S.G., Kilbas, A.A., Marichev, O.I.: Fractional Integrals and Derivatives: Theory and Applications. Gordon and Breach, Yverdon (1993)
31. Seixas, M., Melicio, R., Mendes, V.F.M.: Offshore wind energy system with DC transmission discrete mass: modeling and simulation. Electr. Power Compon. Syst. **44**, 2271–2284 (2016)
32. Seixas, M., Melicio, R., Mendes, V.F.M., Couto, C.: Simulation of OWES with five-level converter linked to the grid: harmonic assessment. In: Proceedings of the 9th IEEE International Conference on Compatibility and Power Electronics, pp. 126–131. Lisbon, Portugal (2015)
33. Seixas, M., Melicio, R., Mendes, V.M.F.: Offshore wind turbine simulation: multibody drive train. Back-to-back NPC (neutral point clamped) converters. Fractional-order control. Energy **69**, 357–369 (2014)
34. Seixas, M., Melicio, R., Mendes, V.M.F.: Simulation of rectifier voltage malfunction on OWECS, four-level converter, HVDC light link: smart grid context tool. Energy Convers. Manag. **97**, 140–153 (2015)
35. Seixas, M., Melicio, R., Mendes, V.M.F., Couto, C.: Blade pitch control malfunction simulation in a wind energy conversion system with MPC five-level converter. Renew. Energy **89**, 339–350 (2016)
36. Seixas, M., Mendes, V., Melicio, R.: Ride through fault on the rectifier controller of an offshore wind system aided by VRFB. In: Proceedings of the IEEE International Symposium on Power Electronics, Electrical Drives and Motion, pp. 925–930. Amalfi (2018)
37. Gryning, S.M.P., Wu, Q., Blanke, M., Niemann, H.H., Andersen, K.P.H.: Wind turbine inverter robust loop-shaping control subject to grid interaction effects. IEEE Trans. Sustain. Energy **7**, 41–50 (2016)
38. Valrio, D., da Costa, J.S.: An Introduction to Fractional Control. IET, Stevenage (2013)
39. Yang Z.C.Y.: A survey of fault diagnosis for onshore grid-connected converter in wind energy conversion systems. IEEE Renew. Sustain. Energy Rev. **66**, 345–359 (2016)
40. Zhao, H., Wu, Q., Wang, J., Lui, Z., Shahidehpour, M., Xue, Y.: Combined active and reactive power control of wind farms based on model predictive control. IEEE Trans. Energy Convers. **16**, 86–803 (2013)

Soft Numerical Algorithm with Convergence Analysis for Time-Fractional Partial IDEs Constrained by Neumann Conditions

Omar Abu Arqub, Mohammed Al-Smadi and Shaher Momani

Abstract Some scientific pieces of research are governed by classes of partial integro-differential equations (PIDEs) of fractional order that are leading to novel challenges in simulation and optimization. In this chapter, a soft numerical algorithm is proposed and analyzed to fitted analytical solutions of PIDEs with appropriate initial and Neumann conditions in Sobolev space. Meanwhile, the solutions are represented in series form with strictly computable components. By truncating n-term approximation of the analytical solution, the solution methodology is discussed for both linear and nonlinear problems based on the nonhomogeneous term. Analysis of convergence and smoothness are given under certain assumptions to show the theoretical structures of the method. Dynamic features of the approximate solutions are studied through an illustrated example. The yield of numerical results indicates the accuracy, clarity, and effectiveness of the proposed algorithm as well as provide a proper methodology in handling such fractional issues.

Keywords Partial integro-differential equations · Reproducing kernel algorithm · Fredholm and Volterra operators · Fractional derivatives

1 Introduction

In the past years, fractional calculus theory has gained a considerable attention in diverse fields of science and engineering according to the enormous range of real-

O. A. Arqub (✉) · S. Momani
Department of Mathematics, Faculty of Science, The University of Jordan,
Amman, 11942, Jordan
e-mail: o.abuarqub@ju.edu.jo

M. Al-Smadi
Department of Applied Science, Ajloun College, Al-Balqa Applied University,
Ajloun 26816, Jordan

S. Momani
Department of Mathematics and Sciences, College of Humanities and Sciences, Ajman
University, Ajman, UAE

© Springer Nature Singapore Pte Ltd. 2019
P. Agarwal et al. (eds.), *Fractional Calculus*, Springer Proceedings
in Mathematics & Statistics 303, https://doi.org/10.1007/978-981-15-0430-3_7

107

world applications and the critical role that it plays to describe the complex dynamical behaviors to such models including fluid dynamic model, traffic flow model,
convection-diffusion model, heat flux model, and so forth [1–4]. As well as, this
topic aids to simplify the controlling design without any shortage of hereditary. The
derivatives of fractional order are powerful to interpret several physical problems,
for instance, electrical circuits, damping laws, and controlled damper. Further, fractional partial differential equations (FPDEs) are constructed due to attention among
the scientists and engineers to exact explanation of nonlinear phenomena which
appear, for example, in fluid mechanics wherever continuum assumption does not
well, and therefore fractional model can be deemed to be the best operator [5–10].
Developing numeric-analytic techniques to the solutions of PIDEs of fractional order
is an essential task. Since it is challenging to get closed-form solutions to fractional
PIDEs in many situations. So a lot of efforts have been made to introduce and improve
analytical techniques that help us to obtain the analytic solution of those fractional
differential equations [11–17].

The reproducing kernel (RK) technique is a well-known systematic approach
in obtaining a feasible solution of both linear and nonlinear differential or integral
operators involving ordinary differential, partial differential, integral and integro-
differential, fractional differential, fuzzy differential, and delay differential equations [18–30]. The RKA is a superb, suitable, and useful tool to provide accurate
and appropriate algorithms for numeric simulations to natural phenomena arising
in physics, chemistry, biology, ecology, and engineering such as diffusive transport,
viscoelastic materials, fluid rheology, intelligent transportation systems, electromag-
netic theory, and probability [31–46]. Inspired by the areas mentioned above, the RK
method has been successfully implemented and utilized directly in solving nonlinear
initial or boundary value problems without the need for unphysically tied hypotheses,
linearization, transformations, discretization, or even perturbation.

In this chapter, we display a soft numerical algorithm, called reproducing kernel,
for handling classes of time-fractional PIDEs restricted by initial and Neumann
functions as follows:

$$\partial_{\xi^a}^{\alpha} \varphi(\eta, \xi) + \mu_1 \partial_{\eta^2}^2 \varphi(\eta, \xi) + \mu_2 \partial_\eta \varphi(\eta, \xi) + \mu_3 \varphi(\eta, \xi) + P\left[\varphi(\eta, \xi)\right] + Q\left[\varphi(\eta, \xi)\right] = F(\eta, \xi),$$

$$P\left[\varphi(\eta, \xi)\right] = \lambda_1 \int_0^1 k_1\left(\eta, \xi, \rho\right) \left(\mu_4 \partial_{\eta^2}^2 \varphi(\eta, \rho) + \mu_5 \partial_\eta \varphi(\eta, \rho) + \mu_6 \varphi(\eta, \rho)\right) d\rho,$$

$$Q\left[\varphi(\eta, \xi)\right] = \lambda_2 \int_0^\eta k_2\left(\eta, \xi, \rho\right) \left(\mu_7 \partial_{\eta^2}^2 \varphi(\eta, \rho) + \mu_8 \partial_\eta \varphi(\eta, \rho) + \mu_9 \varphi(\eta, \rho)\right) d\rho,$$

$$\tag{1}$$

with the initial and Neumann conditions

$$\varphi(\eta, 0) = \omega(\eta),$$
$$\partial_\eta \varphi(0, \xi) = \nu_1(\xi), \partial_\eta \varphi(1, \xi) = \nu_2(\xi),$$

$$\tag{2}$$

where $0 < \alpha \le 1$, $0 \le \eta, \xi, \rho \le 1$, μ_i, $i = 1, 2, ..., 9$ are real finite constant, λ_1 and
λ_2 are constant parameters, $k_1\left(\eta, \xi, \rho\right)$ and $k_2\left(\eta, \xi, \rho\right)$ are arbitrary continuous kernel functions on the cube $[0, 1]^3$, $\omega(\eta)$, $\nu_1(\xi)$, $\nu_2(\xi)$ and $F(\eta, \xi)$ are continuous
functions over the required domain, and $\varphi(\eta, \rho)$ is an analytical solution to be found

numerically. Hereby, we assume that Eq. (1) with conditions (2) has a unique smooth solution. Further, $\partial_{\xi^\alpha}^\alpha$ denotes the time-fractional Caputo derivative of order α, which is given by

$$\partial_{\xi^\alpha}^\alpha \varphi(\eta, \xi) = \frac{1}{\Gamma(1-\alpha)} \int_0^\xi (\xi - \tau)^{-\alpha} \partial_\tau \varphi(\eta, \tau) d\tau. \tag{3}$$

The rest of this chapter is organized as follows. In Sect. 2, the Hilbert spaces required are extended, as well the reproducing kernel functions are provided. The issue is formulated and the computational RK algorithm is presented in Sect. 3. Meanwhile, some necessary theoretical results and convergent validity are studied in the same section. In Sect. 4, numerical results are discussed to show the reliability and efficiency of the proposed algorithm. Finally, this chapter ends with conclusions.

2 Preliminary Spaces

For clarity of presentation, some necessary definitions and preliminary facts are presented. Henceforth, $\|u\|_\Pi^2 = \langle u(\eta), u(\eta) \rangle_\Pi$, $u \in \Pi$, $\eta \in [0, 1]$, and Π is a Hilbert space, and $L^2[0, 1] = \left\{ r \mid \int_0^1 r^2(\rho) d\rho < \infty \right\}$.

Definition 1 [17] If Π is a Hilbert space defined on a nonempty set Ω, then $\mathcal{F}: \Omega \times \Omega \to \mathbb{R}$ is called a reproducing kernel function (RKF) of the space Π when both two conditions are met:

1. For each $\eta \in \Omega$, we have $\mathcal{F}(\rho, \eta) \in \Pi$.
2. For each $\psi \in \Pi$ and each $\eta \in \Omega$, we have $\langle \psi(\rho), \mathcal{F}(\rho, \eta) \rangle = \psi(\eta)$. (*Reprod-ucing Property*)

The space Π that possesses a reproducing kernel is said to be reproducing kernel Hilbert space. The RKF $\mathcal{F}$ of the space Π completely determines the Hilbert space Π.

Next, the required Hilbert spaces $_rH_2^1[0, 1]$ and $_dH_2^2[0, 1]$ will be defined, which are possessing RKFs $_rR_\eta^{\{1\}}(\xi)$ and $_dR_\eta^{\{2\}}(\xi)$, respectively.

Definition 2 [17] Let $r'(\eta)$ be in the space $L^2[0, 1]$. The space $_rH_2^1[0, 1]$ given by $_rH_2^1[0, 1] = \{r = r(\eta) : r$ is absolutely continuous over $[0, 1]\}$. The inner product is equipped by

$$\langle r_1(\eta), r_2(\eta) \rangle_{,H_2^1} = r_1(0) r_2(0) + \int_0^1 r_1'(\rho) r_1'(\rho) d\rho. \tag{4}$$

Remark 1 The space $_rH_2^1[0, 1]$ is complete reproducing kernel and the RKF is obtained as follows:

$$_rR_\eta^{\{1\}}(\xi) = 1 + mim\{\eta, \xi\}. \tag{5}$$

Anyhow, if $[0, 1]$ is the desired domain in direction of τ, then the complete reproducing kernel space $_{\hat{\tau}}H_2^1[0, 1]$ can be defined. Further, the inner product is equipped by

$$\langle r_1(\tau), r_2(\tau) \rangle_{\hat{\tau}H_2^1} = r_1(0)\, r_2(0) + \int_0^1 r_1'(\rho)r_1'(\rho)d\rho.$$

The kernel function is $_{\hat{\tau}}R_\tau^{\{1\}}(\xi) = 1 + mim\{\tau, \xi\}$.

Definition 3 Let $r''(\eta)$ be in the space $L^2[0, 1]$. The space $_dH_2^2[0, 1]$ is given by $_dH_2^2[0, 1] = \{r = r(\eta) : r, r'$ are absolutely continuous over $[0, 1]$, and $r(0) = 0\}$. The inner product is equipped by

$$\langle r_1(\eta), r_2(\eta) \rangle_{dH_2^2} = \sum_{i=0}^2 r_1^{(i)}(0)\, r_2^{(i)}(0) + \int_0^1 r_1''(\rho)r_2''(\rho)d\rho. \tag{6}$$

Remark 2 The space $_dH_2^2[0, 1]$ is complete reproducing kernel and the RKF is obtained as follows:

$$_dR_\eta^{\{2\}}(\xi) = \frac{1}{12}\begin{cases} h_\eta(\xi), & \eta \le \xi, \\ h_\xi(\eta), & \eta > \xi, \end{cases} \tag{7}$$

where $h_\eta(\xi) = 12\eta\xi + 6\eta\xi^2 - 2\xi^3$.

Definition 4 Let $r'''(\eta)$ be in the space $L^2[0, 1]$. The space $_dH_2^3[0, 1]$ is given by $_dH_2^3[0, 1] = \{r = r(\eta) : r, r', r''$ are absolutely continuous over $[0, 1]$, and $r'(0) = r'(1) = 0\}$. The inner product is equipped by

$$\langle r_1(\eta), r_2(\eta) \rangle_{dH_2^3} = \sum_{i=0}^1 r_1^{(i)}(0)\, r_2^{(i)}(0) + r_1(1)r_2(1) + \int_0^1 r_1'''(\rho)r_2'''(\rho)d\rho. \tag{8}$$

Remark 3 The space $_dH_2^3[0, 1]$ is complete reproducing kernel, and the RKF is obtained as follows:

$$_dR_\eta^{\{3\}}(\xi) = \frac{1}{120}\begin{cases} \left((1-\xi)^3\eta^3\right) g_\eta(\xi), & \eta \le \xi, \\ \left((1-\eta)^3\xi^3\right) g_\xi(\eta), & \eta > \xi, \end{cases} \tag{9}$$

where $g_\eta(\xi) = 6\eta^2\xi^2 + 3\eta\xi(\xi - 5\eta) + (10\eta^2 - 5\eta\xi + \xi^2)$.

Next, to extend the novel inner product spaces and to fit its reproducing kernel functions, we construct a reproducing kernel space $W(\Lambda)$, $\Lambda = [0, 1] \otimes [0, 1]$, in which every function satisfies the constraints homogeneous initial and Neumann conditions of the above-mentioned time-fractional PIDEs.

Definition 5 Let be $\partial_{\eta^3}^3 \partial_{\xi^3}^3 \varphi$ in the space $L^2(\Lambda)$. The Hilbert space $W(\Lambda)$ is defined as $W(\Lambda) = \{\varphi = \varphi(\eta, \xi) : \partial_{\eta^2}^2 \partial_{\xi^2}^2 \varphi$ are complete continuous functions in Λ, and $\varphi(\eta, 0) = \partial_\eta \varphi(0, \xi) = \partial_\eta \varphi(1, \xi) = 0\}$. The metric system structure lies in

$$\langle \varphi_1(\eta, \xi), \varphi_2(\eta, \xi) \rangle_W = \sum_{i=0}^{1} \left\langle \partial_{\xi^i}^i \varphi_1(\eta, 0), \partial_{\xi^i}^i \varphi_2(\eta, 0) \right\rangle_{d H_2^3}$$

$$+ \int_0^1 \left[\sum_{i=0}^{1} \partial_{\xi 2}^2 \partial_{\eta^i}^i \varphi_1(0, \xi) \partial_{\xi 2}^2 \partial_{\eta^i}^i \varphi_2(0, \xi) + \partial_{\xi 2}^2 \varphi_1(1, \xi) \partial_{\xi 2}^2 \varphi_2(1, \xi) \right] d\xi \quad (10)$$

$$+ \int_0^1 \int_0^1 \partial_{\xi 3}^3 \partial_{\eta 2}^2 \varphi_1(\eta, \xi) \partial_{\eta 3}^3 \partial_{\xi 2}^2 \varphi_2(\eta, \xi) \, d\eta d\xi.$$

Theorem 1 *The space $W(\Lambda)$ is a complete reproducing kernel, and its RKF is defined by*

$$R_{(x,t)}^W(\eta, \xi) = \left({}_d R_x^{\{3\}}(\eta) \right) \left({}_d R_t^{\{2\}}(\xi) \right), \quad (11)$$

where the functions ${}_d R_t^{\{2\}}(\xi)$ and ${}_d R_x^{\{3\}}(\eta)$ are RKFs of ${}_d H_2^2[0, 1]$ and ${}_d H_2^3[0, 1]$, respectively.

Proof By utilizing the features of $\langle r_1(\eta), r_2(\eta) \rangle_{d H_2^2}$ and $\langle r_1(\eta), r_2(\eta) \rangle_{d H_2^3}$, it follows that

$$\left\langle \varphi(\eta, \xi), {}_d R_x^{\{3\}}(\eta) {}_d R_t^{\{2\}}(\xi) \right\rangle_W = \sum_{i=0}^{1} \left\langle \partial_{\xi^i}^i \varphi(\eta, 0), \partial_{\xi^i}^i \left({}_d R_x^{\{3\}}(\eta) {}_d R_t^{\{2\}}(0) \right) \right\rangle_{d H_2^3}$$

$$+ \int_0^1 \left[\begin{array}{l} \sum_{i=0}^{1} \partial_{\xi 2}^2 \partial_{\eta^i}^i \varphi(0, \xi) \partial_{\xi 2}^2 \partial_{\eta^i}^i \left({}_d R_x^{\{3\}}(0) {}_d R_t^{\{2\}}(\xi) \right) \\ + \partial_{\xi 2}^2 \varphi(1, \xi) \partial_{\xi 2}^2 \left({}_d R_x^{\{3\}}(1) {}_d R_t^{\{2\}}(\xi) \right) \end{array} \right] d\xi$$

$$+ \int_0^1 \int_0^1 \partial_{\xi 3}^3 \partial_{\eta 2}^2 \varphi(\eta, \xi) \partial_{\eta 3}^3 \partial_{\xi 2}^2 \left({}_d R_x^{\{3\}}(\eta) {}_d R_t^{\{2\}}(\xi) \right) d\eta d\xi$$

$$= \sum_{i=0}^{1} \left\langle \partial_{\xi^i}^i \varphi(\eta, 0), {}_d R_x^{\{3\}}(\eta) \partial_{\xi^i d}^i R_t^{\{2\}}(0) \right\rangle_{d H_2^3}$$

$$+ \int_0^1 \left[\begin{array}{l} \sum_{i=0}^{1} \partial_{\xi 2}^2 \partial_{\eta^i}^i \varphi(0, \xi) \partial_{\xi 2 d}^2 R_t^{\{2\}}(\xi) \partial_{\eta^i d}^i R_x^{\{3\}}(0) \\ + \partial_{\xi 2}^2 \varphi(1, \xi) {}_d R_x^{\{3\}}(1) \partial_{\xi 2 d}^2 R_t^{\{2\}}(\xi) \end{array} \right] d\xi$$

$$+ \int_0^1 \left(\int_0^1 \partial_{\xi 3}^3 \partial_{\eta 2}^2 \varphi(\eta, \xi) \partial_{\eta 3 d}^3 R_x^{\{3\}}(\eta) \partial_{\xi 2 d}^2 R_t^{\{2\}}(\xi) \, d\eta \right) d\xi$$

$$= \sum_{i=0}^{1} \left\langle \partial_{\xi^i}^i \varphi(\eta, 0), {}_d R_x^{\{3\}}(\eta) \right\rangle_{d H_2^3} \partial_{\xi^i d}^i R_t^{\{2\}}(0)$$

$$+ \int_0^1 \partial_{\xi 2 d}^2 R_t^{\{2\}}(\xi) \partial_{\xi 2}^2 \left[\begin{array}{l} \sum_{i=0}^{1} \partial_{\eta^i}^i \varphi(0, \xi) \partial_{\eta^i d}^i R_x^{\{3\}}(0) + \varphi(1, \xi) {}_d R_x^{\{3\}}(1) \\ + \int_0^1 \partial_{\xi 3}^3 \varphi(\eta, \xi) \partial_{\eta 3 d}^3 R_x^{\{3\}}(\eta) \end{array} \right] d\xi$$

$$= \sum_{i=0}^{1} \partial_{\xi^i}^i \varphi(x, 0) \partial_{\xi^i d}^i R_t^{\{2\}}(0) + \int_0^1 \partial_{\xi 2 d}^2 R_t^{\{2\}}(\xi) \partial_{\xi 2}^2 \left\langle \varphi(\eta, \xi), {}_d R_x^{\{3\}}(\eta) \right\rangle_{d H_2^3} d\xi$$

$$= \sum_{i=0}^{1} \partial_{\xi^i}^i \varphi(x, 0) \partial_{\xi^i d}^i R_t^{\{2\}}(0) + \int_0^1 \partial_{\xi 2 d}^2 R_t^{\{2\}}(\xi) \partial_{\xi 2}^2 \varphi(x, \xi) d\xi$$

$$= \left\langle \varphi(x, \xi), {}_d R_t^{\{2\}}(\xi) \right\rangle_{d H_2^3} = \varphi(x, t).$$

Hence, $\left\langle \varphi(\eta, \xi), R_{(x,t)}(\eta, \xi) \right\rangle_W = \varphi(x, t)$.   ■

Definition 6 Let be $\partial_\eta \partial_\xi \varphi$ in the space $L^2(\Lambda)$. The space $H(\Lambda)$ is given by $H(\Lambda) = \{\varphi = \varphi(\eta, \xi) : \varphi$ is continuous functions on $\Lambda\}$, and the metric space can be obtained by

$$\langle \varphi_1(\eta, \xi), \varphi_2(\eta, \xi) \rangle_H = \langle \varphi_1(\eta, 0), \varphi_2(\eta, 0) \rangle_{_dH_2^2} \tag{12}$$

$$+ \int_0^1 \partial_\xi \varphi_1(0, \xi) \partial_\xi \varphi_2(0, \xi) \, d\xi + \int_0^1 \int_0^1 \partial_{\eta\xi}^2 \varphi_1(\eta, \xi) \partial_{\eta\xi}^2 \varphi_2(\eta, \xi) \, d\eta d\xi.$$

Theorem 2 *The space $H[\Lambda]$ is complete reproducing kernel, and the RKF is*

$$R_{(x,t)}^H (\eta, \xi) = \left({}_{\hat{\tau}} R_\tau^{\{1\}}(\xi) \right) \left({}_r R_\eta^{\{1\}}(\xi) \right), \tag{13}$$

where the functions ${}_r R_\eta^{\{1\}}(\xi)$ and ${}_{\hat{\tau}} R_\tau^{\{1\}}(\xi)$ are RKFs of ${}_r H_2^1[0, 1]$ and ${}_{\hat{\tau}} H_2^1[0, 1]$, respectively.

3 Numerical RK Algorithm

In this section, the statement of PIDEs (1) and (2) is being redrafted in Hilbert space $W(\Lambda)$. After homogenizing the inhomogeneous restriction conditions using appropriate transformation, the differential operator $T : W(\Lambda) \rightarrow H(\Lambda)$ of fractional order α can be defined, which is invertible, linear, and bounded, such that

$$T\varphi(\eta, \xi) = \partial_{\xi^\alpha}^\alpha \varphi(\eta, \xi) + \mu_1 \partial_{\eta^2}^2 \varphi(\eta, \xi) + \mu_2 \partial_\eta \varphi(\eta, \xi) + \mu_3 \varphi(\eta, \xi) + P\left[\varphi(\eta, \xi)\right] + Q\left[\varphi(\eta, \xi)\right].$$

Therefore, the original FPIED statement will be converted equivalently into the following form:

$$T\varphi(\eta, \xi) = F(\eta, \xi), \tag{14}$$

with respect to the homogeneous initial and Neumann conditions

$$\begin{aligned} \varphi(\eta, 0) &= 0, \\ \partial_\eta \varphi(0, \xi) &= 0, \ \partial_\eta \varphi(1, \xi) = 0. \end{aligned} \tag{15}$$

The orthogonal function systems of $W(\Lambda)$ can be constructed by choosing a countable dense set $\{(\eta_i, \xi_i)\}_{i=1}^\infty$ of Λ, and defining $\omega_i(\eta, \xi) = R_{(\eta_i, \xi_i)}^H (\eta, \xi)$ and $\psi_i(\eta, \xi) = T^* \omega_i(\eta, \xi)$, in which T^* is the adjoint operator of T such tah $T^* : H(\Lambda) \rightarrow W(\Lambda)$. While, the orthonormal basis $\{\bar{\psi}_i(\eta, \xi)\}_{i=1}^\infty$ of the space $W(\Lambda)$ can be obtained using the procedures of G-Schmidt normalization to $\{\psi_i(\eta, \xi)\}_{i=1}^\infty$ such that

$$\overline{\psi}_i(\eta, \xi) = \sum_{k=1}^i \beta_{ik} \psi_k(\eta, \xi). \tag{16}$$

Theorem 3 *The system $\{\psi_i(\eta, \xi)\}_{i=1}^{\infty}$ is complete orthogonal basis of the space $W(\Lambda)$ as follows:*

$$\psi_i(\eta, \xi) = T_{(x,t)}\left(R_{(x,t)}^W(\eta, \xi)\right)\Big|_{(x,t)=(\eta_i,\xi_i)},$$

where $T_{(x,t)}$ indicates that the operator T applies to the function of (x, t).

Proof Clearly that

$$
\begin{aligned}
\psi_i(\eta, \xi) &= T^*\omega_i(\eta, \xi) = \left\langle T^*\omega_i(x, t), R_{(\eta,\xi)}^W(x, t)\right\rangle_W \\
&= \left\langle \omega_i(x, t), T_{(x,t)} R_{(\eta,\xi)}^W(x, t)\right\rangle_H \\
&= T_{(\eta,\xi)}\left(R_{(\eta,\xi)}^W(x, t)\right)\Big|_{(x,t)=(\eta_i,\xi_i)} \\
&= T_{(x,t)}\left(R_{(x,t)}^W(\eta, \xi)\right)\Big|_{(x,t)=(\eta_i,\xi_i)} \in W(\Lambda).
\end{aligned}
\tag{17}
$$

Consequently, for each fixed $\varphi \in W(\Lambda)$, let $\langle\varphi(\eta, \xi), \psi_i(\eta, \xi)\rangle_w = 0$, $i = 1, 2, \dots$. Then, $\langle\varphi(\eta, \xi), \psi_i(\eta, \xi)\rangle_w = \langle\varphi(\eta, \xi), T^*\omega_i(\eta, \xi)\rangle_w = \langle T\varphi(\eta, \xi), \varphi_i(x)\rangle_H = T\varphi(\eta_i, \xi_i) = 0$. Since $\{(\eta_i, \xi_i)\}_{i=1}^{\infty}$ is dense on Λ, therefore $L\varphi(\eta, \xi) = 0$. It follows that $\varphi(\eta, \xi) = 0$ by applying $T^{-1}\varphi$. ∎

Remark 4 The sequence $\left\{R_{(x_i,t_i)}^W(\eta, \xi)\right\}_{i=1}^{\infty}$ is linear independent basis on $W(\Lambda)$.

Theorem 4 *Suppose that $A_m = \sum_{k=1}^{m}\beta_{ik}F(\eta_k, \xi_k)$. Let $\varphi(\eta, \xi) \in W(\Lambda)$ be the the analytical solution of Eqs. (14) and (15), then it has the following form:*

$$\varphi(\eta, \xi) = \sum_{m=1}^{\infty} A_m \bar{\psi}_m(\eta, \xi). \tag{18}$$

Proof By utilizing the features of $\langle\varphi_1(\eta, \xi), \varphi_2(\eta, \xi)\rangle_W$, it follows that

$$
\begin{aligned}
\varphi(\eta, \xi) &= \sum_{m=1}^{\infty}\langle\varphi(\eta, \xi), \bar{\psi}_m(\eta, \xi)\rangle_W \bar{\psi}_m(\eta, \xi) \\
&= \sum_{m=1}^{\infty}\sum_{k=1}^{m}\beta_{mk}\langle\varphi(\eta, \xi), \psi_k(\eta, \xi)\rangle_W \bar{\psi}_m(\eta, \xi) \\
&= \sum_{m=1}^{\infty}\sum_{k=1}^{m}\beta_{mk}\langle\varphi(\eta, \xi), T^*\omega_k(\eta, \xi)\rangle_W \bar{\psi}_m(\eta, \xi) \\
&= \sum_{m=1}^{\infty}\sum_{k=1}^{m}\beta_{mk}\langle T\varphi(\eta, \xi), \omega_k(\eta, \xi)\rangle_H \bar{\psi}_m(\eta, \xi) \\
&= \sum_{m=1}^{\infty}\sum_{k=1}^{m}\beta_{mk}\langle F(\eta, \xi), \omega_k(\eta, \xi)\rangle_H \bar{\psi}_m(\eta, \xi) \\
&= \sum_{m=1}^{\infty}\sum_{k=1}^{m}\beta_{mk}F(\eta_k, \xi_k)\bar{\psi}_m(\eta, \xi) \\
&= \sum_{m=1}^{\infty} A_m \bar{\psi}_m(\eta, \xi). \quad ∎
\end{aligned}
$$

Remark 5 The n-term approximate solution of the analytical solution described in Eq. (18) can be given by

$$\varphi_n(\eta, \xi) = \sum_{i=1}^{n} \sum_{k=1}^{i} \beta_{ik} F(\eta_k, \xi_k) \bar{\psi}_i(\eta, \xi). \tag{19}$$

According to the proposed algorithm, the required domain Λ can be divided into finite $r \times s$ grid points with respect to η space-direction, $\Delta \eta = \frac{1}{r}$, and to ξ time-direction, $\Delta \xi = \frac{1}{s}$ over $[0, 1]$, respectively, $r, s \in \mathbb{N}$, $r, s > 0$. Anyhow, the ordered pair (η_l, ξ_m) of Λ can be given simultaneously by

$$(\eta_l, \xi_m) = (l \Delta \eta, m \Delta \xi), l = 0, 1, ..., r, \ m = 0, 1, ..., s.$$

The following computational RK algorithm is given to summarize the procedures of the proposed method in solving those time-fractional PIDEs.

Algorithm 1 To obtain approximate solution $\varphi_n(\eta, \xi)$ of the analytical solution $\varphi(\eta, \xi)$ of BVPs (14) and (15) on Λ, the following steps can be carried out:

Step A: Divide the required domain Λ into grid points such as $n = rs$, $r, s \in \mathbb{N}$, $r, s > 0$;

Step B: Put $\psi_n(\eta_n, \xi_n) = T\left(R_{(x,t)}^H(\eta, \xi)\right)\Big|_{(x,t)=(\eta_n, \xi_n)}$;

Step C: Find orthonormal coefficients β_{nk}, then let $\bar{\psi}_n(\eta_n, \xi_n) = \sum_{k=1}^{n} \beta_{nk} \psi_n$ $(\eta_n, \xi_n), n = 1, 2, ..., rs$;

Step D: Ues the initial data $\varphi_0(\eta_1, \xi_1)$; then do the following subroutine:

  For $n = 1, n + +$;

  Let $A_n = \sum_{k=1}^{n} \beta_{nk} F(\eta_k, \xi_k)$;

  Set $\varphi_n(\eta_n, \xi_n) = \sum_{k=1}^{n} \sum_{j=1}^{n} A_k \bar{\psi}_k(\eta_k, \xi_k)$;

Step E: If $n < rs$, then let $n = n+1$ and go to Step D; Otherwise, Stop.

Theorem 5 *Let* $\|\varphi(\eta, \xi)\|_W$ *be bounded on* Λ, *then the n-term approximate solution* $\varphi_n(\eta, \xi)$ *in Eq. (19) converges to the analytical solution* $\varphi(\eta, \xi)$ *of Eqs. (14) and (15) in* $W(\Lambda)$ *that is given as* $\varphi(\eta, \xi) = \sum_{i=1}^{\infty} \sum_{k=1}^{i} \beta_{ik} F(\eta_k, \xi_k) \bar{\psi}_i(\eta, \xi)$.

Proof Let $\delta_n = \|\varphi(\eta, \xi) - \varphi_n(\eta, \xi)\|_W$ the nature error at $(\eta, \xi) \in \Lambda$, then $\delta_{n-1}^2 = \left\| \sum_{i=n}^{\infty} \sum_{k=1}^{i} \beta_{ik} F(\eta_k, \xi_k) \bar{\psi}_i(\eta, \xi) \right\|_W^2 = \sum_{i=n}^{\infty} \left(\sum_{k=1}^{i} \beta_{ik} F(\eta_k, \xi_k) \right)^2$ and $\delta_n^2 = $

$\left\| \sum_{i=n+1}^{\infty} \sum_{k=1}^{i} \beta_{ik} F(\eta_k, \xi_k) \bar{\psi}_i(\eta, \xi) \right\|_W^2 = \sum_{i=n}^{\infty} \left(\sum_{k=1}^{i} \beta_{ik} F(\eta_k, \xi_k) \right)^2$. Thus, $\delta_{n-1} \geq \delta_n$.

Consequently $\{\delta_n\}_{n=1}^{\infty}$ is decreasing with respect to the norm of $W(\Lambda)$. If $\sum_{i=1}^{\infty} A_i \bar{\psi}_i$ (η, ξ) is convergent, then, $\|\varphi(\eta, \xi) - \varphi_n(\eta, \xi)\|_W \to 0$ as soon as $n \to \infty$. ∎

4 Numerically Gained Results

In this section, some examples are quantitatively discussed at certain grid points on Λ to demonstrate the ability and performance of the proposed method in solving those fractional PIDEs. For computation, all symbolic and numerical calculations are performed using the Mathematica 9.0.

Example 1 We consider the linear time-fractional PIDE in the following form:

$$\partial_{\xi\alpha}^{\alpha}\varphi(\eta,\xi) + \partial_{\eta^2}^{2}\varphi(\eta,\xi) - \partial_{\eta}\varphi(\eta,\xi) + x\varphi(\eta,\xi) + P\left[\varphi(\eta,\xi)\right] - Q\left[\varphi(\eta,\xi)\right] = F(\eta,\xi),$$

$$P\left[\varphi(\eta,\xi)\right] = \int_{0}^{1} (\xi-\rho)\,e^{\eta+\rho}\left(\partial_{\eta^2}^{2}\varphi(\eta,\rho) + \partial_{\eta}\varphi(\eta,\rho) + \varphi(\eta,\rho)\right)d\rho,$$

$$Q\left[\varphi(\eta,\xi)\right] = \int_{0}^{\eta} \rho^{\alpha+1}e^{\eta-\xi}\left(\partial_{\eta^2}^{2}\varphi(\eta,\rho) + 3\partial_{\eta}\varphi(\eta,\rho) + 4\varphi(\eta,\rho)\right)d\rho,$$

$$\tag{20}$$

with the initial and Neumann conditions

$$\varphi(\eta,0) = 0,$$
$$\partial_{\eta}\varphi(0,\xi) = e^{-1}\xi^{2\alpha},\ \partial_{\eta}\varphi(1,\xi) = 2\xi^{\alpha+1}(\xi^{\alpha-1}+1),\tag{21}$$

where $0 \le \eta, \xi \le 1, 0 < \alpha \le 1$ and $F(\eta,\xi)$ are given functions such that the selected solution will be satisfied in the both left- and right-hand sides of BVPs. Equations (20) and (21) over the domain Λ. Here, the exact solution is $\varphi(\eta,\xi) = \eta e^{\eta-1}\xi^{2\alpha} + \eta^2\xi^{\alpha+1}$.

Example 2 Consider the following nonlinear time-fractional PIDE:

$$\partial_{\xi\alpha}^{\alpha}\varphi(\eta,\xi) - \partial_{\eta^2}^{2}\varphi(\eta,\xi) + e^{\eta}\varphi^3(\eta,\xi) + \sin(\eta)\varphi^2(\eta,\xi) + \partial_{\eta}\varphi(\eta,\xi) - P\left[\varphi(\eta,\xi)\right] - Q\left[\varphi(\eta,\xi)\right] = F(\eta,\xi),$$

$$P\left[\varphi(\eta,\xi)\right] = \int_{0}^{1} (\xi+\eta)\,e^{\eta}\rho^{\alpha}\left(\partial_{\eta^2}^{2}\varphi(\eta,\rho) + \partial_{\eta}\varphi(\eta,\rho) + \varphi(\eta,\rho)\right)d\rho,$$

$$Q\left[\varphi(\eta,\xi)\right] = \int_{0}^{\eta} \sin(\eta\xi)\left(\partial_{\eta^2}^{2}\varphi(\eta,\rho) - \partial_{\eta}\varphi(\eta,\rho) - \varphi(\eta,\rho)\right)d\rho,$$

$$\tag{22}$$

with the initial and Neumann conditions

$$\varphi(\eta,0) = 0,$$
$$\partial_{\eta}\varphi(0,\xi) = \xi^{3\alpha+1},\ \partial_{\eta}\varphi(1,\xi) = \ln(0.5)\xi^{3\alpha+1},\tag{23}$$

where $0 \le \eta, \xi \le 1, 0 < \alpha \le 1$ and $F(\eta,\xi)$ is given function such that the selected solution will be satisfied in the both left- and right-hand sides of BVPs (20) and (21) over the domain Λ. Here, the exact solution is $\varphi(\eta,\xi) = (1-\eta)\ln(1+\eta)\cos^2(3\pi\eta)\,\xi^{3\alpha+1}$.

To demonstrate the effectiveness of the RK solutions, the examples above are tested across the domain Λ. Anyhow, results from numerical analysis are an approximation, in general, which can be made as accurate as desired. Because a computer has a finite word length, only a fixed number of digits are stored and used during computations. Following, absolute errors of approximate solution $\varphi_n(\eta,\xi)$ for both Examples

Table 1 Absolute errors of Example 1 over Λ

x	t	$\alpha = 0.25$	$\alpha = 0.5$	$\alpha = 0.75$	$\alpha = 1$
0	0.25	$1.5202955220 \times 10^{-7}$	$8.5851595338 \times 10^{-7}$	$9.3131586223 \times 10^{-7}$	$1.6088074877 \times 10^{-8}$
	0.5	$3.1310929434 \times 10^{-6}$	$2.9629678339 \times 10^{-7}$	$4.4292618364 \times 10^{-7}$	$9.7271247494 \times 10^{-7}$
	0.75	$4.0499972764 \times 10^{-6}$	$1.6829336033 \times 10^{-6}$	$8.1459842130 \times 10^{-6}$	$5.9375835113 \times 10^{-7}$
	1	$8.9639505999 \times 10^{-6}$	$9.9022004243 \times 10^{-6}$	$2.3384483768 \times 10^{-6}$	$8.1051406404 \times 10^{-6}$
0.25	0.25	$8.2373427553 \times 10^{-7}$	$6.5983172765 \times 10^{-7}$	$9.3054980205 \times 10^{-7}$	$7.6310676740 \times 10^{-7}$
	0.5	$2.7232898676 \times 10^{-6}$	$1.0693336383 \times 10^{-7}$	$3.9973040691 \times 10^{-7}$	$7.7728964454 \times 10^{-7}$
	0.75	$3.2714639390 \times 10^{-6}$	$3.4265168332 \times 10^{-6}$	$8.9924037150 \times 10^{-6}$	$8.1729535402 \times 10^{-6}$
	1	$1.8522344410 \times 10^{-6}$	$8.6625479440 \times 10^{-6}$	$9.7462801838 \times 10^{-6}$	$1.1749765902 \times 10^{-6}$
0.5	0.25	$4.5870682134 \times 10^{-7}$	$3.1634826495 \times 10^{-7}$	$2.7065700155 \times 10^{-7}$	$5.9471694210 \times 10^{-8}$
	0.5	$9.9056258776 \times 10^{-6}$	$4.4738559389 \times 10^{-6}$	$4.6166202647 \times 10^{-7}$	$6.8824054303 \times 10^{-7}$
	0.75	$3.0618500642 \times 10^{-6}$	$6.7163511298 \times 10^{-6}$	$6.9702799521 \times 10^{-6}$	$3.8457050803 \times 10^{-7}$
	1	$2.7560569249 \times 10^{-6}$	$4.2505438452 \times 10^{-6}$	$8.1256012945 \times 10^{-6}$	$8.2173135294 \times 10^{-6}$
0.75	0.25	$1.7835134763 \times 10^{-7}$	$7.3344192438 \times 10^{-7}$	$3.6082013498 \times 10^{-7}$	$6.2822299556 \times 10^{-8}$
	0.5	$3.2605016445 \times 10^{-6}$	$3.4355177778 \times 10^{-7}$	$5.8928801842 \times 10^{-7}$	$8.8955534782 \times 10^{-7}$
	0.75	$1.9684148538 \times 10^{-6}$	$8.5361825235 \times 10^{-6}$	$5.8141127051 \times 10^{-6}$	$3.1743929263 \times 10^{-7}$
	1	$5.3327573861 \times 10^{-6}$	$6.2303305719 \times 10^{-6}$	$2.0146526217 \times 10^{-6}$	$8.0470697269 \times 10^{-6}$
1	0.25	$4.8687548990 \times 10^{-7}$	$1.4518234026 \times 10^{-7}$	$7.3231124692 \times 10^{-7}$	$6.7200224276 \times 10^{-7}$
	0.5	$5.3763575915 \times 10^{-6}$	$8.8579046722 \times 10^{-6}$	$2.5430060005 \times 10^{-7}$	$7.4596696195 \times 10^{-7}$
	0.75	$5.9469042577 \times 10^{-6}$	$7.7876144332 \times 10^{-6}$	$6.8853386269 \times 10^{-6}$	$3.6563750502 \times 10^{-6}$
	1	$1.3313627925 \times 10^{-6}$	$8.9735110839 \times 10^{-6}$	$2.2569043006 \times 10^{-6}$	$4.8476533804 \times 10^{-6}$

Table 2 Absolute errors of Example 2 over Λ

x	t	$\alpha = 0.25$	$\alpha = 0.5$	$\alpha = 0.75$	$\alpha = 1$
0	0.25	$7.9776844684 \times 10^{-7}$	$3.4840685648 \times 10^{-7}$	$2.0885121979 \times 10^{-7}$	$4.1424682191 \times 10^{-8}$
	0.5	$8.3783958819 \times 10^{-7}$	$4.1528820307 \times 10^{-7}$	$7.1702389557 \times 10^{-8}$	$6.1668778742 \times 10^{-8}$
	0.75	$3.3967362498 \times 10^{-7}$	$6.6991474431 \times 10^{-7}$	$8.4884903069 \times 10^{-7}$	$2.0811339500 \times 10^{-7}$
	1	$5.3104194940 \times 10^{-6}$	$5.2032473529 \times 10^{-7}$	$1.8581559833 \times 10^{-7}$	$5.1027116115 \times 10^{-8}$
0.25	0.25	$7.7999626080 \times 10^{-7}$	$2.4640451071 \times 10^{-7}$	$2.8311662609 \times 10^{-7}$	$8.6689276035 \times 10^{-8}$
	0.5	$8.3989967791 \times 10^{-7}$	$5.7091389550 \times 10^{-7}$	$4.8072774473 \times 10^{-7}$	$5.2358140673 \times 10^{-7}$
	0.75	$6.9639358686 \times 10^{-7}$	$2.8087915523 \times 10^{-7}$	$7.0935196083 \times 10^{-7}$	$6.6709448740 \times 10^{-8}$
	1	$5.1420418936 \times 10^{-6}$	$4.0159489813 \times 10^{-7}$	$2.2581468314 \times 10^{-7}$	$3.2212477612 \times 10^{-7}$
0.5	0.25	$6.9540160215 \times 10^{-7}$	$5.6446045032 \times 10^{-7}$	$7.8194176413 \times 10^{-8}$	$1.3560757675 \times 10^{-8}$
	0.5	$1.8873258794 \times 10^{-7}$	$8.7256088711 \times 10^{-7}$	$7.3103863842 \times 10^{-7}$	$6.1796585570 \times 10^{-7}$
	0.75	$9.3946809462 \times 10^{-7}$	$5.5063241201 \times 10^{-7}$	$3.0104629974 \times 10^{-7}$	$1.7518180465 \times 10^{-8}$
	1	$9.4341574440 \times 10^{-6}$	$9.3968995759 \times 10^{-7}$	$8.0664793597 \times 10^{-7}$	$6.1269428722 \times 10^{-8}$
0.75	0.25	$4.7458237332 \times 10^{-7}$	$2.7794343641 \times 10^{-7}$	$1.2112309375 \times 10^{-7}$	$4.8362338856 \times 10^{-8}$
	0.5	$3.4525853552 \times 10^{-7}$	$5.0629795416 \times 10^{-7}$	$4.0379454165 \times 10^{-7}$	$9.4477540589 \times 10^{-7}$
	0.75	$6.6148187244 \times 10^{-7}$	$2.9964614478 \times 10^{-7}$	$2.2900033128 \times 10^{-7}$	$8.9941236417 \times 10^{-8}$
	1	$9.8926059525 \times 10^{-6}$	$2.6698929941 \times 10^{-7}$	$6.7181321202 \times 10^{-8}$	$1.1564777914 \times 10^{-8}$
1	0.25	$2.8313095737 \times 10^{-7}$	$6.6972801239 \times 10^{-7}$	$7.4614502464 \times 10^{-7}$	$3.0846376150 \times 10^{-8}$
	0.5	$6.5203528576 \times 10^{-7}$	$5.8144559231 \times 10^{-7}$	$4.3056435055 \times 10^{-7}$	$7.5776950397 \times 10^{-8}$
	0.75	$3.2788169448 \times 10^{-7}$	$6.3469459479 \times 10^{-7}$	$6.2841762561 \times 10^{-7}$	$7.8177679269 \times 10^{-7}$
	1	$4.9070667887 \times 10^{-6}$	$4.9990426724 \times 10^{-7}$	$9.2211292620 \times 10^{-7}$	$6.7229592233 \times 10^{-7}$

1 and 2 for different values (η, ξ) in Λ with step-size 0.25 are listed in Tables 1 and 2, respectively. From these tables, it can be observed that the error estimate confirm the accuracy of numeric results related to fill time $\xi_m = m \Delta \xi$, $m = 0, 1, ...,r$, $\Delta \xi = \frac{1}{r}$ and distance $\eta_l = l \Delta \eta$, $l = 0, 1, ...,s$, $\Delta \eta = \frac{1}{s}$. Hence, more accurate numeric solutions can be found by utilizing more grid points (η, ξ).

5 Concluding Remarks

In this chapter, the RKM has been applied to obtain approximate solutions for both linear and nonlinear PIDEs of fractional order. The fractional derivative has been described in the Caputo sense. Two examples have been tested to show the efficiency of the proposed method. By comparing our results with the exact solution for integer and non-integer orders derivative, one can observe that the proposed method yields accurate approximations. This adaptive can be used as an alternative technique in solving several nonlinear partial fractional problems arising in diverse engineering, chemistry, biology, and physical sciences.

References

1. Mainardi, F.: Fractional Calculus and Waves in Linear Viscoelasticity. Imperial College Press, London, UK (2010)
2. Zaslavsky, G.M.: Hamiltonian Chaos and Fractional Dynamics. Oxford University Press (2005)
3. Podlubny, I.: Fractional Differential Equations. Academic Press, San Diego, CA, USA (1999)
4. Samko, S.G., Kilbas, A.A., Marichev, O.I.: Fractional Integrals and Derivatives Theory and Applications. Gordon and Breach, New York (1993)
5. Kilbas, A., Srivastava, H., Trujillo, J.: Theory and Applications of Fractional Differential Equations. Elsevier, Amsterdam, Netherlands (2006)
6. Arshed, S.: B-spline solution of fractional integro partial differential equation with a weakly singular kernel. Numer. Methods Part. Differe. Equ. (2017). In Press. https://doi.org/10.1002/num.22153.
7. Rostami, Y., Maleknejad, K.: Numerical solution of partial integro-differential equations by using projection method. Mediterranean J. Math. **14**, 113 (2017). https://doi.org/10.1007/s00009-017-0904-z
8. Huang, L., Li, X.F., Zhao, Y., Duan, X.Y.: Approximate solution of fractional integro-differential equations by Taylor expansion method. Comput. Math. Appl. **62**, 1127–1134 (2011)
9. Mohammed, D.S.: Numerical solution of fractional integro-differential equations by least squares method and shifted Chebyshev polynomial. Math. Problems Eng. **2014**, Article ID 431965, 5 p. (2014). https://doi.org/10.1155/2014/431965
10. Momani, S., Qaralleh, R.: An efficient method for solving systems of fractional integro-differential equations. Comput. Math. Appl. **52**, 459–470 (2006)
11. Tohidi, E., Ezadkhah, M.M., Shateyi, S.: Numerical solution of nonlinear fractional Volterra integro-differential equations via Bernoulli polynomials. Abstract Appl. Anal. **2014**, Article ID 162896, 7 p. (2014). https://doi.org/10.1155/2014/162896
12. Wang, Y., Zhu, L.: Solving nonlinear Volterra integro-differential equations of fractional order by using Euler wavelet method. Adv. Differ. Equ. **2017**, 27 (2017). https://doi.org/10.1186/s13662-017-1085-6

13. Abu Arqub, O., El-Ajou, A., Momani, S.: Constructing and predicting solitary pattern solutions for nonlinear time-fractional dispersive partial differential equations. J. Comput. Phys. **293**, 385–399 (2015)
14. El-Ajou, A., Abu Arqub, O., Momani, S., Baleanu, D., Alsaedi, A.: A novel expansion iterative method for solving linear partial differential equations of fractional order. Appl. Math. Comput. **257**, 119–133 (2015)
15. El-Ajou, A., Abu Arqub, O., Momani, S.: Approximate analytical solution of the nonlinear fractional KdV-Burgers equation: A new iterative algorithm. J. Comput. Phys. **293**, 81–95 (2015)
16. Ray, S.S.: New exact solutions of nonlinear fractional acoustic wave equations in ultrasound. Comput. Math. Appl. **71**, 859–868 (2016)
17. Ortigueira, M.D., Machado, J.A.T.: Fractional signal processing and applications. Signal Process **83**, 2285–2286 (2003)
18. Zaremba, S.: L'equation biharminique et une class remarquable defonctionsfoundamentals harmoniques. Bulletin International de l'Academie des Sciences de Cracovie **39**, 147–196 (1907)
19. Aronszajn, N.: Theory of reproducing kernels. Trans. Am. Math. Soc. **68**, 337–404 (1950)
20. Cui, M., Lin, Y.: Nonlinear Numerical Analysis in the Reproducing Kernel Space. Nova Science, New York, NY, USA (2009)
21. Al-Smadi, M.: Simplified iterative reproducing kernel method for handling time-fractional BVPs with error estimation. Ain Shams Eng. J. **9**(4), 2517–2525 (2018)
22. Daniel, A.: Reproducing Kernel Spaces and Applications. Springer, Basel, Switzerland (2003)
23. Weinert, H.L.: Reproducing Kernel Hilbert Spaces: Applications in Statistical Signal Processing, Hutchinson Ross (1982)
24. Lin, Y., Cui, M., Yang, L.: Representation of the exact solution for a kind of nonlinear partial differential equations. Appl. Math. Lett. **19**, 808–813 (2006)
25. Zhoua, Y., Cui, M., Lin, Y.: Numerical algorithm for parabolic problems with non-classical conditions. J. Comput. Appl. Math. **230**, 770–780 (2009)
26. Abu Arqub, O.: Fitted reproducing kernel Hilbert space method for the solutions of some certain classes of time-fractional partial differential equations subject to initial and Neumann boundary conditions. Comput. Math. Appl. **73**, 1243–1261 (2017)
27. Abu Arqub, O., Shawagfeh, N.: Application of reproducing kernel algorithm for solving Dirichlet time-fractional diffusion-Gordon types equations in porous media. J. Porous Media (2017). In Press
28. Abu Arqub, O., Rashaideh, H.: The RKHS method for numerical treatment for integrodifferential algebraic systems of temporal two-point BVPs. Neural Comput. Appl., 1–12 (2017). https://doi.org/10.1007/s00521-017-2845-7
29. Abu Arqub, O.: The reproducing kernel algorithm for handling differential algebraic systems of ordinary differential equations. Math. Methods Appl. Sc. **39**, 4549–4562 (2016)
30. Abu Arqub, O., Al-Smadi, M., Shawagfeh, N.: Solving Fredholm integro-differential equations using reproducing kernel Hilbert space method. Appl. Math. Comput. **219**, 8938–8948 (2013)
31. Abu Arqub, O., Al-Smadi, M.: Numerical algorithm for solving two-point, second-order periodic boundary value problems for mixed integro-differential equations. Appl. Math. Comput. **243**, 911–922 (2014)
32. Momani, S., Abu Arqub, O., Hayat, T., Al-Sulami, H.: A computational method for solving periodic boundary value problems for integro-differential equations of Fredholm-Voltera type. Appl. Math. Comput. **240**, 229–239 (2014)
33. Abu Arqub, O., Al-Smadi, M., Momani, S., Hayat, T.: Numerical solutions of fuzzy differential equations using reproducing kernel Hilbert space method. Soft Comput. **20**, 3283–3302 (2016)
34. Abu Arqub, O., Al-Smadi, M., Momani, S., Hayat, T. : Application of reproducing kernel algorithm for solving second-order, two-point fuzzy boundary value problems. Soft Comput., 1–16 (2016). https://doi.org/10.1007/s00500-016-2262-3
35. Abu Arqub, O.: Adaptation of reproducing kernel algorithm for solving fuzzy Fredholm-Volterra integrodifferential equations. Neural Comput. Appl., 1–20 (2015). https://doi.org/10.1007/s00521-015-2110-x

36. Abu Arqub, O.: Approximate solutions of DASs with nonclassical boundary conditions using novel reproducing kernel algorithm. Fundamenta Informaticae **146**, 231–254 (2016)
37. Abu Arqub, O., Integral-initial, Z., Al-Smadi, M.: Numerical solutions of time-fractional partial integrodifferential equations of Robin functions types in Hilbert space with error bounds and error estimates. Nonlinear Dyn. **94**(3), 1819–1834 (2018)
38. Abu Arqub, O.: Numerical solutions for the Robin time-fractional partial differential equations of heat and fluid flows based on the reproducing kernel algorithm. Int. J. Numer. Methods Heat Fluid Flow (2017). https://doi.org/10.1108/HFF-07-2016-0278
39. Geng, F.Z., Qian, S.P.: Reproducing kernel method for singularly perturbed turning point problems having twin boundary layers. Appl. Math. Lett. **26**, 998–1004 (2013)
40. Jiang, W., Chen, Z.: A collocation method based on reproducing kernel for a modified anomalous subdiffusion equation. Numer. Methods Part. Differ. Equ. **30**, 289–300 (2014)
41. Geng, F.Z., Qian, S.P., Li, S.: A numerical method for singularly perturbed turning point problems with an interior layer. J. Comput. Appl. Math. **255**, 97–105 (2014)
42. Al-Smadi, M., Abu Arqub, O., Shawagfeh, N., Momani, S.: Numerical investigations for systems of second-order periodic boundary value problems using reproducing kernel method. Appl. Math. Comput. **291**, 137–148 (2016)
43. Jiang, W., Chen, Z.: Solving a system of linear Volterra integral equations using the new reproducing kernel method. Appl. Math. Comput. **219**, 10225–10230 (2013)
44. Geng, F.Z., Qian, S.P.: Modified reproducing kernel method for singularly perturbed boundary value problems with a delay. Appl. Math. Model. **39**, 5592–5597 (2015)
45. Al-Smadi, M., Abu Arqub, O.: Computational algorithm for solving fredholm time-fractional partial integrodifferential equations of dirichlet functions type with error estimates. Appl. Math. Comput. **342**, 280–294 (2019)
46. Abu Arqub, O., Al-Smadi, M.: Atangana-Baleanu fractional approach to the solutions of Bagley-Torvik and Painlevé equations in Hilbert space. Chaos Solitons Fractals **117**, 161–167 (2018)

Approximation of Fractional-Order Operators

Reyad El-Khazali, Iqbal M. Batiha and Shaher Momani

Abstract In order to deal with some difficult problems in fractional-order systems, like computing analytical time responses such as unit impulse and step responses; some rational approximations for the fractional-order operators are presented with satisfying results in simulation and realization. In this chapter, several comparisons in the time response and Bode results between four well-known methods; Oustaloup's method, Matsuda's method, AbdelAty's method, and El-Khazali's method are made for the rational approximation of fractional-order operator (fractional Laplace operator). The various methods along with their advantages and limitations are described in this chapter. Simulation results are shown for different orders of the fractional operator. It has been shown in several numerical examples that the El-Khazali's method is very successful in comparison with Oustaloup's, Matsuda's, and AbdelAty's methods.

Keywords Fractional-Order models · Oustaloup's approximation · Matsuda's approximation · AbdelAty's approximation · El-Khazali's approximation

R. El-Khazali (✉)
ECSE Department, Khalifa University, Abu Dhabi 127788, UAE
e-mail: reyad.elkhazali@ku.ac.ae

I. M. Batiha · S. Momani
Department of Mathematics, The University of Jordan, Amman 11942, Jordan
e-mail: iqbalbatiha22@yahoo.com

S. Momani
e-mail: s.momani@ajman.ac.ae

S. Momani
Nonlinear Analysis and Applied Mathematics (NAAM) Research Group, Faculty of Science, King Abdulaziz University, Jeddah 21589, Kingdom of Saudi Arabia

© Springer Nature Singapore Pte Ltd. 2019
P. Agarwal et al. (eds.), *Fractional Calculus*, Springer Proceedings
in Mathematics & Statistics 303, https://doi.org/10.1007/978-981-15-0430-3_8

1   Introduction

Many attempts have been made by many researchers to obtain different forms of finite-order rational approximation to the fractional-order Laplacian integro-differential operators. Such attempts allow one to develop realizable models of different systems and processes using passive or active electronic devices to mimic the behavior of such operators. For example, a diffusion process in the electrochemical process, which exhibits fractional-order dynamics, can successfully be modeled by finite-order electrical circuits using different approximation algorithms [1, 2].

Fractional-order systems provide more freedom in control theory. For example, fractional-order controllers proved to show superior performance over their integer-order counterparts. It widens the scope of applications in systems and, in some cases, simplifies the design of controllers. The Laplace transform of the input–output relationship provides a powerful tool to investigate the frequency response of linear systems. It is extended to systems that exhibit fractional-order dynamics [1–5]. The frequency response of an integro-differential Laplacian operator, $s^{\pm\alpha}$ can be defined as

$$(j\omega)^{\pm\alpha} = \omega^{\pm\alpha}\left[\cos\left(\frac{\alpha\pi}{2}\right) \pm j\sin\left(\frac{\alpha\pi}{2}\right)\right] \tag{1}$$

It is not possible to estimate the exact time response of a fractional-order transfer function since the analytical inverse Laplace transform does not exist [2], one may compute the system time responses, namely the impulse and the step responses, which may be described by fractional-order transfer function, by constructing rational approximations of the fractional-order operators, $s^{\pm\alpha}$. Such approximation can be used to generate equivalent integer-order transfer functions that describe the original system within a limited frequency band [1].

There are several popular approximation methods that are used to approximate $s^{\pm\alpha}$, such as the Continued Fractional Expansion (CFE), least square method, Oustaloup's, Carlson's, Matsuda's, Chareff's, AbdelAty's et.al., and El-Khazali's approximation methods [1–5]. In the latest paper by the authors in [6], the numerical time-domain solution of fractional-order systems is obtained using two methods; the first one uses the Fourier series representation of a square wave, and the second one uses the inverse Fourier transform. The two methods use the exact numerical data of the system frequency response to obtain accurate representations of the fractional-order dynamics.

In this chapter, some basic concepts about the fractional-order models, the frequency domain analysis, and some rational approximations of fractional-order operators are presented in the first four sections. Section 5, however, includes the main comparison results in both time and frequency domains using four different approximation methods of the Laplacian operators, $s^{\pm\alpha}$, namely, the Oustaloup's, Matsuda's, AbdelAty's et.al, and El-Khazali's approximations [1–5].Two numerical examples are given to provide detailed comparison and to highlight the advantages and disadvantages of each method.

Furthermore, providing modular approximating to $s^{\pm\alpha}, 0 < \alpha \leq 1$, by finite-order rational-transfer functions simplifies the realization of larger class of fractional-order controllers such as fractional-order integrators, I^λ, fractional-order differentiator, D^δ, a combination of PI^λ, PD^δ, to design $PI^\lambda D^\delta$ controllers (FOPID). The properties of these controllers and the effect of their fractional orders on system transient response are briefly highlighted for completeness and compared with their integer-order counterparts.

2 Fractional-Order Models

Figure 1 shows a general classification of LTI systems. It is well known that systems, which exhibit hereditary effect, are described by fractional-order differential equations. The fundamental definitions of fractional-order calculus are used to characterize such systems [7]. The integer-order dynamics, however, is considered as a subset of larger class of fractional-order systems. Furthermore, it was shown in [7–9] that fractional-order controllers outperform their integer-order counterparts due to their flexibility in accommodating more parameters, and due to the constant-phase frequency response, which provides more robustness to the controlled plants.

The class of linear time-invariant systems that will be considered here is described by the following fractional-order differential equations [7]:

$$a_n D^{\alpha_n} y(t) + \cdots + a_0 D^{\alpha_0} y(t) = b_m D^{\beta_m} u(t) + \cdots + b_0 D^{\beta_0} u(t) \qquad (2)$$

where $u(t)$ and $y(t)$ are the control and output variables, and D^α defines a fractional-order differential operator (according to Caputo definition) of different fractional orders $\alpha_k; k = 1, 2, 3, \ldots, n$, and $\beta_l; l = 1, 2, 3, \ldots, m$, are arbitrary constants and $n, m \in \mathbb{N}$.

Fig. 1 Classification of LTI systems

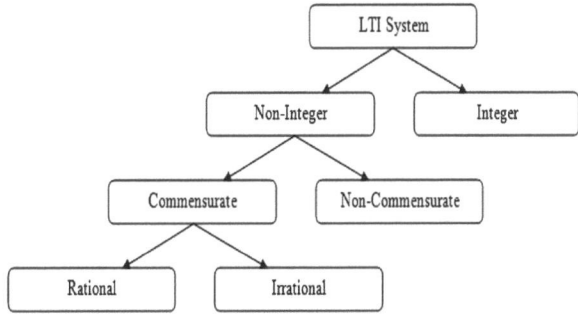

The system of (2) is said to be of commensurate order if all its orders are integer multiples of a base of a fractional order, q, such that α_k; $\beta_k = k_q$; $q \in \mathbb{R}^+$. The system can then be expressed as [4]:

$$\sum_{k=0}^{n} a_k D^{k_q} y(t) = \sum_{k=0}^{m} b_k D^{k_q} u(t) \tag{3}$$

Obviously, if one sets $q = \frac{1}{n}$, where $n > 1$; then (3) defines a typical fractional-order system of commensurate order. A fractional-order linear time-invariant (FOLTI) system is mathematically equivalent to an infinite-dimensional LTI filter. In order to model such systems, one has to realize finite-order approximate models that best describe the original fractional-order systems. Obviously, the size of approximation depends on the type of the numerical algorithms used to implement such models.

Discrete-time fractional-order systems are usually approximated by higher order FIR or IIR filters of integer orders [10]. The size of the FIR filters needed to model such systems are usually much larger than those of IIR filters. Therefore, most researchers focus on developing a set of rational-transfer functions of much lower orders. The presence of the feedback action imbedded in the rational transfer functions compensate for the need to realize larger size of either continuous or discrete FIR filters for the FOLTI systems [11, 12]. Figure 1 shows a classification of FOLTI that is of interest in this chapter.

The frequency response of systems is usually carried out by using transfer functions. The transfer function of a linear time-invariant system is defined as the ratio of the Laplace transform of the output (system output response) to the Laplace transform of the input (system input) under the assumption that all initial conditions are zero [11], i.e.,

$$G(s) = \left. \frac{\mathscr{L}(\text{output})}{\mathscr{L}(\text{input})} \right|_{\text{zero initial conditions}} = \frac{Y(s)}{U(s)} \tag{4}$$

A typical form of a fractional-order LTI (FoLTI) system can be described by the following transfer function [11]:

$$G(s) = \frac{Y(s)}{U(s)} = \frac{b_m s^{\beta_m} + b_{m-1} s^{\beta_{m-1}} + \cdots + b_1 s^{\beta_1} + b_0 s^{\beta_0}}{a_n s^{\alpha_n} + a_{n-1} s^{\alpha_{n-1}} + \cdots + a_1 s^{\alpha_1} + a_0 s^{\alpha_0}} \tag{5}$$

The next sections outline a complete comparison between the different types of approximations used to evaluate the time and frequency response of the approximated integro-differential operators. In addition, the finite-order rational approximations will be used to model and design fractional-order PID (FOPID) controllers, and compared with the ideal controllers.

3 Frequency Response of FOLTI Systems

The frequency response of $FOLTI$ systems (5) may be obtained by replacing, s, in (5) by a generating function, $s = (\omega(z^{-1}))$ [11] to yield

$$G(z) = \frac{b_m(\omega(z^{-1}))^{\beta_m} + b_{m-1}(\omega(z^{-1}))^{\beta_{m-1}} + \cdots + b_1(\omega(z^{-1}))^{\beta_1} + b_0(\omega(z^{-1}))^{\beta_0}}{a_n(\omega(z^{-1}))^{\alpha_n} + a_{n-1}(\omega(z^{-1}))^{\alpha_{n-1}} + \cdots + a_1(\omega(z^{-1}))^{\alpha_1} + a_0(\omega(z^{-1}))^{\alpha_0}}$$

(6)

where $s = (\omega(z^{-1}))$ denotes the discrete equivalence of the Laplace operator s, expressed as a function of the complex variable z or the shift operator z^{-1}. System (5) defines an irrational continuous transfer function in the Laplace domain or/and an infinite-dimensional discrete-time transfer function in the z-domain [5]. In both cases, FOLTI systems have unlimited memory size, while integer-order ones are described by finite-dimensional models [1].

4 Rational Approximation of Fractal-Order Operators

The rational approximation of fractional-order Laplacian operators simplifies the realization of fractal elements of real orders, which can also be characterized as Constant-Phase Elements (CPE) [13]. Such elements provide a good description of these operators over a limited frequency range. In this study, our interest is limited to fractional-order operators of real order. The electronic circuit realizations of different fractional-order operators are left for future consideration [13].

The following section describes briefly the different approximation methods of the fractional-order integro-differential Laplacian operators, $s^{\pm\alpha}$, that are used in many applications [1, 3–5, 14].

4.1 Oustaloup's Approximation

Oustaloup's approximation is a popular one and generates rational-transfer functions of odd order only. The bandwidth over which the approximation is considered can be customized to yield a good fitting to the fractional-order elements $s^{\pm\alpha}$ within a predefined frequency band. Thus, for geometrically distributed frequencies over the frequency range of interest (ω_b, ω_h), the following rational function is used for approximating s^α [1]:

$$s^\alpha \cong K \prod_{k=-N}^{N} \frac{s + \omega_k'}{s + \omega_k} = \frac{B_n s^n + B_{n-1} s^{n-1} + \cdots + B_1 s + B_0}{A_n s^n + A_{n-1} s^{n-1} + \cdots + A_1 s + A_0}$$

(7)

where the poles, zeros, and gain are evaluated from

$$\omega_k = \omega_b \left(\frac{\omega_h}{\omega_b} \right)^{\frac{k+N+0.5(1+\alpha)}{2N+1}} \tag{8a}$$

$$\omega_k' = \omega_b \left(\frac{\omega_h}{\omega_b} \right)^{\frac{k+N+0.5(1-\alpha)}{2N+1}} \tag{8b}$$

$$K = \left(\frac{\omega_h}{\omega_b} \right)^{\frac{-\alpha}{2}} \prod_{k=-N}^{N} \frac{\omega_k}{\omega_k'} \tag{8c}$$

Due to the geometrical distribution of frequencies, the unity gain geometric frequency ω_u is calculated from

$$\omega_u = \sqrt{\omega_b \cdot \omega_h} \tag{9}$$

where the approximation depends on the order of the filer N and the upper and the lower frequency range (ω_b, ω_h). Observe that the order of the transfer function (7) will always be $n = 2N + 1$, i.e., only odd-order approximations are possible through the Oustaloup's method. In the special case where the limited frequencies ω_b and ω_h are symmetrical around the center frequency $\omega_u = 1 rad/sec$, (i.e., $\omega_b = 1/\omega_h$), then the coefficients of (7) are correlated to each other as follows [12]:

$$A_{n-i} = B_i, i = 0, 1, 2, \ldots, N. \tag{10}$$

One can define the fractional-order derivative using the a fractional-order integral as follows:

$$D_a^\alpha f(t) = D^m J_a^{m-\alpha} f(t) \tag{11}$$

where m is a positive integer such that $m - 1 < \alpha \leq m$.

Consequently, for the case when the fractional orders $\alpha \geq 1$, one may rewrite s^α as

$$s^\alpha = s^n s^\gamma \tag{12}$$

where $n = \alpha - \gamma$ denotes the integer part of α and s^γ is obtained by Oustaloup's approximation using (7).

For digital implementations, the obtained approximation may be discretized to using suitable discretization methods [5].

4.2 Matsuda's Approximation

This method provides continuous approximation by calculating gain at logarithmically spaced frequencies. If the value of a function $F(j\omega)$ is known over a set of frequencies $\omega_0, \omega_1, \omega_2, \ldots, \omega_N$, then the following set of functions is recursively defined by [15]:

$$d_0(\omega) = |F(j\omega)| \tag{13}$$

$$d_{k+1}(\omega) = \frac{\omega - \omega_k}{d_k(\omega) - d_k(\omega)}, k = 0, 1, 2, \ldots, N \tag{14}$$

Then, an $(N + 1)(N + 1)$ superior upper triangular matrix is formed as follows:

$$D = \begin{bmatrix} d_0(\omega_0) & d_0(\omega_1) & d_0(\omega_2) & \ldots & d_0(\omega_N) \\ & d_1(\omega_1) & d_1(\omega_2) & \ldots & d_1(\omega_N) \\ & & d_2(\omega_2) & \ldots & d_2(\omega_N) \\ & & & \ddots & \vdots \\ & & & & d_N(\omega_N) \end{bmatrix} \tag{15}$$

The desired approximation is then given by the following continued fraction:

$$F(s) = a_0 + \cfrac{s - \omega_0}{a_1 + \cfrac{s-\omega_1}{a_2 + \cfrac{s-\omega_2}{a_3 + \cdots}}} = a_0 + \cfrac{s - \omega_0}{a_1 + \cdots} \cfrac{s - \omega_1}{a_2 + \cdots} \cfrac{s - \omega_2}{a_3 + \cdots} \tag{16}$$

such that the set of coefficients is defined as

$$a_k = \begin{cases} |F(j\omega_0)|, & if\ k = 0 \\ \\ \frac{\omega_k - \omega_{k-1}}{d_{k-1}(\omega_k) - d_{k-1}(\omega_{k-1})}, & if\ k = 1, 2, 3, \ldots, N \end{cases} \tag{17}$$

4.3 AbdelAty's Approximation

The approximation is based on using a weighted sum of 1st-order high-pass filters. The parameters of the filters are obtained using a flower pollination algorithm (FPA) for each fractional order [16], which can be synthesized using Forster II RC realization [4] and given by

$$s^\alpha = \frac{1}{\Gamma(\alpha)\Gamma(1 - \alpha)} \sum_{i=1}^{n} k_i \frac{s}{\tau_i^\alpha(1 + \tau_i s)} \tag{18}$$

where τ_i and k_i are two constants found using the FPA optimization algorithm.

Notice that the high-pass structure of each filter section of (20) creates larger phase errors for fractional orders less than 0.5 than other approximation methods, since the phase value of each section is $\pi/2$ at low frequency.

4.4 El-Khazali's Integro-Differential Approximation

Fractional-order integro-differential Laplacian operators, $s^{\pm\alpha}$, can be approximated using a biquadratic approximation algorithm introduced in [17], where s^α defines a differential operator, while $s^{-\alpha}$ defines a fractional-order integrator. The orders of both the numerator and denominator are equal, where the reciprocal of one approximation yields the other one. Thus, realizing fractal elements (i.e., fractional-order capacitors or inductors) is straightforward and only depends on the order of differentiation or integration. For a single module, it enjoys a flat phase response at its center frequency, but with a narrower bandwidth than its counterparts. It consists of cascaded several 2nd-order biquadratic transfer functions of the form [13, 17]:

$$\left(\frac{s}{\omega_g}\right)^\alpha = \prod_{i=1}^n H_i(s/\omega_i) = \prod_{i=1}^n \frac{N_i\left(\frac{s}{(\omega_i/\omega_g)}\right)}{D_i\left(\frac{s}{(\omega_i/\omega_g)}\right)} \tag{19}$$

where ω_i, $i = 1, 2, \ldots, n$, is the center frequency of each biquadratic module and $\omega_g = \sqrt[n]{\prod_{i=1}^n \omega_i}$ is their geometric mean.

If one selects the first center frequency, ω_1, of the first section, then to obtain a constant-phase element, the subsequent center frequencies of each section can be calculated from the following recursive formula [17]:

$$\omega_i = \omega_x^{2(i-1)}\omega_1; \quad i = 2, 3, \ldots, n \tag{20}$$

where ω_x is the maximum real solution of the following polynomial:

$$a_0 a_2 \eta \gamma^4 + a_1(a_2 - a_0)\gamma^3 + (a_1^2 - a_2^2 - a_0^2)\eta\gamma^2 + a_1(a_2 - a_0)\gamma + a_0 a_2 \eta = 0 \tag{21}$$

and where $\eta = tan(\alpha\pi/4)$. Each biquadratic module in (21) is given by

$$\left(\frac{s}{\omega_g}\right)^\alpha = H_i\left(\frac{s}{\omega_i}\right) = \frac{N_i\left(\frac{s}{\omega_i}\right)}{D_i\left(\frac{s}{\omega_i}\right)} \simeq \frac{a_0\left(\frac{s}{\omega_i}\right)^2 + a_1\left(\frac{s}{\omega_i}\right) + a_2}{a_2\left(\frac{s}{\omega_i}\right)^2 + a_1\left(\frac{s}{\omega_i}\right) + a_0}, \quad i = 1, 2, 3, \ldots \tag{22}$$

where

$$\left.\begin{array}{l} a_0 = \alpha^\alpha + 2\alpha + 1 \\ a_2 = \alpha^\alpha - 2\alpha + 1 \\ a_1 = (a_2 - a_0)tan\left(\frac{(2+\alpha)\pi}{4}\right) = -6\alpha\, tan\left(\frac{(2+\alpha)\pi}{4}\right) \end{array}\right\} \tag{23}$$

Observe that (24) is the only approximation that yields $H_i(\frac{s}{\omega_i}) = (\frac{s}{\omega_i})$ as $\alpha \to 1$; i.e.,

$$H_i\left(\frac{s}{\omega_i}\right) = \frac{a_0(\frac{s}{\omega_i})^2 + a_0(\frac{s}{\omega_i})}{a_0(\frac{s}{\omega_i}) + a_0} = \left(\frac{s}{\omega_i}\right) \tag{24}$$

Moreover, the reciprocal of (21) approximates a fractional-order integrator [17], or simply:

$$(s/\omega_g)^{(-\alpha)} = \prod_{i=1}^{n} \hat{H}_i(s/\omega_g) = \prod_{i=1}^{n} \frac{D_i\left(\frac{s}{(\omega_i/\omega_g)}\right)}{N_i\left(\frac{s}{(\omega_i/\omega_g)}\right)}. \tag{25}$$

5 Comparison Results

In this section, we introduce a comparison simulation result between the four different approximation algorithms discussed in section (4) in both time and frequency domains. Two numerical examples are investigated to highlight the main differences of these methods; the first one approximates a fractional-order Laplacian differentiator $s^{0.4}$, and the second example is the approximation of a closed-loop transfer function of a FOLTI system.

Example 1 The integer-order of a rational transfer function that approximates $s^{0.4}$ using Oustaloup's, Matsuda's, AbdelAty's, and El-Khazali's approximation, respectively, are given by

– Oustaloup's approximation

$$s^{0.4} = \frac{6.31s^3 + 77.14s^2 + 41.74s + 1}{s^3 + 41.74s^2 + 77.14s + 6.31}, \quad 0.1 \leq \omega \leq 10 \tag{26}$$

– Matsuda's approximation

$$s^{0.4} = \frac{10.01s^3 + 163.6s^2 + 83.01s + 1}{s^3 + 83.01s^2 + 163.6s + 10.01} \tag{27}$$

– AbdelAty's approximation

$$s^{0.4} = \frac{\begin{aligned}&412.48s^8 + 206415397.69s^7 + 11956108280000.56s^6 + 94706047639064780s^5 \\ &+ 104739477770482730000s^4 + 1.616e22s^3 + 3.399e23s^2 + 8.429e23s\end{aligned}}{\begin{aligned}&s^8 + 1443327.34s^7 + 190377994705.8s^6 + 3320443435965385.5s^5 \\ &+ 8064411171562614000s^4 + 2.737e21s^3 + 1.285e23s^2 + 7.801e23s + 3.550e23\end{aligned}} \tag{28}$$

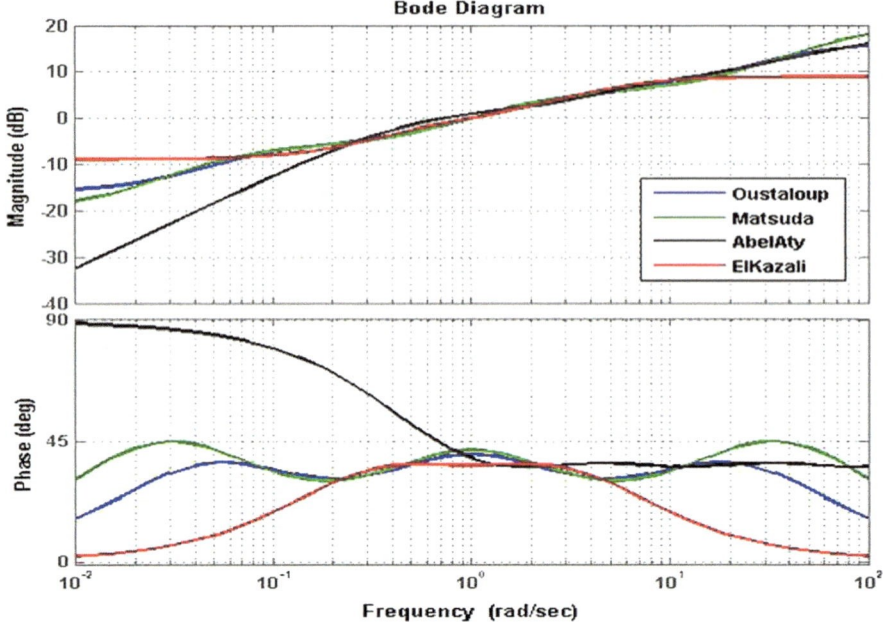

Fig. 2 Bode plot of the approximations (26), (27), (28), and (29)

– El-Khazali's approximation

$$s^{0.4} = \frac{2.493s^2 + 4.924s + 0.8931}{0.8931s^2 + 4.924s + 2.493} \tag{29}$$

The Bode diagrams and the step response of (26)–(29) are shown in Figs. 2 and 3, respectively. Figure 2 shows the frequency response of four different approximations. They all show similar frequency response, except for AbdelAty's approximation since the phase error at low frequency is larger than the rest of approximations. This error would increase for lower values of α. Figure 3, however, shows an almost identical response of the fractional-order derivative of the unit-step function using the four approximations given by (26)–(29). However, the 8th-order approximation of AbdelAty does not give the right time response.

Example 2 Consider the following fractional-order transfer function reported in [6]:

$$H(s) = \frac{1}{2s^{2.1} + 0.8s^{0.8} + 1} \tag{30}$$

If one wishes to sufficiently simplify (30) by a lowest integer-order transfer function using the four approximations discussed in Sect. 4, one needs to replace $s^{0.1}$ and $s^{0.8}$ by their integer-order equivalence to be able to develop integer-order approximation to (30). Using Oustaloup's 5th-order approximations for $s^{0.1}$ and $s^{0.8}$ yields [1]:

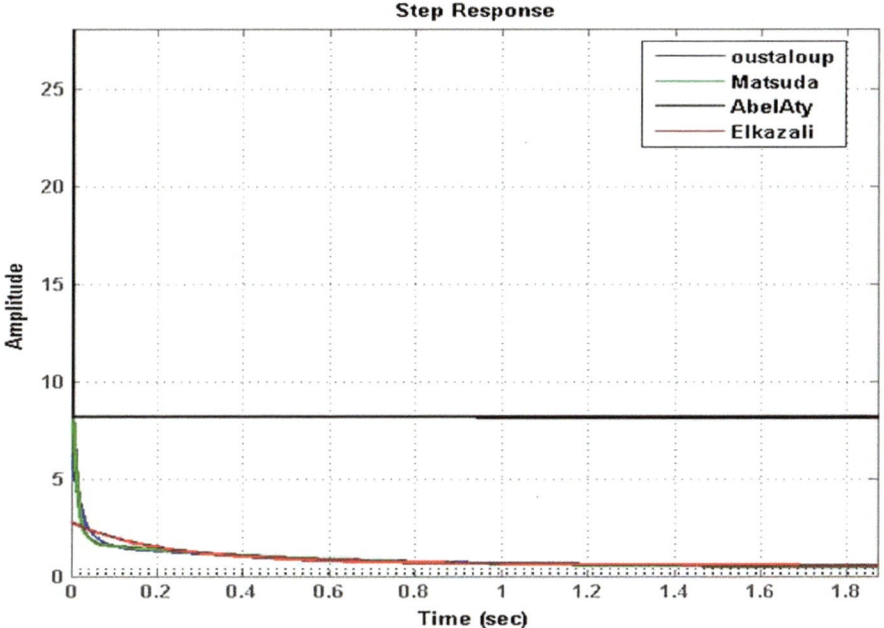

Fig. 3 Fractional-order derivative of order 0.4 of the unit-step function using (26), (27), (28), and (29)

$$s^{0.1} = \frac{1.585s^5 + 68.37s^4 + 403.3s^3 + 367.9s^2 + 51.87s + 1}{s^5 + 51.87s^4 + 367.9s^3 + 403.3s^2 + 68.37s + 1.585} \tag{31}$$

$$s^{0.8} = \frac{39.81s^5 + 901.4s^4 + 2790s^3 + 1336s^2 + 98.83s + 1}{s^5 + 98.83s^4 + 1336s^3 + 2790s^2 + 901.4s + 39.81} \tag{32}$$

Substituting (31) and (32) into (30) yields a 10th-order transfer function of the form:

$$\boldsymbol{H}_{Ous}(s) = \frac{\begin{aligned}&s^{10} + 150.7s^9 + 6830s^8 + 1.088e05s^7 + 6.769e05s^6 + 1.619e06s^5 \\ &\quad + 1.551e06s^4 + 5.711e05s^3 + 8.211e04s^2 + 4151s + 63.1\end{aligned}}{\begin{aligned}&3.17s^{12} + 450s^{11} + 1.859e04s^{10} + 2.745e05s^9 + 1.593e06s^8 + 3.871e06s^7 + 4.771e06s^6 \\ &\quad + 3.953e06s^5 + 2.293e06s^4 + 6.858e05s^3 + 8.961e04s^2 + 4331s + 64.36\end{aligned}} \tag{33}$$

Repeating the same procedure using the following Matsuda's 5th-order approximations of $s^{0.1}$ and $s^{0.8}$:

$$s^{0.1} = \frac{1.891s^5 + 175.3s^4 + 1329s^3 + 1197s^2 + 126.3s + 1}{s^5 + 126.3s^4 + 1197s^3 + 1329s^2 + 175.3s + 1.891} \tag{34}$$

$$s^{0.8} = \frac{313.3s^5 + 1.188e04s^4 + 4.261e04s^3 + 1.826e04s^2 + 809.3s + 1}{s^5 + 809.3s^4 + 1.826e04s^3 + 4.261e04s^2 + 1.188e04s + 313.3} \tag{35}$$

yields the following 10th-order rational-transfer function of (30) of the form:

$$
H_{Mat}(s) = \frac{\begin{aligned} &s^{10} + 935.6s^9 + 1.216e05s^8 + 3.317e06s^7 + 2.831e07s^6 + 7.689e07s^5 \\ &+ 7.406e07s^4 + 2.366e07s^3 + 2.579e06s^2 + 7.739e04s + 592.3 \end{aligned}}{\begin{aligned} &3.78s^{12} + 3411s^{11} + 3.557e05s^{10} + 8.757e06s^9 + 6.709e07s^8 + 1.806e08s^7 + 2.219e08s^6 \\ &+ 1.815e08s^5 + 1.041e08s^4 + 2.725e07s^3 + 2.722e06s^2 + 7.875e04s + 593.8 \end{aligned}} \tag{36}
$$

In similar manner, the 8th-order AbdelAty's approximations of $s^{0.1}$ and $s^{0.8}$ are given by

$$
s^{0.1} = \frac{\begin{aligned} &4.51s^8 + 3722617.69s^7 + 390344733034.4s^6 + 6445483209685290s^5 \\ &+ 17462315946623164000s^4 + 7.739e21s^3 + 5.437e23s^2 + 5.054e24s \end{aligned}}{\begin{aligned} &s^8 + 1053025.15s^7 + 133752991203.54s^6 + 2649165710288196s^5 \\ &+ 8597754455256356000s^4 + 4.569e21s^3 + 3.872e23s^2 + 4.503e24s + 1.386e24 \end{aligned}} \tag{37}
$$

$$
s^{0.8} = \frac{\begin{aligned} &228721.34s^8 + 56105355367.14s^7 + 1671424211082485s^6 + 7107665167043685000s^5 \\ &+ 4.418e21s^4 + 4.005e23s^3 + 5.176e24s^2 + 7.867e24s \end{aligned}}{\begin{aligned} &s^8 + 3327245.95s^7 + 518122071483.89s^6 + 10400295711020872s^5 \\ &+ 30071114886647075000s^4 + 1.273e22s^3 + 7.848e23s^2 + 6.793e24s + 6.216e24 \end{aligned}} \tag{38}
$$

Consequently, the 16th-order approximation of (30) is

$$
H_{Abd}(s) = \frac{\begin{aligned} &s^{16} + 4380271.1s^{15} + 4155548728273.07s^{14} + 1003674131746307100s^{13} + 8.911e22s^{12} \\ &+ 2.824e27s^{11} + 3.606e31s^{10} + 1.732e35s^9 + 3.401e38s^8 + 2.529e41s^7 + 7.661e43s^6 \\ &+ 8.707e45s^5 + 3.923e47s^4 + 6.210e48s^3 + 3.408e49s^2 + 3.741e49s + 8.618e48 \end{aligned}}{\begin{aligned} &9.02s^{18} + 37456993.85s^{17} + 30226279998733.27s^{16} + 6561788513256727000s^{15} \\ &+ 5.251e23s^{14} + 1.514e28s^{13} + 1.758e32s^{12} + 7.688e35s^{11} + 1.376s^{10} + 9.315e41s^9 \\ &+ 2.574e44s^8 + 2.655e46s^7 + 1.090e48s^6 + 1.558e49s^5 + 7.889e49s^4 + 9.057e49s^3 \\ &+ 6.816e49s^2 + 4.613e49s + 8.618e48 \end{aligned}} \tag{39}
$$

In addition, the biquadratic approximation method of El-Khazali yields the following rational transfer function for $s^{0.1}$ and $s^{0.8}$, respectively.

$$
s^{0.1} = \frac{1.994s^2 + 5.082s + 1.594}{1.594s^2 + 5.082s + 1.994} \tag{40}
$$

$$
s^{0.8} = \frac{3.436s^2 + 4.404s + 0.236}{0.2365s^2 + 4.404s + 3.436} \tag{41}
$$

Substituting (40) and (41) into (30) gives a 6th-order approximation of the form:

$$
H_k(s) = \frac{0.377s^4 + 8.222s^3 + 28.333s^2 + 26.247s + 6.852}{0.943s^6 + 19.968s^5 + 63.984s^4 + 76.78s^3 + 62.978s^2 + 34.234s + 7.229} \tag{42}
$$

The frequency and the unit-step responses of all four approximations given by (33), (36), (39), and (42) are shown in Figs. 4 and 5, respectively. Obviously, El-Khazali's approximation yields a competitive frequency response to those of Oustaloup's and Matsuda's approximations, but gives larger steady-state errors than its counterparts as depicted in Fig. 5. This is due to using a 2nd-order approximation to the fractional-order derivatives in (40) and (41). Increasing their order of approximation to a 4th-one, for example, would improve the rational approximation of $s^{0.1}$ and $s^{0.8}$ and consequently, it would improve the steady-state step response. Furthermore, the parameter values of El-Khazali's approximation are smaller than its competitive counterparts and that would imply less expensive circuit design.

The effectiveness of the approximation methods in designing different types of controllers is further investigated in the next section.

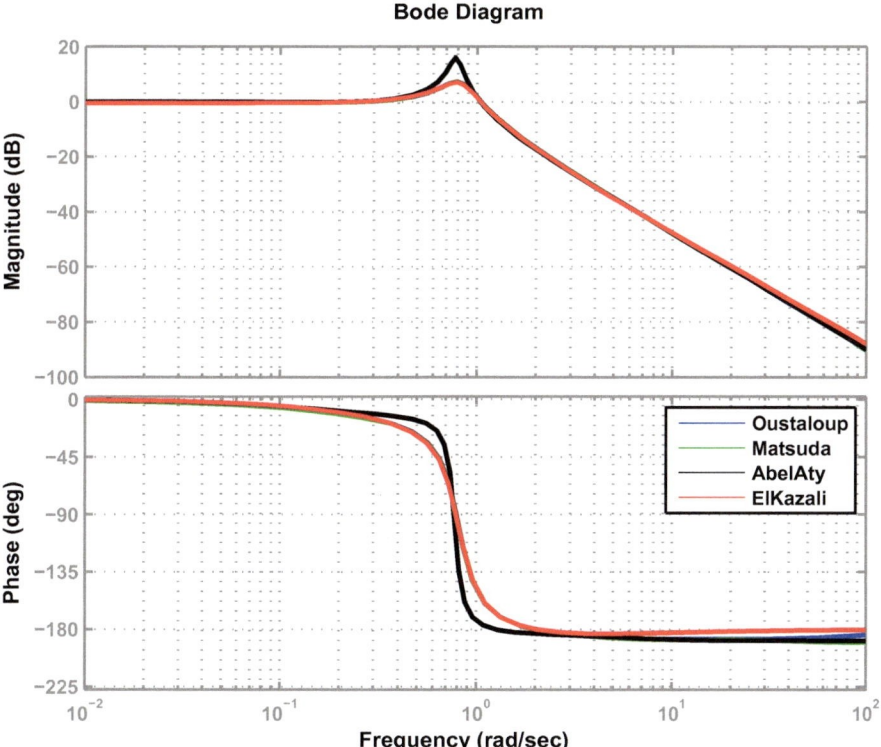

Fig. 4 Bode diagrams of (33), (36), (39) and (42)

Step Response

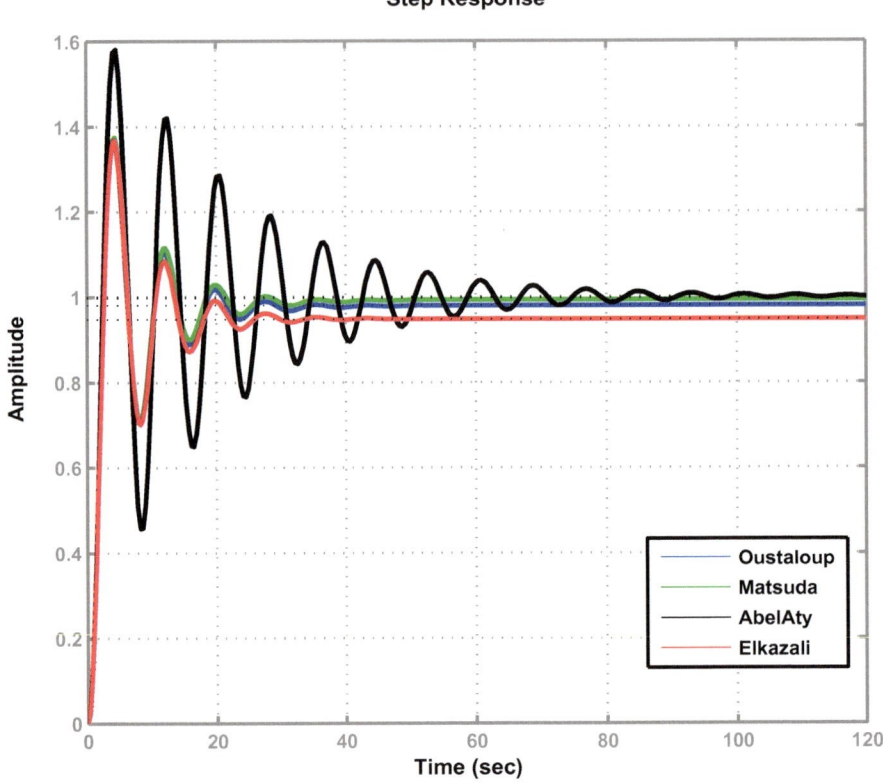

Fig. 5 Step response of (33), (36), (39) and (42)

6 Approximation of Fractional-Order Controllers

The commonly used proportional–integral–derivative (PID) controller, which consists of three parameters has been successfully used in industrial applications for several decades. The popularity of the PID controller lies in the simplicity of the design procedures and in the effectiveness of its system performance [18]. Some applications of PID controller show undesirable system performance, which may be enhanced by using fractional-order PID (FOPID) controllers [7–9, 19–21].

A FOPID controller was introduced in [7]. it is a generalization of the conventional integer-order PID controller and denoted by $PI^\lambda D^\delta$, where λ and δ are the fractional orders of the integral and the derivative components, respectively. Thus, adding two more parameters to tune the original three parameters of the classical integer-order PID controllers, which add more degrees of freedom [7]. The PID controller processes the present, past, and future values of the error signal, $e(t)$. A digital implementation of such controllers grew rapidly and overcomes the difficulties embedded in the realizations of the continuous versions.

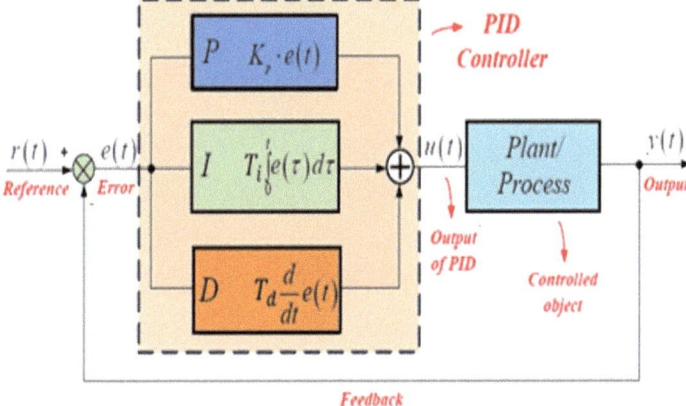

Fig. 6 Block diagram of a controlled system using integer-order PID controller

Figure 6 shows a typical block diagram of an integer-order PID controller. The proportional controller amplifies the present value of the error signal. The integral part refers to the accumulation of the past errors, while its derivative part predicts the future values of the error; i.e., acts on the anticipated value of the error signal. One may use the weighted sum of these three actions to make a final adjustment to tune its parameters [7].

6.1 Proportional-Order Controllers

Proportional controllers are widely used and are simple to design. They simply form a direct scaling of the error signal to alter the transient and steady-state system responses, i.e.,

$$y(t) = K_p e(t) \tag{43}$$

where $e(t) = r(t) - y(t)$ is the error signal between a reference signal, $r(t)$, and the output signal, $y(t)$. Thus,

$$Y(s) = K_p E(s) \tag{44}$$

The steady-state error in a proportional controller, $C(s) = K_p$, is inversely proportional to the controller gain [22, 23]. It is well known that increasing it makes the system response faster and minimizes the steady-state errors, but increases the system overshoot [15].

6.2   *Fractional-Order Integral Controllers (I^{λ}-Controllers)*

The fractional-order integral controllers (I^{λ}- controllers) produce the control signal $y(t)$, which is proportional to the fractional integral of the error signal $e(t)$. It can be described by the following relation [23]:

$$y(t) = T_i.I^{\lambda}e(t) \equiv T_i.D^{-\lambda}e(t) \tag{45}$$

which is described by the following transfer function:

$$\frac{Y(s)}{E(s)} = \frac{T_i}{s^{\lambda}} \equiv C(s, \lambda) \tag{46}$$

where $C(s, \lambda)$ denotes a fractional-order controller, (I^{λ}-controller) of order λ.

Obviously, increasing λ reduces the steady-state error. The fractional-order integrator, $s^{-\lambda}$, eliminates the offset error when $\lambda > 1$ without increasing the integral constant T_i. It is possible to reduce the offset error when $\lambda < 1$ by adding a pole and zero at the origin. Therefore, $s^{-\lambda}$ in this case can be expressed as [22, 24]:

$$\frac{1}{s^{\lambda}} = \frac{1}{s^{\lambda}}.\frac{s}{s} = \frac{1}{s}.s^{(1-\lambda)}, \quad 0 < \lambda < 1 \tag{47}$$

Equation (47) shows that $s^{-\lambda}$ can be expressed as a product of a pure integral ($1/s$) and a fractional-order differentiator $s^{(1-\lambda)}$.

Now, we briefly summarize the effects of extending the integral control actions to the fractional case. Let us first explore $s^{-\lambda}$ with $\lambda = 0.4$ and $T_i = 1$. The transfer function of the $I^{0.4}$-controller is then given by

$$C(s) = \frac{1}{s^{0.4}} \tag{48}$$

Using the approximations discussed in example (1), the $I^{0.4}$-controller, (48), can be expressed by the following transfer functions:

$$C_{Ous}(s) = \frac{s^3 + 41.74s^2 + 77.14s + 6.31}{6.31s^3 + 77.14s^2 + 41.74s + 1} \tag{49}$$

$$C_{Mat}(s) = \frac{s^3 + 83.01s^2 + 163.6s + 10.01}{10.01s^3 + 163.6s^2 + 83.01s + 1} \tag{50}$$

$$C_{Abd}(s) = \frac{\begin{array}{c} s^8 + 1443327.34s^7 + 190377994705.8s^6 + 3320443435965385.5s^5 \\ + 8064411171562614000s^4 + 2.737e21s^3 + 1.285e23s^2 + 7.801e23s + 3.550e23 \end{array}}{\begin{array}{c} 412.48s^8 + 206415397.69s^7 + 11956108280000.56s^6 + 94706047639064780s^5 \\ + 1047394777704827300000s^4 + 1.616e22s^3 + 3.399e23s^2 + 8.429e23s \end{array}} \tag{51}$$

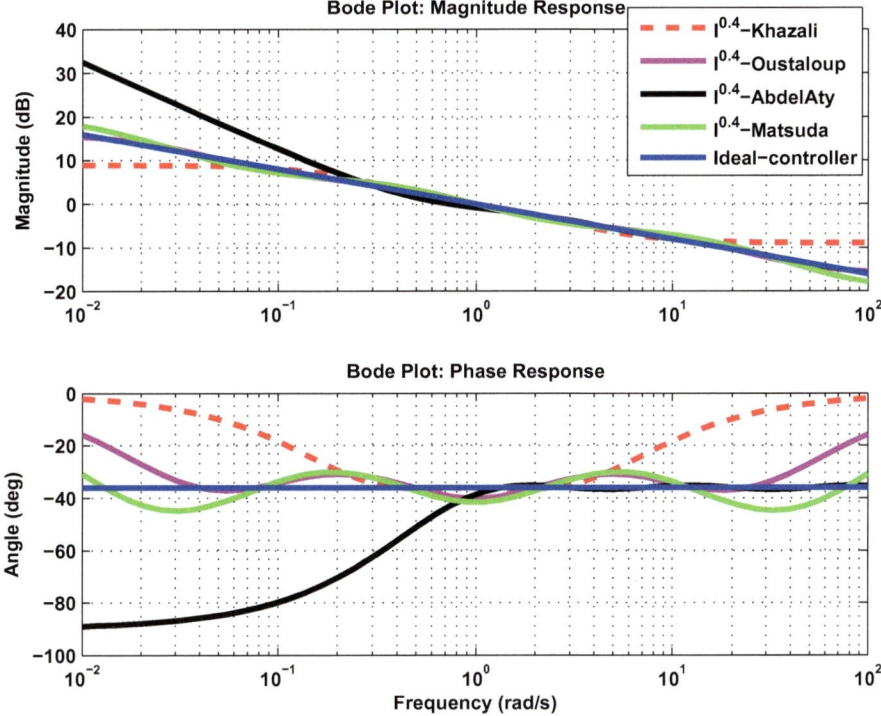

Fig. 7 Bode diagram of an integer I- and a fractional-order $I^{0.4}$-controller approximated by Oustaloup's, Matsuda's, AbdelAty's, and El-Khazali's approximations for $T_i = 1$

and

$$C_k(s) = \frac{0.8931s^2 + 4.924s + 2.493}{2.493s^2 + 4.924s + 0.8931} \tag{52}$$

Figure 7 shows a numerical simulation of the ideal $I^{0.4}$-controller's four of its approximations using Oustaloup's, Matsuda's, AbdelAty's, and El-Khazali's methods.

6.3 Fractional-Order Differential Controllers (D^δ-Controllers)

The fractional-order differential controllers of order δ, denoted by D^δ-controllers, produce a control signal, $y(t)$, that is proportional to the fractional-order derivative of the error signal $e(t)$. It described by the following expression [23]:

$$y(t) = T_d.D^\delta e(t) \tag{53}$$

Hence, the corresponding transfer function of (53) is given by

$$\frac{Y(s)}{E(s)} = T_d.s^\delta \tag{54}$$

where T_d is a constant of the differential controller, $C(s) = T_d.s^\delta$.

The fractional differentiator, s^δ, acts as a δ-predictor that predicts the future value of the error signal. It reduces the rate of change of error, which improves the control performance. Notice that s^δ cannot be implemented alone since it does not eliminate the steady-state errors. Thus, it must be augmented at least with a proportional controller, P, to form a PD controller, known as lead controller, to shape the frequency response of the controlled system. Notice that the δth-differential controller amplifies noise signals when δ increases and has no effect on the steady-state error [9, 23].

For comparison, let us consider the case when $\delta = 0.4$ and $T_d = 1$, then, the $D^{0.4}$-controller is given by $C(s) = s^{0.4}$. Figure 8 shows the frequency response of the rational approximation of $C(s) = s^{0.4}$ using (7), (16), (18), and (19). Obviously, the low-frequency deviation of (18) cannot be avoided due to the presence of the

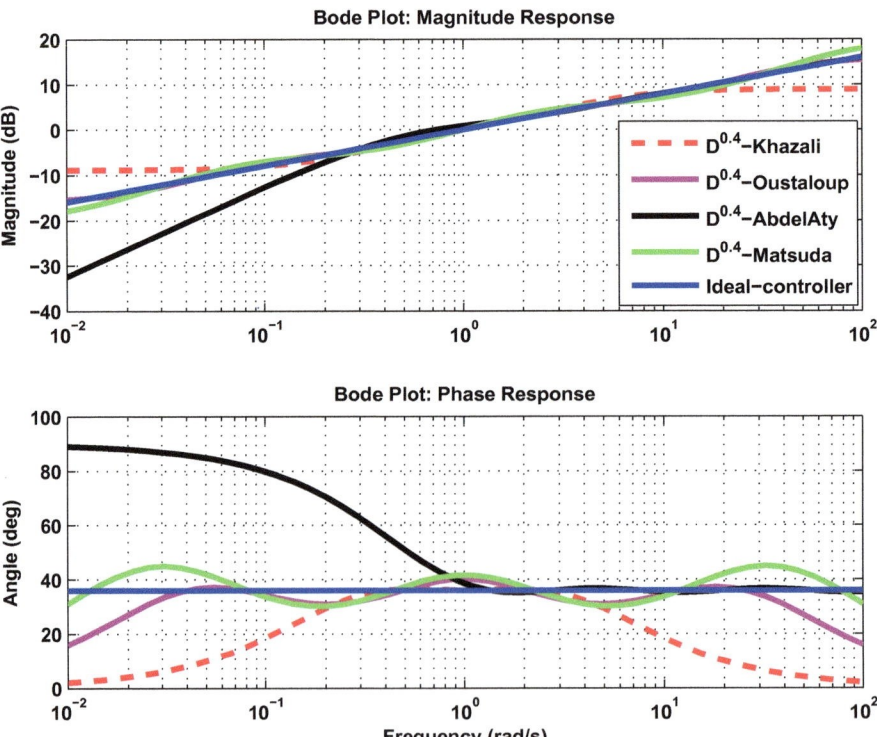

Fig. 8 Bode diagram of a classical D- and $D^{0.4}$-controller approximated by Oustaloup's, Matsuda's, AbdelAty's, and El-Khazali's approximations with $T_d = 1$

differential effect of the high-pass filters approximations. However, the biquadratic approximation of (19) does behave like a proportional controller for both the low- and high-frequency bands, thus suppressing signal noise.

6.4 Fractional-Order PI Controllers (PI^λ-Controller)

The fractional-order PI controller (PI^λ-Controller), or lag controllers, combines both the proportional and the fractional-order integral action and is defined as [23]

$$y(t) = K_p e(t) + T_i . I^\lambda e(t) \tag{55}$$

which yields a lag controller of order λ of the form:

$$\frac{Y(s)}{E(s)} = C(s) = K_p + \frac{T_i}{s^\lambda} \tag{56}$$

As λ increases, this controller has the following features [23]:

- It reduces the steady-state error.
- It decreases the rise time.
- It filters out the noise at high frequencies.
- It increases bandwidth of the system.
- It increases the order and type of the system.

Consider the case when $\lambda = 0.4$, $K_p = 1$, and $T_i = 1$, the following transfer functions approximate $PI^{0.4} = 1 + 1/s^{0.4}$ using (7), (16), (18), and (19), respectively:

$$C_{Ous}(s) = \frac{731s^3 + 11888s^2 + 11888s + 731}{631s^3 + 7714s^2 + 4174s + 100} \tag{57}$$

$$C_{Mat}(s) = \frac{1101s^3 + 24661s^2 + 24661s + 1101}{1001s^3 + 16360s^2 + 8301s + 100} \tag{58}$$

$$C_{Abd}(s) = \frac{\begin{array}{c} s^8 + 502710.5s^7 + 29376520704.24s^6 + 237079034998476.75s^5 \\ + 272818468162711600s^4 + 45699052676123180000s^3 \\ + 1.133e21s^2 + 3.925e21s + 3.859e05 \end{array}}{\begin{array}{c} 0.9976s^8 + 499219.78s^7 + 28916087705.21s^6 + 229048475943137.66s^5 \\ + 253314527978607520s^4 + 39078673785758040000s^3 \\ + 822012098401221200000s^2 + 2.039e21s \end{array}} \tag{59}$$

and

$$C_k(s) = \frac{s^2 + 2.908s + 1}{0.7362s^2 + 1.454s + 0.2638} \tag{60}$$

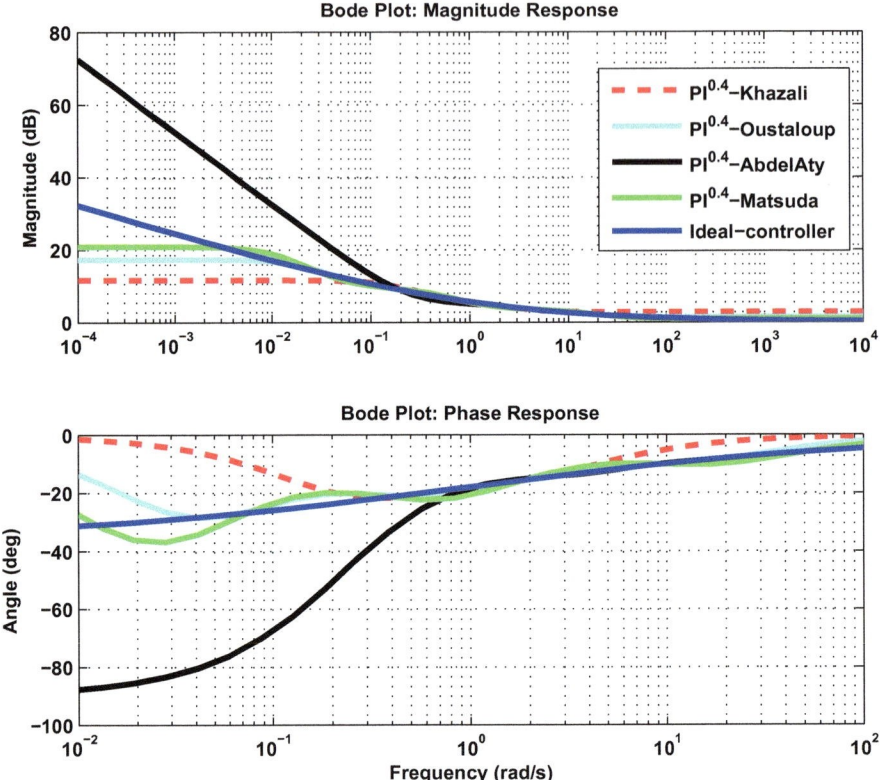

Fig. 9 Bode diagram of an approximated $PI^{0.4}$-controller using Oustaloup's, Matsuda's, Abdel-lAty's, and 2nd-order El-Khazali's biquadratic form with $K_p = T_i = 1$

Graphically, we make the following comparison in the frequency domain, shown in Fig. 9, between a classical PI-controller and $PI^{0.4}$-controller.

Figure 10 shows the same approximation shown in Fig. 9, except for El-Khazali's approximation, where the following 4th-order biquadratic form is used to approximate $s^{0.4}$ instead of a 2nd-order one:

$$PI^{0.4} = \frac{70137s^4 + 1374500s^3 + 3422000s^2 + 1374500s + 70137}{7977s^4 + 362500s^3 + 1711000s^2 + 1012000s + 62160} \quad (61)$$

Clearly, the frequency response of all approximations match that of the ideal controller except for AbdelAty's one, which has large deviations in both the magnitude and phase responses for low frequency. If one considers the simplicity of the realization and construction, one would choose the biquadratic approximation given by (61) since it depends only on a single parameter, which could be the order of the integrators or the differentiator.

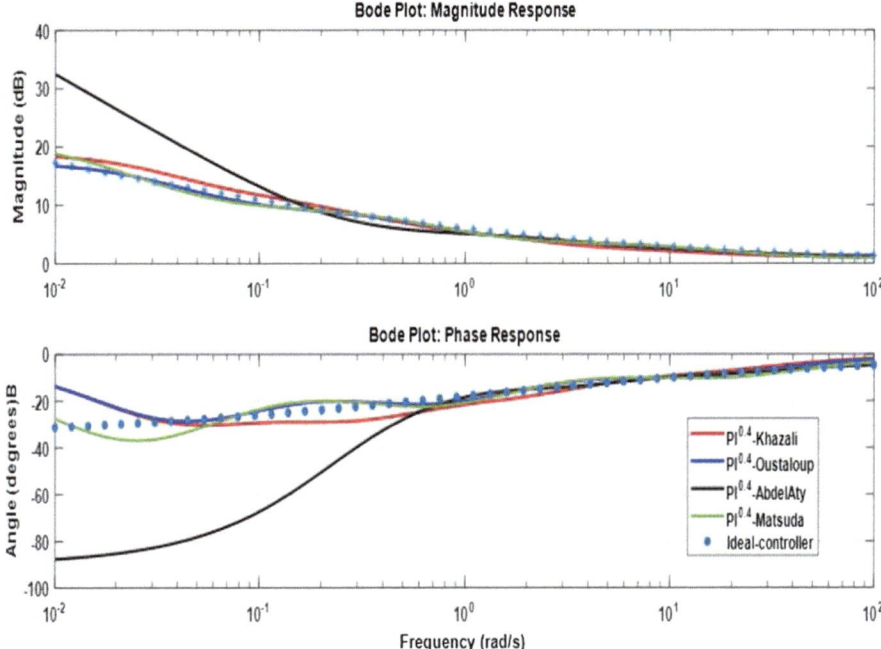

Fig. 10 Bode diagram of an approximated $PI^{0.4}$-controller using (7), (16), (18), and (61) with $K_p = T_i = 1$

6.5 *Fractional-Order PD Controller (PD^δ-Controller)*

The transfer function of the fractional-order controller (lead controller) of order δ is given by

$$C(s) = K_p + T_d.s^\delta \tag{62}$$

As δ increases, the PD^δ or the fractional-order lead controller enjoys the following properties [19]:

- It reduces the overshoot.
- It improves transient response.
- It reduces the settling time.
- It improves the bandwidth of the system.
- It may make noises at high frequencies.
- It does not affect steady-state errors.

Now, to summarize the effects of extending the derivative control actions to the fractional case; let us explore the case when $\delta = 0.4$, $K_p = 1$, and $T_d = 1$. Then, the $PD^{0.4}$-controller becomes $C(s) = 1 + s^{0.4}$. Now using (26), (27), (28), and (29),

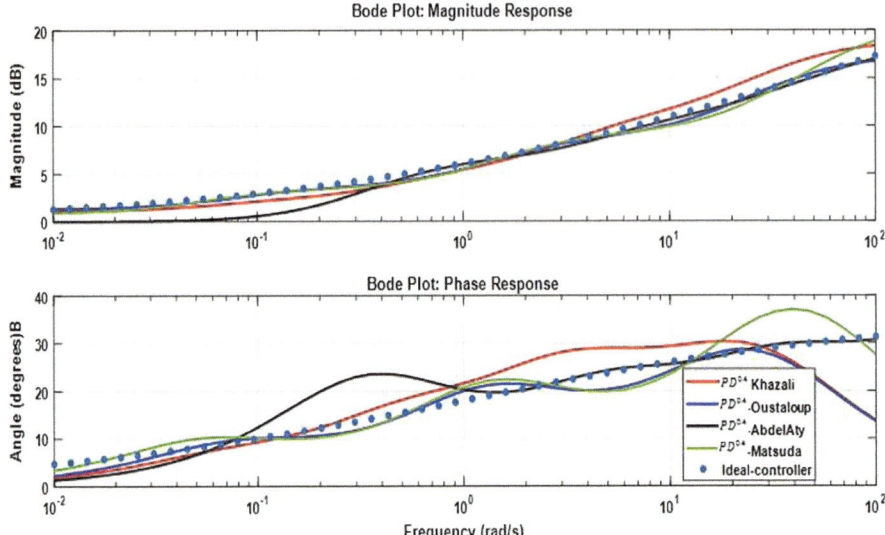

Fig. 11 Bode diagram of PD^{δ}-controller and compared with the ideal one for $\delta = 0.4$, $K_p = T_d = 1$

the frequency response of the $PD^{0.4}$ is shown in Fig. 11. All approximations yield a good match with the ideal lead integrator, however, the AbdelAty's one gives a perfect match with the ideal lead controller at high frequency only.

7 FOPID Controllers

It is well known that the time response of the PID controllers output is given by the following expression [7]:

$$u(t) = K_p e(t) + T_i \int_0^t e(\tau).d\tau + T_d \frac{d}{dt} e(t) \tag{63}$$

and its transfer function is given by

$$C(s) = \frac{U(s)}{E(s)} = K_p + \frac{T_i}{s} + T_d.s \tag{64}$$

where $E(s) = \mathscr{L}\{e(t)\}$, and $U(s) = \mathscr{L}\{u(t)\}$ are the Laplace transforms of the error and control signals, respectively.

The Zeigler–Nichols design method is a popular one that can be used to design integer-order PID controllers for most systems [15, 20, 21]. The integer-order PID controller is applicable for many control problems and it often yields satisfactory performance and, in some cases, it requires parameter tuning [21].

The concept of the fractional-order PID (FOPID or $PI^\lambda D^\delta$) controller was introduced by Podlubny in [23]. This controller has an integrator of an order λ and a differentiator of order δ. It was shown that the fractional-order controller outperforms its integer counterpart [19, 25]. When controlling industrial plants, they require a complete satisfaction of a wide range of specification, where wide ranges of techniques are needed. Recently, FOPID controllers are used for industrial applications to improve system performance. They provide extra degrees of freedom by adding two more parameters to tune (namely, λ and δ) to the original three parameters, (K_p, T_i, T_d), thus increasing the complexity of tuning its parameters [20].

7.1 FOPID Controllers

The fractional-order integro-differential equation that describes the FOPID controllers is given by [23]

$$u(t) = K_p e(t) + T_i . I^\lambda e(t) + T_d . D^\delta e(t) \tag{65}$$

The Laplace transform of (65) is given by

$$C(s) = K_p + T_i . s^{-\lambda} + T_d . s^\delta \tag{66}$$

Obviously, one can get the classical integer-order PID controller by taking $\lambda = \delta = 1$. With more freedom in tuning the controller, the four-point PID diagram can be seen as a PID controller plane, which is depicted in Fig. 12 [7].

Notice that the integer-order controllers are classified as particular cases of the more general FOPID controller, which provides more flexibility and robustness and gives the capability for better adjustment of the dynamical properties of fractional-order control system [21, 26]. For example, by assuming $\delta = 0$ and $T_d = 0$, then a PI^λ- controller is derived, and so on.

The FOPID controller can be considered as an infinite-dimensional linear filter due to the fractional orders of the differentiators and the integrators. Since PID controllers are ubiquitous in industry process control, then fractional-order PID control will also be ubiquitous when tuning and implementation techniques are well devel-

Fig. 12 Generalization of
the FOPID controllers

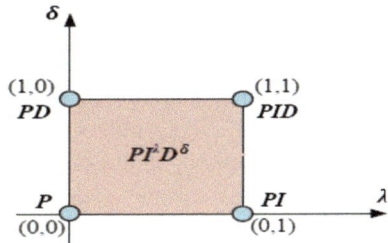

oped [27–35]. As compared to PID controller, a FOPID is supposed to offer the following advantages [30]:

- If the parameter of a controlled system changes, a FOPID controller is less sensitive than the classical PID controller.
- Fractional-order controller has two extra variables to tune. This provides extra degrees of freedom to the dynamic properties of fractional-order system.

Now, we will study the effects of extending the integral and derivative control actions on the fractional-order case. Let us first explore s^λ and s^δ with $\lambda = \delta = 0.4$ and $K_p = T_i = T_d = 1$. Then, the $PI^{0.4}D^{0.4}$-controller is given by

$$C(s) = 1 + \frac{1}{s^{0.4}} + s^{0.4} \tag{67}$$

Moreover, the $PI^{0.4}D^{0.4}$-controller expressed in (67) can be approximated by Oustaloup's, Matsuda's, AbdelAty's, and El-Khazali's approximations using Eqs. (26), (27), (28) and (29), respectively, as follows:

$$C_{Ous}(s) = \frac{\begin{array}{c} 471261s^6 + 13975062s^5 + 121221630s^4 + 206381577s^3 \\ + 121221630s^2 + 13975062s + 471261 \end{array}}{\begin{array}{c} 63100s^6 + 3405194s^5 + 37483170s^4 + 77336233s^3 \\ + 37483170s^2 + 3405194s + 63100 \end{array}} \tag{68}$$

$$C_{Mat}(s) = \frac{\begin{array}{c} 1112101s^6 + 44358221s^5 + 509457623s^4 + 881186042s^3 \\ + 509457623s^2 + 44358221s + 1112101 \end{array}}{\begin{array}{c} 100100s^6 + 9945301s^5 + 153010820s^4 + 337568202s^3 \\ + 153010820s^2 + 9945301s + 100100 \end{array}} \tag{69}$$

$$C_{Abd}(s) = \frac{\begin{array}{c} s^{16} + 1003151.62s^{15} + 309946795646.82s^{14} + 29742326890849164s^{13} + 1.086e21s^{12} \\ + 1.389e25s^{11} + 6.992e28s^{10} + 1.256e32s^9 + 8.960e34s^8 + 2.301e37s^7 + 2.360e39s^6 \\ + 8.778e40s^5 + 1.303e42s^4 + 6.770e42s^3 + 1.283e43s^2 + 5.003e42s + 7.391e41 \end{array}}{\begin{array}{c} 0.0024s^{16} + 4700.99s^{15} + 2277382680.28s^{14} + 340180953977873.3s^{13} + 3.491e23s^{11} \\ + 18186425387598578000s^{12} + 2.530e27s^{10} + 6.727e30s^9 + 6.797e33s^8 + 2.523e36s^7 \\ + 3.548e38s^6 + 1.815e40s^5 + 3.437e41s^4 + 2.223e42s^3 + 4.563e42s^2 + 1.755e42s \end{array}} \tag{70}$$

and

$$C_k(s) = \frac{s^4 + 5.4134s^3 + 9.5961s^2 + 5.41399s + 1}{0.241s^4 + 1.8047s^3 + 3.3834s^2 + 1.8047s + 0.241} \tag{71}$$

A direct comparison in the frequency domain is shown in Fig. 14 for the approximated $PI^{0.4}D^{0.4}$-controller given by (68), (69), (70), and (71). Observe that for AbdelAty's method, an 8th-order approximation was used for simulation purposes. In spite of using a higher order approximation, a larger magnitude and phase error is demonstrated for low frequencies.

The following example introduces further comparison in both time and frequency domains between the previous three methods; Oustaloup's, Matsuda's, and El-Khazali's methods by considering a closed-loop controlled system.

Example 3 Let us consider Fig. 13 with FOPID controller and integer-order plant transfer function below:

$$C(s) = 18 + \frac{13}{s^{0.8}} + 6s^{1.4} \tag{72}$$

and

$$G(s) = \frac{1}{s(s+1)(s+5)} \tag{73}$$

Then, the open-loop transfer function of the controlled system is given by

$$L(s) = C(s)G(s) = \frac{6s^{2.2} + 18s^{0.8} + 13}{s^{3.8} + 6s^{2.8} + 5s^{1.8}} \tag{74}$$

and the closed-loop transfer function for a unity feedback system is equal to

$$P(s) = \frac{L(s)}{1 + L(s)} = \frac{6s^{2.2} + 18s^{0.8} + 13}{s^{3.8} + 6s^{2.8} + 6s^{2.2} + 5s^{1.8} + 18s^{0.8} + 13} \tag{75}$$

That is,

$$P(s) = \frac{6s^2(s^{0.2}) + 18s^{0.8} + 13}{s^3(s^{0.8}) + 6s^2(s^{0.8}) + 6s^2(s^{0.2}) + 5s(s^{0.8}) + 18s^{0.8} + 13} \tag{76}$$

Obviously, (76) includes $s^{0.2}$ and $s^{0.8}$ such that the Oustaloup's 5th-order approximation of $s^{0.8}$ was given previously in equation (32), while the Oustaloup's approximation of $s^{0.2}$ is given by

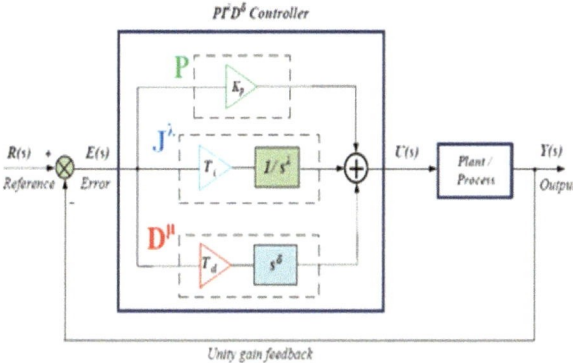

Fig. 13 Block diagram of a FOPID controller with unity gain feedback

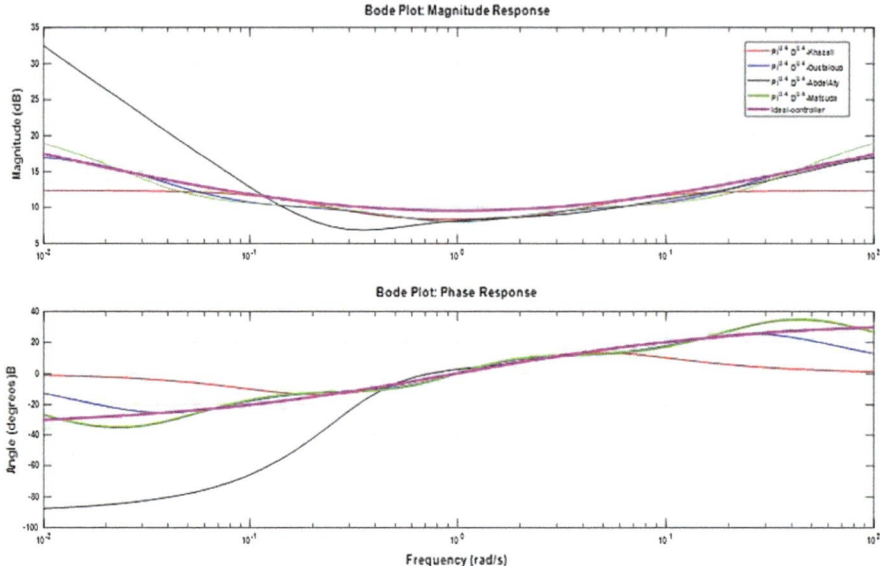

Fig. 14 Bode diagram of a frequency response of a classical PID controller and $PI^\lambda D^\delta$ controller approximated by Oustaloup's, Matsuda's, AbdelAty's, and El-Khazali's approximations with $\lambda = \delta = 0.4$, $K_p = T_i = T_d = 1$

$$s^{0.2} = \frac{2.512s^5 + 98.83s^4 + 531.7s^3 + 442.3s^2 + 56.87s + 1}{s^5 + 56.87s^4 + 442.3s^3 + 531.7s^2 + 98.83s + 2.512} \tag{77}$$

One can obtain an approximate closed-loop transfer function $P(s)$ by substituting (32) and (77) into (76). Similarly, the Matsuda's approximation of $s^{0.2}$ is given by

$$s^{0.2} = \frac{3.357s^4 + 161s^3 + 453.9s^2 + 95s + 1}{s^4 + 95s^3 + 453.9s^2 + 161s + 3.357} \tag{78}$$

The approximated closed-loop transfer function can now be obtained by using (35), (78) and (76). On the other hand, El-Khazali's approximation for $s^{0.2}$ is given by

$$s^{0.2} = \frac{2.125s^2 + 5.051s + 1.325}{1.325s^2 + 5.051s + 2.125} \tag{79}$$

The closed-loop transfer functions of the three approximations are, respectively, given as follows:

$$P_{Ous}(s) = \frac{\begin{matrix} 0.3786s^{12} + 52.31s^{11} + 2076.37s^{10} + 30424.88s^{9} + 190374.23s^{8} \\ + 629279.32s^{7} + 1343379.50s^{6} + 1723054.78s^{5} + 1136020.90s^{4} \\ + 340700.72s^{3} + 44471.77s^{2} + 2181.14s + 33.79 \end{matrix}}{\begin{matrix} s^{13} + 85.89s^{12} + 2334.45s^{11} + 27839.99s^{10} + 171865.80s^{9} \\ + 587969.75s^{8} + 1207365.50s^{7} + 1776199.71s^{6} + 1881971.42s^{5} \\ + 1162206.67s^{4} + 342468.13s^{3} + 44515.75s^{2} + 2181.46s + 33.79 \end{matrix}} \quad (80)$$

$$P_{Mat}(s) = \frac{\begin{matrix} 0.0643s^{11} + 55.11s^{10} + 3695.99s^{9} + 68507.40s^{8} + 371769.57s^{7} \\ + 1075594.57s^{6} + 2020717.97s^{5} + 1873642.24s^{4} + 711212.26s^{3} \\ + 102240.19s^{2} + 3913.16s + 43.839 \end{matrix}}{\begin{matrix} s^{12} + 138.98s^{11} + 5049.83s^{10} + 59864.84s^{9} + 344954.58s^{8} \\ + 1012528.04s^{7} + 1745850.16s^{6} + 2331027.67s^{5} + 1932393.81s^{4} \\ + 714332.34s^{3} + 102286.18s^{2} + 3913.21s + 43.83 \end{matrix}} \quad (81)$$

and

$$P_k(s) = \frac{0.6623s^6 + 13.91s^5 + 58.25s^4 + 142.3s^3 + 202s^2 + 118s + 22.84}{s^7 + 11.76s^6 + 56.03s^5 + 125.4s^4 + 189.1s^3 + 214.3s^2 + 118.6s + 22.84} \quad (82)$$

The exact step and frequency response of the systems are shown in Figs. 15 and 16. Obviously, the three approximations yield almost identical step response. However, using the 2nd-order biquadratic approximation of El-Khazali reduces the size of the controllers needed for similar cases.

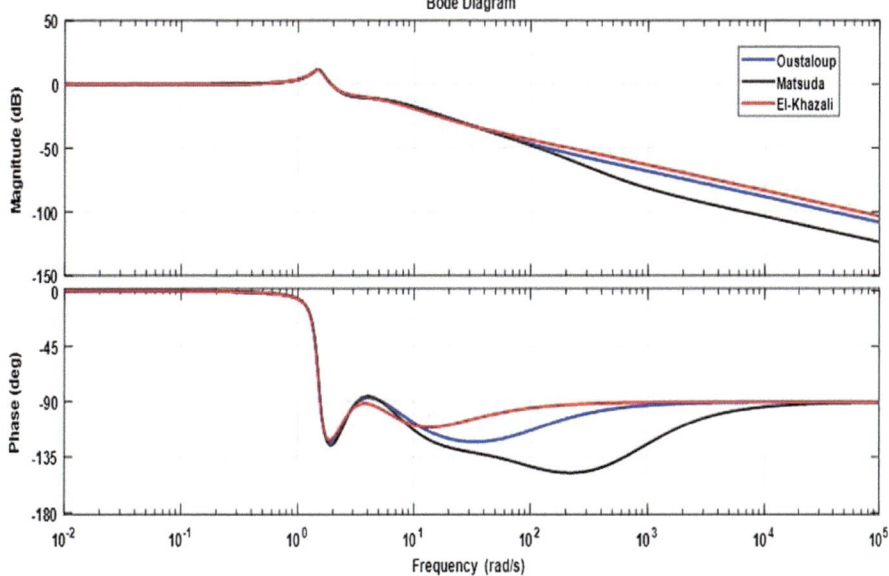

Fig. 15 Bode diagram of the closed-loop controlled system using FPID controllers

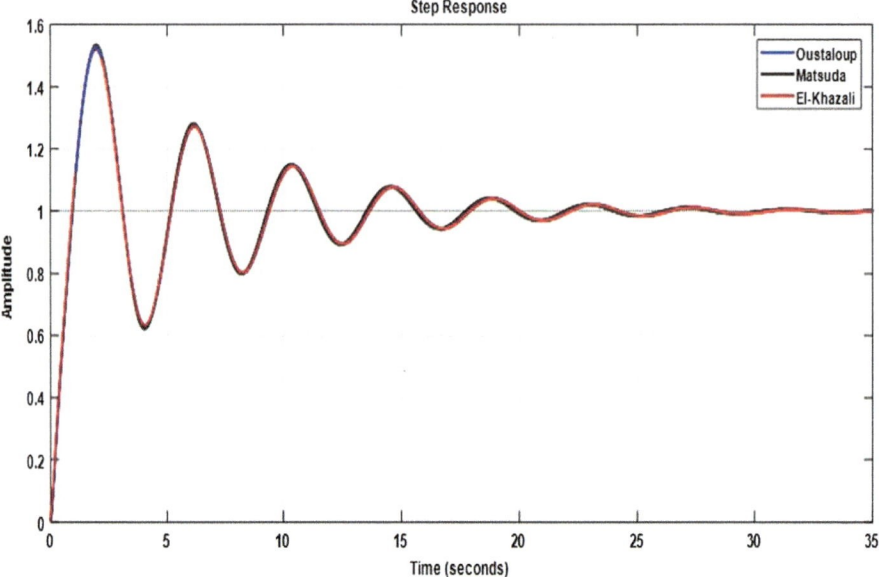

Fig. 16 Step responses for the closed-loop controlled system FOPID controllers

8 Conclusions

Four different approximation methods were used to approximate fractional-order Laplacian operators. These are Oustaloup's, Matsuda's, AbdelAty's, and EL-Khazali's methods. Oustaloup yields only odd order of rational-transfer functions, while Matsuda's method yields an unrealizable model if the sum of poles and zeros is odd. Both methods require high order of approximation with large coefficients to generate constant phases in the frequency domain. AbdelAty's method, however, is a sum of high-pass filter that yields large phase error for small fractional orders (say below 0.5). It depends on using optimum phase algorithm (OPA) to calculate the parameters of the approximations. It yields large parameter values than expected, which is expensive to design. The fourth method is the biquadratic approximation of El-Khazali. It yields an exact phase response at the center frequency of the approximation with unity gain. In most cases, lower order approximation of El-Khazali is very competitive to higher order approximations of other methods, which yields, most of the time, reasonable parameter values. Notice that one may approximate fractional-order integrators using El-Khazali's method by simply using the reciprocal approximation of the fractional-order differentiator. This feature is not possible by the other three approximations. It is worth mentioning that when discretizing these approximations using bilinear transformation, which is left for further study, it showed that EL-Khazali's method gave the most accurate and stable discrete-time version of the given approximation out of all four methods.

Acknowledgements We thank the sponsors of the International Conference on Fractional Differentiation and its Applications (ICFDA 2018), who provided insight and expertise that greatly assisted the research to be in hand. We would also like to thank the international editors, Praveen Agarwal, Dumitru Baleanu, YangQuan Chen, and Tenreiro Machado, for their valuable comments and remarks.

Appendix: Springer-Author Discount

1. Oustaloup's Approximation

```
clear all
alf=input ('Enter value of ALFA = ')
% frequency range
w_L=0.1; w_H=10.
 r=alf;  NN=2;
%
[v1,v2,D_N_K,sysO_tf]=ora_foc(r,NN,w_L,w_H);
%%
% Oustaloup-Recursive-Approximation for fractional order differentiator
%
% Input variables:
%        r: the fractional order as in s^r, r is a real number
%        N: order of the finite TF approximation for both (num/den)
%           (Note: 2N+1 recursive z-p pairs)
%        w_L: low frequency limit of the range of the frequency of interest
%        w_H: upper freq. limit of the range of the frequency of interest
% Output:
% sys_foc: continuous time transfer function system object (TF)
% Sample values: w_L=0.1;w_H=1000;  r=0.5;  N=4;
% Existing problem: Be careful when doing "c2d", I met some problems.
function [v1,v2,D_N_K,sys_N_tf]=ora_foc(r,N,w_L,w_H)
w_L=w_L*0.1;w_H=w_H*10; % enlarge the freq. range of interest for % goodness
mu=w_H/w_L;                  %
w_u=sqrt(w_H*w_L);
alpha=(mu)^(r/(2*N+1));
eta=(mu)^((1-r)/(2*N+1));
k=-N:N;
w_kp=(mu).^( (k+N+0.5-0.5*r)/(2*N+1) )*w_L;
w_k=(mu).^( (k+N+0.5+0.5*r)/(2*N+1) )*w_L;
D_N_K=(w_u/w_H)^r * prod(w_k) / prod(w_kp);
D_N_P=-w_k;D_N_Z=-w_kp;
[num,den]=zp2tf(D_N_Z',D_N_P',D_N_K);
sys_N_tf=tf(num,den);
v1=num;  v2=den;
sys_foc=tf(num,den);
end
```

2. El-Khazali's Approximation

```
%
alf=input ('Enter value of ALFA = ')
n= input ('No. of Biquadratic Modules = ')
% wc is the initial selection of the center frequency of the first %biquadratic form
     wc=1;
% This is a call statement of the function "Biquad_K" that generates
% a modular structure of biquadratic approximations using EL-Khazali % approximation.
[Nk,Dk,sysk]= Biquad_K (alf,n,wc)
```

```
%%
function [Numk,Denk,sysk]=Biquad_K(alf,n,w)
% This function generates a normalized modular biquadratic structure
% structures of order 2*n.
%
et=tan(alf*pi/4);
wc(1)=w;
     ao=alf^alf+2*alf+1;
     a2=alf^alf-2*alf+1;
     a1=(a2-ao)*tan((2+alf)*pi/4);
if ( n > 1 )
% The solution of the following polynomial is used to generate
% a recursive formula to select the next center frequency for the
% next modular structure.
Y=roots([ao*a2*et a1*(a2-ao) (a1^2-a2^2-ao^2) a1*(a2-ao) ao*a2*et ]);
 wx=(abs(max(Y)))
 for k=2:n
  wc(k)=(wx^(2*(k-1)))*wc(1)
   end
% normalizing by the geometric mean;
wm=(prod(wc))^(1/n);
sysk=1;
  for l=1:n;
    N=[ao  a1*wc(l)/wm  a2*(wc(l)/wm)^2];
    D=[a2  a1*wc(l)/wm  ao*(wc(l)/wm)^2];
sysk=sysk*(tf(N,D));
   end
   [Numk,Denk]=tfdata(sysk);
else
    Numk=[ao  a1*w  a2*w^2];
    Denk=[a2  a1*w  ao*w^2];
    sysk=tf(Numk,Denk);
end
end
```

References

1. Oustaloup, A., Levron, F., Mathieu, B., Nanot, F.M.: Frequency band complex noninteger differentiator: characterization and synthesis. IEEE Trans. Circuit. Syst. I, Fund. Theory and Appl. **47**, 25–39 (2000)
2. Vinagre, B.M., Podlubny, I., Hernandez, A., Feliu, V.: Some approximations of fractional order operators used in control theory and applications. Fract. Calculus Appl. Anal. **3**, 231–248 (2000)
3. Yce, A., Deniz, F.N., Tan, N.: A new integer-order approximation table for fractional order derivative operators. Int. Federation Autom. Control **50**, 9736–9741 (2017)
4. AbdelAty, A.M., AbdelAty, A.S., Radwan, A.G., Psychalinos, C., Maundy, B.J.: Approximation of the fractional-order Laplacian s^α As a weighted sum of first-order high-pass filters. IEEE Trans. Circuit. Sys. II: Express Briefs **65** (2018)
5. El-Khazali, R.: Discretization of Fractional-Order Laplacian Operators. 19th IFAC Congress, Cape Town, 25/8/2014
6. Atherton, D.P., Tan, N., Yce, A.: Methods for computing the time response of fractional-order systems. IET Control Theory Appl. **9**, 817–830 (2015)
7. Podlubny, I.: Fractional-order systems and $PI^\lambda D^\mu$ controllers. IEEE Trans. Automat. Control **44**, 208–214 (1999)
8. Lin, W., Chongquan, Z.: Design of optimal fractional-order PID controllers using particle swarm optimization algorithm for DC motor system. In: IEEE Advanced Information Technology, Electronic and Automation Control Conference (IAEAC), pp. 175–179 (2015)

9. El-Khazali, R.: Fractional-order $PI^\lambda D^\mu$ controller design. Comp. Math. App. **66**, 639–646 (2013)
10. El-Khazali, R.: Discretization of Fractional-Order Differentiators and Integrators. 19th IFAC Congress (2014)
11. Tepljakov, A.: Fractional-order Modeling and Control of Dynamic Systems. Tallin University of Technology, Tallinn (2015)
12. Petras, I., Podlubny, I., OLeary, P.: Analogue Realization of Fractional Order Controllers. Fakulta BERG, TU Kosice (2002)
13. Anis Kharisovich Gilmutdinov: Pyotr Arkhipovich Ushakov, Reyad El-Khazali: Fractal Elements and their Applications. Springer, Switzerland (2017)
14. Matsuda, K., Fujii, H.: Optimized wave-absorbing control: analytical and experimental results. J. Guidance Control Dyn. **16**, 1153–1164 (1993)
15. Ogata, K.: Modern Control Engineering. Prentice Hall, New Jersey (2010)
16. Yang, X.S.: Flower pollination algorithm for global optimization in Unconventional computation and natural computation, pp. 240–249. Springer (2012)
17. El-Khazali, R.: On the biquadratic approximation of fractional-order Laplacian operators. Analog Int. Circuits and Sig. Proc. **82**, 503–517 (2015)
18. Astrom, K., Hagglund, T.: PID Controllers; Theory. Design and Tuning. Instrument Society of America, Research Triangle Park (1995)
19. Saleh, K.: Fractional Order PID Controller Tuning by Frequency Loop-Shaping: Analysis and Applications. Arizona State University, Arizona (2017)
20. Podlubny, I.: Fractional-order systems and $PI^\lambda D^\mu$-controllers. IEEE Trans. Autom. Control **44**, 208–214 (1999)
21. Cech, M., Schlegel, M.: The Fractional-order PID Controller Outperforms the Classical One. Process control 2006: Pardubice Technical University, pp. 1–6 (2006)
22. Podlubny, I., Dorcak, L., Kostial, I.: On fractional derivatives, fractional-order dynamic systems and $PI^\lambda D^\mu$ controllers. In: Proceedings of the 36th Conference on Decision and Control, San Diego, California, USA (1997)
23. Podlubny, I.: Fractional Differential Equations. Academic Press, New York (1999)
24. Rinu, P.S.: Design and analysis of optimal fractional-order PID controller. Int. J. Appl. Eng. Res. **10**, 23000–23002 (2015)
25. Xue, D., Chen, Y.Q., Atherton, D.P.: Linear Feedback Control: Analysis and Design with MATLAB. Society for Industrial and Applied Mathematics, Philadelphia (2007)
26. Luo, Y., Chen, Y.Q.: Fractional-order [proportional derivative] controller for robust motion control: tuning procedure and validation. Am. Control Conf., 1412–1417 (2009)
27. Kumar, A.: Controller Design for Fractional Order Systems. National Institute of Technology-Rourkela, Odisha (2013)
28. Maione, G., Lino, P.: New tuning rules for fractional PI^α controllers. Nonlinear Dyn. **49**, 251–257 (2007)
29. Luo, Y., Chen, Y.Q.: Fractional Order Motion Controls. Wiley, Odisha (2012)
30. Monje, C.A., Chen, Y.Q., Vinagre, B.M., Xue, D., Feliu, V.: Fractional-Order Systems and Controls: Fundamentals and Applications. Springer, London, New York (2010)
31. Dimeas, I.: Design of an Integrated Fractional-Order Controller. University of Patras, Patras (2017)
32. Oldham, K.B., Spanier, J.: The fractional calculus: Theory and Applications of Differentiation and Integration to Arbitrary Order. Academic Press, New York (1974)
33. Caponetto, R., Dongola, G., Fortuna, L., Petras, I.: Fractional order Systems: Modelling and Control Applications. World Scientific, Singapore (2010)
34. Chen, Y.Q.: Ubiquitous fractional order controls. In: Proceedings of the Second IFAC Symposium on Fractional Derivatives and Applications. Porto (2006)
35. Monje, C.A., Vinagre, B.M., Feliu, V., Chen, Y.Q.: Tuning and auto-tuning of fractional order controllers for industry application. Control Eng. Pract. **16**, 798–812 (2008)

Multistep Approach for Nonlinear Fractional Bloch System Using Adomian Decomposition Techniques

Asad Freihat, Shatha Hasan, Mohammed Al-Smadi, Omar Abu Arqub and Shaher Momani

Abstract In this chapter, a superb multistep approach, based on the Adomian decomposition method (ADM), is successfully implemented for solving nonlinear fractional Bolch system over a vast interval, numerically. This approach is demonstrated by studying the dynamical behavior of the fractional Bolch equations (FBEs) at different values of fractional order α in the sense of Caputo concept over a sequence of the considerable domain. Further, the numerical comparison between the proposed approach and implicit Runge–Kutta method is discussed by providing an illustrated example. The gained results reveal that the MADM is a systematic technique in obtaining a feasible solution for many nonlinear systems of fractional order arising in natural sciences.

Keywords Multistep approach · Fractional system · Bolch equations · Adomian decomposition method

1 Introduction

During the past few decades, a growing interest in the study of complex systems has been observed including glasses, amorphous systems, microemulsions, polymers, biopolymers, and so forth. The term "complex" is due to a broad distinction of the elementary units, intense interactions between the units, or an irregular evolution of units over time. Nowadays, such complex systems are investigated on all structural levels from microscopic to macroscopic and in all fields of physics, biophysics, engi-

A. Freihat · S. Hasan · M. Al-Smadi
Department of Applied Science, Ajloun College, Al-Balqa Applied University, Ajloun 26816, Jordan

O. A. Arqub · S. Momani
Department of Mathematics, Faculty of Science, Jordan University, Amman 11942, Jordan

S. Momani (✉)
Department of Mathematics and Sciences, College of Humanities and Sciences, Ajman University, Ajman, UAE
e-mail: S.Momani@ju.edu.jo

© Springer Nature Singapore Pte Ltd. 2019
P. Agarwal et al. (eds.), *Fractional Calculus*, Springer Proceedings in Mathematics & Statistics 303, https://doi.org/10.1007/978-981-15-0430-3_9

neering, chemistry, and medicine. For instance, nuclear magnetic resonance (NMR), magnetic resonance imaging (MRI), or electron spin resonance (ESR) are physical phenomena that are widely utilized to study complex systems. Indeed, these systems have nonlocal interaction and a long memory due to the disorder that appears in the magnetic relaxation in complex environments. In this aspect, Bloch equations are a set of macroscopic equations that are used to calculate the nuclear magnetization $M = (M_x, M_y, M_z)$ as a function of time in the presence of the relaxation times T_1 and T_2, where the components $M_x(t)$, $M_y(t)$, and $M_z(t)$ are the system magnetization, T_1 is the spin–lattice relaxation time that characterizes the rate of which the longitudinal Mz component recovers exponentially toward the thermodynamic equilibrium, and T_2 represents the spin–spin relaxation time that characterizes the signal decay in the NMR and MRI systems. For more details, see [1–3] and the references therein.

The standard Bloch equations are a set of first-order ordinary differential equations that describe the magnetization behavior in static, varying magnetic fields, and relaxation. However, to study the heterogeneity, complex structure, and memory effects in the relaxation process, the classical Bloch equations were generalized to fractional order by extending the first-order time derivative to a derivative of non-integer order [4]. From this point of view, the fractional operator is considered as a robust framework to account for anomalous diffusion in structurally heterogeneous tissues, porous and composite materials. This is due to the nonlocal nature of fractional derivatives. On the other hand, the utility of generalizing the FBEs with their contributions have recently appeared in the literature. For example, numerical and simulation models of integer and fractional orders of the BEs have been proposed in [5]. In [6], the FBEs have been used to describe anomalous NMR relaxation phenomena. Also, these equations have been considered with time delays, and different stability behaviors for T_1 and T_2 processes were analyzed [7].

Our motivation for this chapter has been devoted to studying approximate solutions for nonlinear Bloch models of fractional order utilizing a numerical multistep approach based on the Adomian decomposition method (ADM). The problem under consideration is subjected to appropriate initial conditions over a vast domain. Indeed, the ADM is efficiently used to provide approximate solutions for many nonlinear fractional problems in convergent series formula with accurately computable structures. Unfortunately, such approximations are found to be not valid for the large values of t for some systems. So, the multistep approach is needed that offers an accurate solution over a longer time frame compared to standard ADM.

Consider the following fractional-modified transformed model of nonlinear BEs that govern the evolution of the magnetization:

$$D^{\alpha_1} x(t) = \delta y(t) + \gamma z(t)\left(x(t)\sin(c) - y(t)\cos(c) - \tfrac{x(t)}{\Gamma_2}\right),$$
$$D^{\alpha_2} y(t) = -\delta x(t) - z(t) + \gamma z(t)\left(x(t)\cos(c) + y(t)\sin(c) - \tfrac{y(t)}{\Gamma_2}\right), \qquad (1)$$
$$D^{\alpha_3} z(t) = y(t) + \gamma \sin(c)\left(x^2(t) + y^2(t) - \tfrac{z(t)-1}{\Gamma_1}\right),$$

subject to the initial conditions

$$x(0) = x_0, \; y(0) = y_0, \; z(0) = z_0, \tag{2}$$

where $t \geq 0$, x_0, y_0, z_0 are real finite constants, δ, γ, c, Γ_1, and Γ_2 are physical parameters, D^{α_i}, $i = 1, 2, 3$, are the fractional derivatives of order α_i in Caputo sense that will be introduced in the next section, and $x(t)$, $y(t)$, and $z(t)$ are analytical unknown functions to be determined.

Anyhow, some of the well-known analytic and numeric techniques were modified for solving the FBEs such as the finite difference method [8], the homotopy perturbation method [9], the predictor-corrector method [10], the Chebyshev polynomials method [3], the operational matrix methods based on Legendre scaling and Laguerre polynomials [11], and the multistep generalized differential transform method [12].

This chapter introduces MADM for fractional nonlinear problems and contains the following sections:

1. Introduction
2. Preliminaries for fractional calculus
3. Principle of Adomian decomposition method
4. Multistep Adomian decomposition method
5. Nonlinear fractional Bloch equations and its modification
6. Multistep approach for modified fractional Bloch equations.

2 Preliminaries for Fractional Calculus

The subject of fractional calculus is not new. It is a generalization of classical calculus that deals with the ordinary differentiation and integration of arbitrary order. The basic idea of fractional calculus goes back to Leibniz in a letter to L'Hospital in 1695 *"Can the meaning of derivatives with integer order be generalized to derivatives with non-integer orders?"*. This concept was developed almost in tandem with the evolution of the classical ones. Anyhow, the fractional operators highlight the intermediate behaviors that cannot be modeled by traditional theory [13]. Nowadays, fractional calculus has become an effective instrument in theoretical and applied fields including physics, bioengineering, finance, signal processing, and so forth [14–21]. Moreover, fractional models can be used to describe the memory and transmissibility for multiple types of materials. So, it plays a vital role in modeling many scientific issues, especially in the anomalous transport process and Hamiltonian chaos.

Unlike the classical calculus, which has unique definitions and clear geometrical and physical interpretations, there are numerous definitions for the operations of differentiation and integration of fractional order. Riemann–Liouville, Riesz, Grünwald–Letnikov, and Caputo are some examples of these definitions. In this chapter, the Caputo concept was preferred due to the facts that the derivative of any constant is equal to zero, and the initial conditions are treated in similar form to those for integer order [22–31]. Next, some main definitions and results concerned with fractional calculus theory are briefly mentioned.

It is well known that the Cauchy's formula for $n \in \mathbb{N}$, $a, t \in R, t > a$ holds such that

$$J_a^n f(t) = \int_a^t \int_a^{\tau_1} \dots \int_a^{\tau_{n-1}} f(\tau_n) d\tau_n \dots d\tau_2 d\tau_1 = \frac{1}{(n-1)!} \int_a^t (t-\tau)^{n-1} f(\tau) d\tau.$$

Thus, if n replaced by a positive real number α and $(n-1)!$ by Gamma function $\Gamma(n)$, then a formula of fractional integration can be obtained as in the following definition:

Definition 1 The fractional operator J_a^α of order α for a function $f(t)$

$$J_a^\alpha f(t) = \frac{1}{\Gamma(\alpha)} \int_a^t (t-\tau)^{1-\alpha} f(\tau) d\tau, \ 0 \leq \tau < t, \ \alpha > 0,$$

is called the Riemann–Liouville fractional integral operator.

The following are some of the interesting properties of the operator J_a^α:

1. For $\alpha = 0$, J_a^α is the identity operator.
2. The operator J_a^α is linear, that is, $J_a^\alpha(cf(t) \pm g(t)) = cJ_a^\alpha f(t) \pm J_a^\alpha g(t)$, for any $c \in \mathbb{R}$.
3. If $f(t)$ is continuous for $t \geq 0$, then $\lim_{\alpha \to 0} J_a^\alpha f(t) = f(t)$.
4. $J_a^{\alpha_1}\left(J_a^{\alpha_2} f(t)\right) = J_a^{\alpha_1+\alpha_2}(f(t)) = J_a^{\alpha_2}\left(J_a^{\alpha_1} f(t)\right)$, $\alpha_1, \alpha_2 > 0$.

Definition 2 The fractional operator $D_a^{*\alpha}$ of order α for a function $f(t)$

$$D_a^{*\alpha} f(t) = \frac{1}{\Gamma(n-\alpha)} \frac{d^n}{dt^n} \int_a^t \frac{f(\tau)}{(t-\tau)^{\alpha-n+1}} d\tau, \ n-1 < \alpha < n, \ n \in \mathbb{N},$$

is called the Riemann–Liouville fractional derivative operator.

Here, it should be observed that if $\alpha \in \mathbb{N}$, then the operator $D_a^{*\alpha}$ is reduced to the standard integer-order differential operator $D^n = \frac{d^n}{dt^n}$. In 1967, an alternative operator to the above Riemann–Liouville fractional derivative has been presented by Caputo as follows:

Definition 3 The fractional operator D_0^α of order α for a function $f(t)$

$$D^\alpha f(t) = \frac{1}{\Gamma(n-\alpha)} \int_a^t \frac{f^{(n)}(\tau)}{(t-\tau)^{\alpha-n+1}} d\tau, \ n-1 < \alpha < n, \ n \in \mathbb{N},$$

is called the Caputo-fractional derivative operator.

The following are some of the interesting properties of the operator D^α:

1. For $\alpha = n$, we have $D^\alpha f(t) = \frac{d^n}{dt^n} f(t)$.
2. The operator D^α is linear, that is, $D^\alpha(cf(t) \pm g(t)) = cD^\alpha f(t) \pm D^\alpha g(t)$, for any $c \in \mathbb{R}$.
3. $D^\alpha c = 0$ for any constant $c \in \mathbb{R}$.
4. For $\gamma > n - 1$, we have $D^\alpha t^\gamma = \frac{\Gamma(\gamma+1)}{\Gamma(\gamma+1-\alpha)} t^{\gamma-\alpha}$ for $n - 1 < \alpha < n$, and is equal to zero otherwise.

3 Principle of Adomian Decomposition Method

The Adomian decomposition method (ADM) is an alternative systematic technique for providing a robust algorithm for analytically approximate solutions and numerical optimization of several fractional applications in physics and engineering. The main features of ADM lie in that it can be directly applied for solving nonlinear fractional problems without the need for unphysical restrictive assumptions such as linearization, discretization, perturbation, guessing the initial data, etc. [32–35]. Indeed, the ADM concept evolved to deal with the linear, nonlinear, stochastic, and deterministic operator problems of Taylor's analytical series with an easily computable, easily verifiable, and rapidly convergent sequence of analytic approximate functions.

For understanding the ADM concept, consider the following nonlinear problem in general form:

$$u = Nu + f, \tag{3}$$

where f is the system input, u is the system output, and N is the nonlinear operator which is assumed to be analytic.

The ADM decomposes the solution into a series

$$u = \sum_{n=0}^{\infty} u_n,$$

and decomposes the nonlinear term Nu into a series

$$Nu = \sum_{n=0}^{\infty} A_n,$$

where A_n's are called the Adomian polynomials, which are depending on the values of $u_0, u_1, ..., u_n$.

In the first approach given by Adomian, A_n's are obtained from the following equalities:

$$q = \sum_{n=0}^{\infty} \lambda^n u_n,$$

$$Nq = N\left(\sum_{n=0}^{\infty} \lambda^n u_n\right) = \sum_{n=0}^{\infty} \lambda^n A_n,$$

where λ is a grouping parameter of convenience.

Formally, the Adomian polynomials A_n's for the nonlinearity are obtained by the following formula:

$$A_n = \frac{1}{n!} \frac{d^n}{d\lambda^n} \left[N \left(\sum_{k=0}^{\infty} \lambda^k u_k \right) \right]_{\lambda=0}, \quad n = 0, 1, 2, \dots.$$

Consequently, the above process leads to the equality

$$\sum_{n=0}^{\infty} u_n = \sum_{n=0}^{\infty} A_n + f,$$

in which the Adomian polynomials A_n can be listed, inclusively, by

$$
\begin{aligned}
A_0 &= f, \\
A_1 &= A_0(u_0), \\
A_2 &= A_1(u_0, u_1), \\
&\vdots \\
A_n &= A_{n-1}(u_0, u_1, \dots, u_{n-1}).
\end{aligned}
$$

Therefore, the solution u can be written as a series of functions u_n such that

$$\sum_{n=0}^{\infty} |u_n| < +\infty.$$

4 Multistep Adomian Decomposition Method

Multistep Adomian decomposition method (MADM) is effectively utilized due to many advantages in the scientific application. Indeed, since it is based on ADM, so there is no need for unphysical restrictive assumptions or small and auxiliary parameters. However, the approximate solution obtained by ADM is usually converged in a small interval but it is not valid or completely divergent over the broader term. So, the MADM is needed to partition the domain of interest into small time steps, which offers a powerful accuracy, especially for the nonlinear problems [36–40].

For perception of Ms-DTM basic idea, consider a general system of fractional differential equations

$$
\begin{aligned}
D^{\alpha_1} x_1(t) &= F_1(t, x_1(t), \dots, x_n(t)), \\
D^{\alpha_2} x_2(t) &= F_2(t, x_1(t), \dots, x_n(t)), \\
&\vdots \\
D^{\alpha_n} x_n(t) &= F_n(t, x_1(t), \dots, x_n(t)),
\end{aligned}
\tag{4}
$$

subject to the initial conditions

$$x_i(0) = c_i, \ i = 1, 2, ..., n, \tag{5}$$

where $0 \leq t \leq T$, $c_i \in \mathbb{R}$ $(i = 1, 2, ..., n)$, D^{α_i}'s are the Caputo-fractional derivative of order α_i, $0 < \alpha_i \leqslant 1$, for $i = 1, 2, \ldots, n$, and F_i's, $i = 1, 2, ..., n$ are linear or nonlinear functions in terms of $x_1(t), ..., x_n(t)$.

To illustrate the MADM for solving such fractional system, the main ideas of the multistep technique are introduced as follows:

Suppose that the interval $[0, T]$ can be divided into m-subintervals of equal length Δt, such as $[t_0, t_1]$, $[t_1, t_2]$, $[t_2, t_3]$, ..., $[t_{m-1}, t_m]$ with $t_0 = 0$, $t_m = T$. Let t^* be the initial value for each subintervals and let $x_{i,j}(t)$, $i = 1, 2, ..., n$, $j = 1, 2, ..., m$ be approximate solutions in each subinterval $[t_{j-1}, t_j]$, $j = 1, 2, ..., m$.

Consequently, system (4) can be converted equivalently into

$$\begin{aligned}
D^{\alpha_1} x_{1,j}(t) &= F_{1,j}(t, x_{1,j}(t), ..., x_{n,j}(t)), \\
D^{\alpha_2} x_{2,j}(t) &= F_{2,j}(t, x_{1,j}(t), ..., x_{n,j}(t)), \\
&\vdots \\
D^{\alpha_n} x_{n,j}(t) &= F_{n,j}(t, x_{1,j}(t), ..., x_{n,j}(t)),
\end{aligned} \tag{6}$$

where $0 < \alpha_i \leq 1$, $i = 1, 2, \ldots, n$, and $j = 1, 2, ..., m$.

By applying J^{α_i} on both the sides of (6) for $j = 1, 2, ..., m$, it follows that

$$\begin{aligned}
x_{1,j}(t) &= x_{1,j}(t^*) + J^{\alpha_1} F_{1,j}(t, x_{1,j}(t), ..., x_{n,j}(t)), \\
x_{2,j}(t) &= x_{2,j}(t^*) + J^{\alpha_2} F_{2,j}(t, x_{1,j}(t), ..., x_{n,j}(t)), \\
&\vdots \\
x_{n,j}(t) &= x_{n,j}(t^*) + J^{\alpha_n} F_{n,j}(t, x_{1,j}(t), ..., x_{n,j}(t)),
\end{aligned} \tag{7}$$

Here, by employing the ADM to system (7), we have the following equalities:

$$x_{i,j}(t) = x_{i,j}(t^*) + \sum_{k=1}^{\infty} x_{i,j,k}(t), \ i = 1, 2, \ldots, n, \ j = 1, 2, ..., m, \tag{8}$$

$$F_{i,j}(t, x_{1,j}(t), \ldots, x_{n,j}(t)) = \sum_{k=0}^{\infty} A_{i,j,k}, \ i = 1, 2, \ldots, n, \ j = 1, 2, ..., m, \tag{9}$$

where $A_{i,j,k}$'s are Adomian polynomials, which are depending on the values of $x_{1,j,0}, ..., x_{1,j,k}, x_{2,j,0}, ..., x_{2,j,k}, ..., x_{n,j,0}, ..., x_{n,j,k}$.

Consequently, the above process leads to the equality

$$\sum_{k=0}^{\infty} x_{i,j,k}(t) = x_{i,j}(t^*) + J^{\alpha_i}\left(\sum_{k=0}^{\infty} A_{i,j,k}(x_{1,j,0}, ..., x_{1,j,k}, ..., x_{n,j,0}, ..., x_{n,j,k})\right),$$
(10)

$i = 1, 2, \ldots, n, \ j = 1, 2, ..., m.$

Now, for $i = 1, 2, \ldots, n, \ j = 1, 2, ..., m,$ and $k = 0, 1, 2, ...,$ set

$$x_{i,j,0}(t) = x_{i,j}(t^*),$$
$$x_{i,j,1}(t) = J^{\alpha_i} A_{i,j,1}(x_{1,j,0}, x_{1,j,1}, ..., x_{n,j,0}, x_{n,j,1}),$$
$$\vdots$$
(11)
$$x_{i,j,k+1}(t) = J^{\alpha_i} A_{i,j,k}(x_{1,j,0}, ..., x_{1,j,k}, ..., x_{n,j,0}, ..., x_{n,j,k}).$$

To determine the Adomian polynomials $A_{i,j,k}$ introduce a parameter q into (9) such that

$$F_{i,j}(t, \sum_{k=0}^{\infty} x_{1,j,k}q^k, \ldots, \sum_{k=0}^{\infty} x_{n,j,k}q^k) = \sum_{k=0}^{\infty} A_{i,j,k}q^k,$$
(12)

thus by letting $x_{i,j,q}(t) = \sum_{k=0}^{\infty} x_{i,j,k}q^k,$ one can get that

$$A_{i,j,k} = \frac{1}{k!} D_q^{k\alpha}\left[F_{i,j,q}(t, x_{1,j,q}, ..., x_{n,j,q})\right]_{q=0}, \ q = 0, 1, 2, ...$$

$$= \frac{1}{k!} D_q^{k\alpha}\left[F_{i,j,q}(t, \sum_{k=0}^{\infty} x_{1,j,k}q^k, ..., \sum_{k=0}^{\infty} x_{n,j,k}q^k)\right]_{q=0},$$
(13)

$$= \left[\frac{1}{k!} D_q^{k\alpha} F_{i,j,q}(t, \sum_{k=0}^{\infty} x_{1,j,k}q^k, ..., \sum_{k=0}^{\infty} x_{n,j,k}q^k)\right]_{q=0}.$$

Hence, for $q = 0, 1, 2, ..., j = 1, 2, ..., m,$ the following recurrence relations are satisfied:

$$x_{i,j,0}(t) = x_{i,j}(t^*),$$
$$x_{i,j,1}(t) = J^{\alpha_i}\left[F_{i,j,q}(t, \sum_{k=0}^{\infty} x_{1,j,1}q, ..., \sum_{k=0}^{\infty} x_{n,j,1}q)\right]_{q=0,}$$
$$\vdots$$
(14)
$$x_{i,j,k+1}(t) = J^{\alpha_i}\left[\frac{1}{k!} D_q^{k\alpha} N_{i,j}(t, \sum_{k=0}^{\infty} x_{1,j,k}q^k, ..., \sum_{k=0}^{\infty} x_{n,j,k}q^k)\right]_{q=0,}.$$

The N-term approximate solution $x_{i,j}^N(t)$ can be given by

$$x_{i,j}^N(t) = \sum_{k=0}^{N} x_{i,j,k}(t),$$

such as $\lim_{N \to \infty} x_{i,j}^N = x_{i,j}$.

For the convergence of MADM, if the system (4) admits a unique solution, then the MADM will produce a unique solution, while if the system (4) does not possess a unique solution, then the MADM will give a solution among many (possible) other solutions [41].

The solution of system (4) in each subinterval $[t_{j-1}, t_j]$, $j = 1, 2, ..., n$, has the following form:

$$\widehat{x}_{i,j}(t) = \sum_{k=0}^{\infty} x_{i,j,k}(t - t_{j-1}), \ i = 1, 2, \ldots, n, \ j = 1, 2, ..., m, \tag{15}$$

while the solution of system (4) in the interval $[0, T]$ can be given as

$$x_i(t) = \begin{cases} \widehat{x}_{i,1}(t) & t \in [t_0, t_1], \\ \widehat{x}_{i,2}(t) & t \in [t_1, t_2], \\ \vdots & \vdots \\ \widehat{x}_{i,m}(t) & t \in [t_{m-1}, t_m], \end{cases} \tag{16}$$

subject to the initial guesses

$$x_{i,1}(t^*) = c_i,$$
$$x_{i,j}(t^*) = \widehat{x}_{i,j}(t_{j-1}) = \widehat{x}_{i,j-1}(t_{j-1}), \ i = 1, 2, ..., n, \ j = 2, 3, ..., m.$$

5 Nonlinear Fractional Bloch Equations and its Modification

MR experiments are mostly performed with a large number of electron spins' resonance and nuclear spins that measure the behaviors and quantities of identical spins. The BEs describe the spin systems of electronic and nuclear resonance in arbitrary magnetic fields over the time–space from transient processes to steady states. The classical BEs is derived from a magnetization M processing in the magnetic induction field with the presence of a constant radio frequency and given in the following form:

$$\frac{dM_x(t)}{dt} = \omega_0 M_y(t) - \frac{M_x(t)}{T_2},$$

$$\frac{dM_y(t)}{dt} = -\omega_0 M_x(t) - \frac{M_y(t)}{T_2}, \tag{17}$$

$$\frac{dM_z(t)}{dt} = \frac{M_0 - M_z(t)}{T_1},$$

subject to the initial conditions

$$M_x(0) = 0, \ M_y(0) = 100, \ M_z(0) = 0, \tag{18}$$

where M_0 is the equilibrium magnetization, ω_0 is the resonant frequency in terms of the static magnetic field B_0 (z-component) given by the Larmor relationship $\omega_0 = \gamma B_0$, $\gamma/2\pi$ is the gyromagnetic ratio, T_1 is the spin–lattice relaxation time that characterizes the rate of which the longitudinal M_z-component recovers exponentially toward the thermodynamic equilibrium, T_2 is the spin–spin relaxation time that characterizes the signal decay in the NMR and MRI systems, and $M_x(t)$, $M_y(t)$, and $M_z(t)$ represent the system magnetization in x, y, and z component, respectively. Here, the set of analytical solution is given by

$$M_x(t) = e^{-t/T_2} \left(M_x(0) \cos \omega_0 t + M_y(0) \sin \omega_0 t \right),$$

$$M_y(t) = e^{-t/T_2} \left(M_y(0) \cos \omega_0 t - M_x(0) \sin \omega_0 t \right), \tag{19}$$

$$M_z(t) = M_z(0)e^{-t/T_1} + M_0 \left(1 - e^{-t/T_1} \right).$$

Some fractional models have been proposed for the BEs, for instance, the following model is investigated in [42] by utilizing the Caputo sense with fractional order $0 < \alpha \le 1$ to study the spin dynamics and magnetization relaxation, in the simple case of a single spin particle at resonance in a static magnetic field B_0:

$$D^\alpha M_x(t) = \omega_0' M_y(t) - \frac{M_x(t)}{T_2'},$$

$$D^\alpha M_y(t) = -\omega_0' M_x(t) - \frac{M_y(t)}{T_2'}, \tag{20}$$

$$D^\alpha M_z(t) = \frac{M_0 - M_z(t)}{T_1'},$$

where $\omega_0' = \omega_0/\tau_2^{\alpha-1}$, $T_1' = \tau_1^{1-\alpha} T_1$, $T_2' = \tau_2^{1-\alpha} T_2$, and τ_1 and τ_2 are fractional time constants.

The next anomalous model is investigated in [43] by utilizing the Riemann–Liouville with fractional orders $0 < \alpha, \beta \le 1$ to fit the derived spin–spin relaxation T_2 decay curves to relaxation data from normal and trypsin-digested bovine nasal cartilage:

$$\frac{dM_x(t)}{dt} = \omega_0 M_y(t) - \frac{D^{1-\alpha} M_x(t)}{T_2},$$

$$\frac{dM_y(t)}{dt} = -\omega_0 M_x(t) - \frac{D^{1-\alpha} M_y(t)}{T_2}, \tag{21}$$

$$\frac{dM_z(t)}{dt} = D^{1-\beta} \frac{M_0 - M_z(t)}{T_1},$$

where $D^{1-\alpha}$ and $D^{1-\beta}$ are the time-fractional Riemann–Liouville derivative. For more details about these models, see [10] and the references therein.

Furthermore, the following modified model of nonlinear BEs governs the evolution of the magnetization M:

$$\frac{dM_x}{dt} = \rho M_y + G M_z(M_x \sin \psi - M_y \cos \psi) - \frac{M_x}{T_2},$$

$$\frac{dM_y}{dt} = -\rho M_x - \omega_1 M_z + G M_z(M_x \cos \psi + M_y \sin \psi) - \frac{M_y}{T_2}, \tag{22}$$

$$\frac{dM_z}{dt} = \omega_1 M_y - G \sin \psi \left((M_x)^2 + (M_y)^2\right) - \frac{M_z - M_0}{T_1},$$

where $\rho = \omega_{rf} - \omega_0$, ω_{rf} is the frequency of a constant radio frequency field with intensity ω_1/γ, $\gamma/2\pi$ is the gyromagnetic ratio, G is the enhancement factor with respect to the magnitude of the transverse magnetization, ψ is the feedback field, T_1 and T_2 are the longitudinal time and transverse relaxation time, respectively. This model can be transformed by introducing

$$t \to \omega_1 t, \ G \to \frac{G}{\omega_1} = \gamma, \ \delta \to \frac{\rho}{\omega_1}, \ \Gamma_i \to \omega_1 T_i, \ i = 1, 2,$$

and

$$M_x \to \frac{M_x}{M_0} = x, \ M_y \to \frac{M_y}{M_0} = y, \ M_z \to \frac{M_z}{M_0} = z,$$

into dimensionless variables model in the following form:

$$\frac{dx}{dt} = \delta y + \gamma z(x \sin \psi - y \cos \psi) - \frac{x}{\Gamma_2},$$

$$\frac{dy}{dt} = -\delta x - z + \gamma z(x \cos \psi + y \sin \psi) - \frac{y}{\Gamma_2}, \tag{23}$$

$$\frac{dz}{dt} = y - \gamma \sin \psi \left(x^2 + y^2\right) - \frac{z - 1}{\Gamma_1}.$$

In this chapter, we consider the modified transformed model by utilizing the Caputo-fractional derivative of order α_i ($0 < \alpha_i \leq 1, i = 1, 2, 3$) described in Eqs. (1) and (2).

6 Multistep Approach for Modified Fractional Bloch Equations

The objective of the section is to obtain the approximate solution of the fractional-modified transformed model (1) and (2) using the MADM. To perform so, set the values of the magnetization parameters as follows:

$$\gamma = 35, \ \delta = -1.26, \ c = 0.173, \ \Gamma_1 = 5 \text{ and } \Gamma_2 = 2.5, \tag{24}$$

and set the initial conditions as

$$x(0) = 0.5, \ y(0) = -0.5 \text{ and } z(0) = 0. \tag{25}$$

For $j = 1, 2, 3, ..., n$, define the nonlinear terms by

$$N_{1,j}(q) = \delta y_j(q) + \gamma z_j(q)(x_j(q)\sin(c) - y_j(q)\cos(c)) - \frac{x_j(q)}{\Gamma_2}$$

$$= \sum_{m=0}^{\infty} A_{1,j,m},$$

$$N_{2,j}(q) = -\delta x_j(q) - z_j(q) + \gamma z_j(q)(x_j(q)\cos(c) + y_j(q)\sin(c)) - \frac{y_j(q)}{\Gamma_2}$$

$$= \sum_{m=0}^{\infty} A_{2j,m}, \tag{26}$$

$$N_{3,j}(q) = y_j(q) - \gamma \sin(c)(x_j^2(q) + y_j^2(q)) - \frac{z_j(q) - 1}{\Gamma_1}$$

$$= \sum_{m=0}^{\infty} A_{3,j,m},$$

where

$$x_j(q) = \sum_{m=0}^{K} x_{j,m}(t)q^m, \ y_j(q) = \sum_{m=1}^{K} y_{j,m}(t)q^m, \ z_j(q) = \sum_{m=1}^{K} z_{j,m}(t)q^m,$$

and the Adomian polynomials $A_{i,j,m}, \ i = 1, 2, 3$, is given by

$$A_{i,j,m} = \frac{1}{m!}\left[D_q^{m\alpha}N_{i,j}(q)\right]_{q=0}, \ j = 1, 2, ..., n, \ m = 1, 2, ..., K.$$

So in this case, we have to satisfy the initial conditions at each subinterval $[t_{j-1}, t_j], \ j = 1, 2, 3, ..., n$, such that

$$x_1(t^*) = 0.5, \ x_j(t^*) = x_j(t_{j-1}) = x_{j-1}(t_{j-1}),$$
$$y_1(t^*) = -0.5, \ y_j(t^*) = y_j(t_{j-1}) = y_{j-1}(t_{j-1}),$$
$$z_1(t^*) = 0, \ z_j(t^*) = z_j(t_{j-1}) = z_{j-1}(t_{j-1}),$$

where t^* is the initial value for each subinterval.

For $j = 1, 2, 3, ..., n, \ m = 0, 1, 2, ...K$, the Adomain decomposition series (11) leads to the following scheme:

$$x_{j,0} = x_j(t^*), \ x_{j,m+1} = J^{\alpha_1} A_{1,j,m},$$
$$y_{j,0} = y_j(t^*), \ y_{j,m+1} = J^{\alpha_2} A_{2,j,m},$$
$$z_{j,0} = z_j(t^*), \ z_{j,m+1} = J^{\alpha_3} A_{3,j,m}.$$

The solutions of system (1), (2) in each subinterval $[t_{j-1}, t_j], \ j = 1, 2, ..., n$, has the following form:

$$\widehat{x}_j(t) = \sum_{m=0}^{K} x_{j,m}(t - t_{j-1}),$$

$$\widehat{y}_j(t) = \sum_{m=0}^{K} y_{j,m}(t - t_{j-1}),$$

$$\widehat{z}_j(t) = \sum_{m=0}^{K} z_{j,m}(t - t_{j-1}),$$

and the solution in the interval $[0, T]$ is given by

$$x(t) = \begin{cases} \widehat{x}_1(t) & t \in [t_0, t_1], \\ \widehat{x}_2(t) & t \in [t_1, t_2], \\ \vdots & \vdots \\ \widehat{x}_n(t) & t \in [t_{n-1}, t_n], \end{cases}$$

$$y(t) = \begin{cases} \widehat{y}_j(t) & t \in [t_0, t_1], \\ \widehat{y}_j(t) & t \in [t_1, t_2], \\ \vdots & \vdots \\ \widehat{y}_j(t) & t \in [t_{n-1}, t_n], \end{cases}$$

$$z(t) = \begin{cases} \widehat{z}_1(t) & t \in [t_0, t_1], \\ \widehat{z}_2(t) & t \in [t_1, t_2], \\ \vdots & \vdots \\ \widehat{z}_n(t) & t \in [t_{n-1}, t_n]. \end{cases}$$

Table 1 Numerical results of $x(t)$ at fractional order $\alpha = 1$

t	MADM	IRKM	Absolute error	Relative error
0.5	0.0900899	0.09009	2.63948×10^{-8}	2.92983×10^{-7}
1.0	0.0100309	0.0100309	1.28886×10^{-8}	1.28489×10^{-6}
1.5	0.0362608	0.0362609	1.63743×10^{-8}	4.51568×10^{-7}
2.0	0.0289491	0.0289491	1.28465×10^{-8}	4.43762×10^{-7}
2.5	0.0289752	0.0289751	5.42791×10^{-9}	1.87330×10^{-7}
3.0	-0.00164103	-0.00164104	1.23807×10^{-8}	7.54445×10^{-6}
3.5	0.0399933	0.0399933	1.02814×10^{-8}	2.57077×10^{-7}
4.0	-0.0277855	-0.0277855	5.37176×10^{-9}	1.93330×10^{-7}
4.5	0.0100665	0.0100664	1.23103×10^{-8}	1.22291×10^{-6}
5.0	0.0967854	0.0967853	5.44949×10^{-8}	5.63049×10^{-7}

Table 2 Numerical results of $y(t)$ at fractional order $\alpha = 1$

t	MADM	IRKM	Absolute error	Relative error
0.5	-0.139658	-0.139658	1.54080×10^{-8}	1.10327×10^{-7}
1.0	-0.0178745	-0.0178746	2.15256×10^{-8}	1.20426×10^{-6}
1.5	-0.0000591291	-0.0000591296	5.18577×10^{-10}	8.77018×10^{-6}
2.0	0.0142948	0.0142948	3.69011×10^{-11}	2.58144×10^{-9}
2.5	0.0231906	0.0231906	1.42967×10^{-8}	6.16489×10^{-7}
3.0	0.020148	0.020148	3.34537×10^{-9}	1.66039×10^{-7}
3.5	-0.0321054	-0.0321054	1.83989×10^{-9}	5.73080×10^{-8}
4.0	0.0174674	0.0174673	6.89852×10^{-9}	3.94938×10^{-7}
4.5	0.100417	0.100417	4.78751×10^{-9}	4.76763×10^{-8}
5.0	0.187399	0.187399	5.27830×10^{-9}	2.81661×10^{-8}

For numerical simulation, we have compared the MAD results from the implicit Runge–Kutta method (IRKM) at the fractional order $\alpha_i = 1$, $i = 1, 2, 3$, over the interval [0, 5] with step size 0.5, $K = 5$, $n = 800$, and the numeric results are listed in Tables 1, 2, and 3. From these tables, it is observed that the accuracy by the MADM is compatible with the IRKM at the value of fractional order α. All the results are calculated by using the Mathematica software package.

The graphical results and parametric plots of the MADM and IRKM are given in Figs. 1, 2, and 3 at fractional order $\alpha_i = 1$, $i = 1, 2, 3$, over the interval [0, 5]. While Figs. 4 and 5 are given for the MAD solutions at fractional order $\alpha_i = 0.9$, $i = 1, 2, 3$, over the interval [0, 5].

Table 3 Numerical results of $z(t)$ at fractional order $\alpha = 1$

t	MADM	IRKM	Absolute error	Relative error
0.5	−0.519828	−0.519828	1.75095×10^{-8}	3.36833×10^{-8}
1.0	−0.398178	−0.398178	1.51377×10^{-8}	3.80175×10^{-8}
1.5	−0.263365	−0.263365	1.44217×10^{-8}	5.47591×10^{-8}
2.0	−0.142461	−0.142461	8.71885×10^{-9}	6.12018×10^{-8}
2.5	−0.0286219	−0.0286219	1.27388×10^{-8}	4.45072×10^{-7}
3.0	0.0786463	0.0786463	1.34023×10^{-8}	1.70413×10^{-7}
3.5	0.158251	0.158251	7.90765×10^{-9}	4.99689×10^{-8}
4.0	0.237731	0.237731	8.78612×10^{-9}	3.69582×10^{-8}
4.5	0.288485	0.288485	1.02007×10^{-8}	3.53596×10^{-8}
5.0	0.285525	0.285525	2.49369×10^{-8}	8.73371×10^{-8}

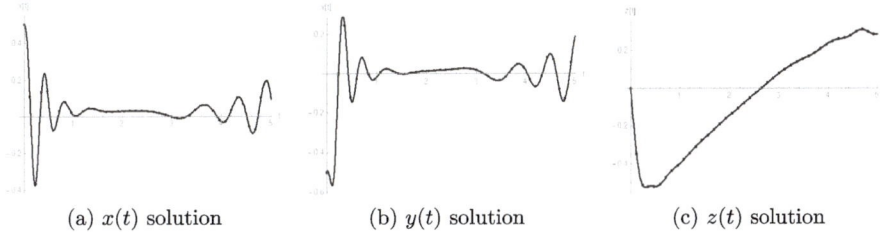

(a) $x(t)$ solution $\qquad$ (b) $y(t)$ solution $\qquad$ (c) $z(t)$ solution

Fig. 1 The MAD solutions and the corresponding IRKM at $\alpha = 1$ and $t \in [0, 5]$

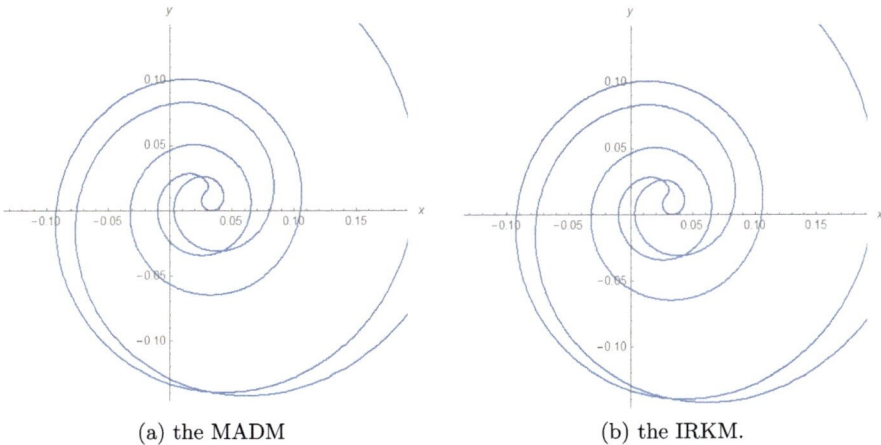

(a) the MADM $\qquad\qquad$ (b) the IRKM.

Fig. 2 The parametric plots of the solution x versus y at $\alpha = 1$

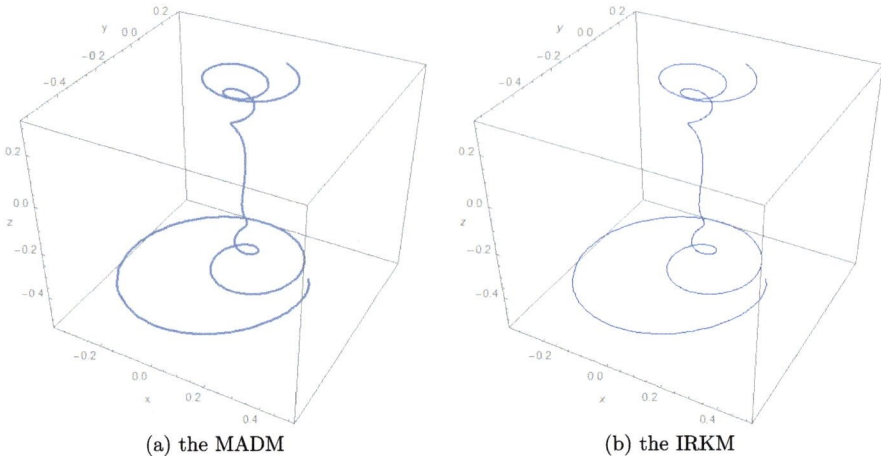

(a) the MADM   (b) the IRKM

Fig. 3 The 3D parametric plots of the solutions at $\alpha = 1$

Fig. 4 The parametric plots
of the MADM solutions x
versus y at $\alpha = 0.9$

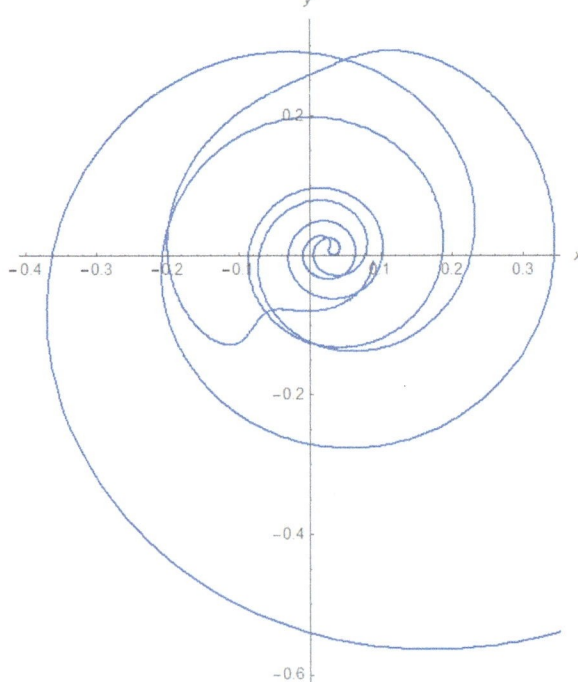

Fig. 5 The 3D parametric plots of the MADM solutions at $\alpha = 0.9$

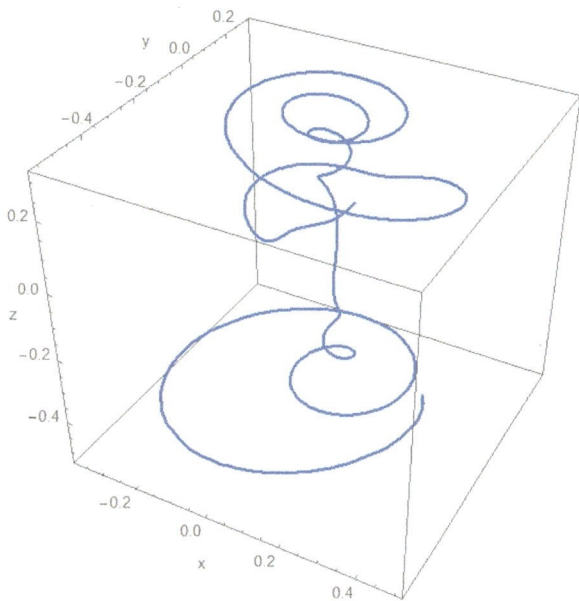

References

1. Matuszak, Z.: Fractional Bloch's equations approach to magnetic relaxation. Curr. Topics Biophys. **37**, 9–22 (2014)
2. Petras, I.: Modeling and numerical analysis of fractional order Bloch equations. Comp. Math. Appl. **61**, 341–56 (2011)
3. Singh, H.: A new numerical algorithm for fractional model of Bloch equation in nuclear magnetic resonance. Alexandria Eng. J. **55**, 2863–2869 (2016)
4. Ascher, U.M., Mattheij, R.M., Russell, R.D.: Numerical solution of boundary value problems for ordinary differential equations. Classics Appl. Math. **13** (1995)
5. Bhalekar, S., Daftardar-Gejji, V., Baleanu, D., Magin, R.L.: Transient chaos in fractional Bloch equations. Comput. Math. Appl. **64**, 3367–3376 (2012)
6. Magin, R., Li, W., Velasco, M., Trujillo, J., Reiter, D., Morgenstern, A., Spencer, R.: J. Magn. Reson. **210**(2), 184 (2011)
7. Bhalekar, S., Daftardar-Gejji, V., Baleanu, D., Magin, R.L.: Fractional Bloch equation with delay. Comput. Math. Appl. **61**(5), 1355–1365 (2011)
8. Hajipour, M., Jajarmi, A., Baleanu, D.: An efficient nonstandard finite difference scheme for a class of fractional chaotic systems. J. Comput. Nonlinear Dyn. **13**, (2018). In press
9. Kumar, S., Faraz, N., Sayevand, K.: A fractional model of Bloch equation in NMR and its analytic approximate solution. Walailak J. Sci. Technol. **11**(4), 273–285 (2014)
10. Yu, Q., Liu, F., Turner, I., Burrage, K.: Numerical simulation of the fractional Bloch equations. J. Comput. Appl. Math. **255**, 635–651 (2014)
11. Singh, H., Singh, C.S.: A reliable method based on second kind Chebyshev polynomial for the fractional model of Bloch equation. Alexandria Eng. J. (2017). In press
12. Abuteen, E., Momani, S., Alawneh, A.: Solving the fractional nonlinear Bloch system using the multi-step generalized differential transform method. Comput. Math. Appl. **68**, 2124–2132 (2014)
13. Herrmann, R.: Fractional calculus: an introduction for physicists. World Scientific (2014)

14. Al-Smadi, M., Abu Arqub, O.: Computational algorithm for solving fredholm time-fractional partial integrodifferential equations of dirichlet functions type with error estimates. Appl. Math. Comput. **342**, 280–294 (2019)

15. Abu Arqub, O., Al-Smadi, M.: Atangana-Baleanu fractional approach to the solutions of Bagley-Torvik and Painlevé equations in Hilbert space, Chaos Solitons & Fractals (2018). In press

16. Abu Arqub, O., Odibat, Z., Al-Smadi, M.: Numerical solutions of time-fractional partial inte-grodifferential equations of Robin functions types in Hilbert space with error bounds and error estimates. Nonlinear Dyn., 1–16 (2018)

17. Moaddy, K., Freihat, A., Al-Smadi, M., Abuteen, E., Hashim, I.: Numerical investigation for handling fractional-order Rabinovich-Fabrikant model using the multistep approach. Soft Comput. **22**(3), 773–782 (2018)

18. Agarwal, P., Chand, M., Baleanu, D., O'Regan, D., Jain, S.: On the solutions of certain fractional kinetic equations involving k-Mittag-Leffler function. Adv. Differ. Equ. **2018**(1), 249 (2018)

19. Khalil, H., Khan, R.A., Al-Smadi, M., Freihat, A.: Approximation of solution of time fractional order three-dimensional heat conduction problems with Jacobi Polynomials. Punjab Univ. J. Math. **47**(1), 35–56 (2015)

20. Agarwal, P., Al-Mdallal, Q., Cho, Y.J., Jain, S.: Fractional differential equations for the gener-alized Mittag-Leffler function. Adv. Differ. Equ. **2018**(1), 58 (2018)

21. Agarwal, P., El-Sayed, A.A.: Non-standard finite difference and Chebyshev collocation meth-ods for solving fractional diffusion equation. Phys. A: Stat. Mech. Appl. **500**, 40–49 (2018)

22. Alaroud, M., Al-Smadi, M., Ahmad, R.R., Salma Din, U.K.: Computational optimization of residual power series algorithm for certain classes of fuzzy fractional differential equations. Int. J. Differ. Equ. **2018**, 1–11 (2018). Art. ID 8686502

23. Al-Smadi, M.: Simplified iterative reproducing kernel method for handling time-fractional BVPs with error estimation. Ain Shams Eng. J. (2017). In press. https://doi.org/10.1016/j.asej.2017.04.006

24. Abu Arqub, O., Al-Smadi, M.: Numerical algorithm for solving time-fractional partial inte-grodifferential equations subject to initial and Dirichlet boundary conditions. Numer. Methods Partial Differ. Equ., 1–21 (2017). https://doi.org/10.1002/num.22209

25. Al-Smadi, M., Abu Arqub, O., Shawagfeh, N., Momani, S.: Numerical investigations for sys-tems of second-order periodic boundary value problems using reproducing kernel method. Appl. Math. Comput. **291**, 137–148 (2016)

26. Agarwal, P., Choi, J., Jain, S., Rashidi, M.M.: Certain integrals associated with generalized Mittag-Leffler function. Commun. Korean Math. Soc **32**(1), 29–38 (2017)

27. Agarwal, P., Berdyshev, A., Karimov, E.: Solvability of a non-local problem with integral transmitting condition for mixed type equation with Caputo fractional derivative. Results Math. **71**(3–4), 1235–1257 (2017)

28. Agarwal, P., Jain, S., Mansour, T.: Further extended Caputo fractional derivative operator and its applications. Russian J. Math. Phys. **24**(4), 415–425 (2017)

29. Abuteen, E., Freihat, A., Al-Smadi, M., Khalil, H., Khan, R.A.: Approximate series solution of nonlinear, fractional Klein-Gordon equations using fractional reduced differential transform method. J. Math. Stat. **12**(1), 23–33 (2016)

30. Khalil, H., Khan, R.A., Al-Smadi, M., Freihat, A., Shawagfeh, N.: New operational matrix for Shifted Legendre polynomials and fractional differential equations with variable coefficients. Punjab Univ. J. Math. **47**(1), 81–103 (2015)

31. Abu-Gdairi, R., Al-Smadi, M., Gumah, G.: An expansion iterative technique for handling fractional differential equations using fractional power series scheme. J. Math. Stat. **11**(2), 29–38 (2015)

32. Adomian, G., Rach, R.C., Meyers, R.E.: Numerical integration, analytic continuation, and decomposition. Appl. Math. Comput. **88**, 95–116 (1997)

33. Wazwaz, A.M.: The modified Adomian decomposition method for solving linear and nonlinear boundary value problems of tenth-order and twelfthorder. Int. J. Nonlinear Sci. Numer. Simul. **1**, 17–24 (2000)

34. Wazwaz, A.M.: A note on using Adomian decomposition method for solving boundary value problems. Found. Phys. Lett. **13**, 493–498 (2000)
35. Ebadi, G., Rashedi, S.: The extended Adomian decomposition method for fourth order boundary value problems. Acta Univ. Apulensis **22**, 65–78 (2010)
36. Al-Smadi, M., Freihat, A., Khalil, H., Momani, S., Khan, R.A.: Numerical multistep approach for solving fractional partial differential equations. Int. J. Computat. Methods **14**(1750029), 1–15 (2017). https://doi.org/10.1142/S0219876217500293
37. Al-Smadi, M., Freihat, A., Abu Hammad, M., Momani, S., Abu Arqub, O.: Analytical approximations of partial differential equations of fractional order with multistep approach. J. Comput. Theor. Nanosci. **13**(11), 7793–7801
38. Momani, S., Abu Arqub, O., Freihat, A., Al-Smadi, M.: Analytical approximations for Fokker-Planck equations of fractional order in multistep schemes. Appl. Comput. Math. **15**(3), 319–330 (2016)
39. Odibat, Z.M., Bertelle, C., Aziz-Alaoui, M.A., Duchamp, G.H.E.: A multi-step differential transform method and application to non-chaotic or chaotic systems. Comput. Math. Appl. **59**, 1462–1472 (2010)
40. Gokdoğan, A., Merdan, M., Yıldırım, A.: Adaptive multi-step differential transformation method for solving nonlinear equations. Math. Comput. Model. **55**(3–4), 761–769 (2012)
41. Jafari, H., Daftardar-Gejji, V.: Solving a system of nonlinear fractional differential equations using Adomian decomposition. J. Comput. Appl. Math. **196**, 644–651 (2006)
42. Magin, R., Feng, X., Baleanu, D.: Solving the fractional order Bloch equation. Concepts Magn. Reson. Part A **34A**(1), 16–23 (2009)
43. Velasco, M., Trujillo, J., Reiter, D., Spencer, R., Li, W., Magin, R.: Anomalous fractional order models of NMR relaxation, In: Proceedings of FDA'10, the 4th IFAC Workshop Fractional Differentiation and its Applications, Vol. 1, 2010. Paper number FDA10-058

Simulation of the Space–Time-Fractional Ultrasound Waves with Attenuation in Fractal Media

E. A. Abdel-Rehim and A. S. Hashem

Abstract In this paper, we are interested in studying the propagation of the over diagnostic ultrasound waves through complex biological vascular networks such as the tumor tissue. Evidence shows that the over diagnostic wave propagates through complex media with power law of non-integer order $t^{-\nu}$, $1 < \nu < 2$. Evidence shows also that the vascular morphology of the tumor is non-smooth and is a complex media that means it is a fractal media. The wave propagates through this fractal media which exhibits with extremely long jumps whose length is distributed according to the Lévy long tail $\sim |x|^{-1-\alpha}$, $0 < \alpha < 2$. Therefore, the space–time-fractional forced wave equation with attenuation, or the so-called multi-term wave equation, mathematically models this medicine problem. This equation mathematically models many other physical, biological, chemical, and environmental problems. We get the approximate solution of this model to study the time evolution of the propagated wave by adopting the backward Grünwald–Letnikov scheme joining with the common finite difference method. We investigate numerically the effect of the time fractional on the propagation of the wave as well as the effect of the space-fractional order for the three cases as: $0 < \alpha < 1$, $1 < \alpha < 2$, and $\alpha = 1$. The stability condition of each approximate solution is also discussed separately. Finally, we prove that the space-fractional order α has no effect on the stability condition.

Keywords Attenuation · Over diagnostic wave equation · Explicit scheme · Caputo time derivative · Stability · Space–fractional derivative · Backward grünwald–Letnikov scheme · memory process · vascular morphology of tumor

Mathematics Subject Classification: 26A33 · 35L05 · 35J05 · 45K05 · 47G30 · 33E20 · 65N06 · 80-99

E. A. Abdel-Rehim (✉) · A. S. Hashem
Department of Mathematics, Faculty of Science, Suez Canal University, Ismailia, Egypt
e-mail: entsarabdelrehim@gmail.com; entsarabdelrehim@yahoo.com

© Springer Nature Singapore Pte Ltd. 2019
P. Agarwal et al. (eds.), *Fractional Calculus*, Springer Proceedings
in Mathematics & Statistics 303, https://doi.org/10.1007/978-981-15-0430-3_10

1 Introduction

The propagation of the ultrasound waves with attenuation in a smooth biological tissue is mathematically modeled by the forced wave equation with attenuation. Whereas attenuation means loss of wave amplitude due to all mechanisms including absorption, scattering through the tissues, and mode conversion [1]. This equation is defined as

$$\frac{\partial^2 u(x,t)}{\partial t^2} = k \frac{\partial u(x,t)}{\partial t} + a \frac{\partial^2 u(x,t)}{\partial x^2} - \frac{\partial}{\partial x}(F(x)\,u(x,t)), \quad -R \le x \le R,\ t > o\ ,$$

(1.1)

here $u(x,t)$ is the pressure amplitude, $a > 0$ is the general positive constant, and $0 < k < 1$ is the attenuation coefficient. $F(x)$ represents the external force, which supplies energy to the ultrasound wave.

It is known that the vascular morphology of tumor is significantly different from normal tissues. Vascular networks are developed and ordered with a hierarchical vessel arrangement. While the tumor vascular networks randomly consisted of a disorderly tangle of vessels. In other words, tumor vasculature has long been known to be more chaotic in appearance than normal vasculature. Complexity, irregularities, and poorly regulated growth are some of the known characteristics of cancer. Tumor vasculature, in particular, defines the optimized growth patterns of healthy vasculature and is known to contain many tortuous vessels, shunts, vascular loops, widely variable inter-vascular distances, and large vascular areas. This complex structure represents a fractal media. That means, when the over diagnostic ultrasound wave propagates in tumor vasculature, it propagates in fractal medium. For more information, see [2–8]. In other words, the over diagnostic ultrasound wave propagates with large deviation from the stochastic process of Brownian motion. The Lévy stable motion is the natural generalization of the Brownian motion but it is different because of the occurrence of the extremely long jumps whose length is proportional to the Lévy tail $\sim |x|^{-1-\alpha}, 0 < \alpha < 2$. This long jump requires not finite velocity. Evidence shows that the velocity of propagation in fractal media has a power law of non-integer frequency of order $t^{-\nu}$, $1 < \nu < 2$, see [9–11, 18]. Then to study the space–time-fractional forced diagnostic ultrasound wave propagation with attenuation, Eq. (1.1) should be modified to

$$\underset{t\,*}{D^{\beta}} u(x,t) = k \frac{\partial u(x,t)}{\partial t} + a \underset{x\,0}{D^{\alpha}} u(x,t) - \frac{\partial}{\partial x}(F(x)\,u(x,t))\ ,$$

(1.2)

where $1 < \beta \le 2$ and $0 < \alpha < 2$. The used time-fractional derivative operator $\underset{t\,*}{D^{\beta}} u(x,t)$ is called the Caputo-fractional operator, see [12] for more information about the relation between Caputo-fractional derivative and the Riemann–Liouville-fractional integral operators. The used $\underset{x\,0}{D^{\alpha}} u(x,t)$ is the Riesz–Feller space-fractional differentiation operator, see [13]. Time-fractional wave equations with attenuation term have been studied by Szabo [1], Caputo [14], and Treeby and Cox [15]. The recent works [16] proposed transient-wave propagation in porous materials using fractional modeling to take into account the frequency variability of some dynamic coefficients of the medium like tortuosity and compressibility. Other authors, like Tarasov [11], gave a space-fractional formulation of the hydrodynamic equations to describe fluid flow in fractal media. Casasanta and Garra [17] studied the space-fractional wave equation in relation to the propagation of acoustic waves with space-dependent sound speed. We have given the approximate solution of the time-fractional wave, forced wave (shear wave), and damped wave equations, see [9, 18].

In this paper, we are interested in finding the approximate solution of the studied model (1.2). The approximate solution is given by adopting the backward Grünwald–Letnikov scheme joined with the common finite difference method. We investigate numerically the effect of the time fractional on the propagation of the wave as well as the effect of the space- fractional order for the three cases as: $0 < \alpha < 1$, $1 < \alpha < 2$, and $\alpha = 1$. The paper is organized as follows: Section 1 is devoted to the introduction. Section 2 is to introduce the approximate solution of the space–time-fractional forced wave equation with attenuation term for all the fractional-order values. In Sect. 3, the proof of the stability conditions are studied. Finally, Sect. 4 is devoted to simulate the propagation of the waves of the previous model for different values of the parameters β, α, $f(x)$, and t. The interpretation of the numerical results with investigations of the effect of the memory is also given.

2 Approximate Solutions

In this section, we discuss the approximate solution of Eq. (1.2) for all values of α and β. To do so, identify first the used external force $F(x)$. There are many forms of $F(x)$, which can be used depending on the kind of the model. In this paper, we consider $F(x) = -bx$, $b > 0$, i.e., $F(x)$ is a linear attractive force. So far, Eq. (1.2) takes the form

$$\underset{t\,*}{D^{\beta}} u(x, t) = k \, \frac{\partial u(x, t)}{\partial t} + a \, \underset{x\,0}{D^{\alpha}} u(x, t) + b \, \frac{\partial}{\partial x}(x \, u(x, t)) \,, \qquad (2.1)$$

we solve this equation with the initial conditions

$$u(x, 0) = f(x) \,, \qquad u_t(x, 0) = 0 \,, \qquad (2.2)$$

and the boundary conditions

$$u(-R, t) = u(R, t) = 0 \,. \qquad (2.3)$$

Discretize x and t by the grid $\{(x_j, t_n) : -R \le j \le R, n \ge 0\}$ with $x_j = jh$, $t_n = n\tau$. Where $h > 0$, and $\tau > 0$ are the steps in space and time, respectively. Introduce the clump $y^{(n)} = \{y_{-R}^{(n)}, y_{-R+1}^{(n)}, \cdots, y_{R-1}^{(n)}, y_R^{(n)}\}^T$ to approximate the integral of the pressure function $u(x,t)$ over the small interval h. The initial value $y^{(0)}$ is obtained by the aid of initial condition $u(x,0) = f(x)$. To discretize $\underset{t\ *}{D^\beta}\, u(x,t)$, we utilize the backward Grünwald–Letnikov scheme as

$$\underset{\tau\ *}{D^\beta}\, y_j(t_{n+1}) = \sum_{m=0}^{n+1}(-1)^m \binom{\beta}{m} \frac{y_j(t_{n+1-m}) - y_j(t_0)}{\tau^\beta}, \quad 1 < \beta \le 2 ,\ \forall n \in N_0. \quad (2.4)$$

This scheme has been successfully utilized at [9, 18–20] for modeling and simulation the time-fractional diffusion processes. The used space-fractional operator $\underset{x\ 0}{D^\alpha}$, is the symmetric Riesz–Feller operator, see [21]. We use here the Zaslavski notation [13] to define the inverse Riesz–Feller, see also Oldham and Spanier [22], Ross and Miller [23] and Samko [24], as

$$\underset{0}{D^\alpha} = \frac{-1}{2cos(\alpha\pi/2)}\,[I_+^{-\alpha} + I_-^{-\alpha}] , 0 < \alpha \le 2 ,\ \alpha \ne 1 , \quad (2.5)$$

where $I_\pm^{-\alpha}$ are the inverse of the operators $I_\pm^\alpha$, being called the *Weyl* integrals. So far, the discretization of the Riesz potential operator is as follows:

$$\underset{h\ 0}{D^\alpha}\, y_j(t_n) = \frac{-1}{2cos\frac{\alpha\pi}{2}}\,(\underset{h\ +}{I}^{-\alpha} + \underset{h\ -}{I}^{-\alpha})\, y_j(t_n) , \quad 0 < \alpha \le 2\ \alpha \ne 1\ j \in \mathbb{Z} . \quad (2.6)$$

This scheme has been effectively used for finding the approximate solutions of the space-fractional diffusion processes, see also [20] and the references therein. $\underset{x\ 0}{D^\alpha}$ has three different discretization schemes depending on the values of α as follows:

$$\underset{h\ \pm}{I}^{-\alpha}\, y_j(t_n) = \frac{1}{h^\alpha}\sum_{s=0}^\infty (-1)^s \binom{\alpha}{s} y_{j\pm1\mp s}, \quad 1 < \alpha < 2 , \quad (2.7)$$

while

$$\underset{h\ \pm}{I}^{-\alpha}\, y_j(t_n) = \frac{1}{h^\alpha}\sum_{s=0}^\infty (-1)^s \binom{\alpha}{s} y_{j\mp s}, \quad 0 < \alpha < 1 . \quad (2.8)$$

Finally, for the case $\alpha = 1$, which is related to the Cauchy distribution and because of Eq. (2.6) one cannot use the Grünwald–Letnikov to discretize it, instead we use the method introduced by Gorenflo and Mainardi [25].

Now joining the discretization of the time- and space-fractional differential operators, one can give the discretization of the space–time-fractional forced wave equation with damping term for each case. Beginning by $1 < \alpha < 2$, $1 < \beta < 2$, see for more detail [26], one gets after minor mathematical manipulating

$$y_j^{(n+1)} = \frac{b_n}{1 - k\tau^{\beta-1}} y_j^{(0)} + \frac{1}{1 - k\tau^{\beta-1}} \sum_{m=2}^{n} c_m y_j^{(n+1-m)}$$

$$+ \left(\frac{\beta - k\tau^{\beta-1}}{1 - k\tau^{\beta-1}} + \frac{a\mu}{\cos\frac{\alpha\pi}{2}(1 - k\tau^{\beta-1})} \binom{\alpha}{1} \right) y_j^{(n)}$$

$$- \frac{1}{1 - k\tau^{\beta-1}} \left(\frac{b\tau^{\beta}(j-1)}{2} + \frac{a\mu}{2\cos\frac{\alpha\pi}{2}} \left(1 + \binom{\alpha}{2}\right) \right) y_{j-1}^{(n)}$$

$$+ \frac{1}{1 - k\tau^{\beta-1}} \left(\frac{b\tau^{\beta}(j+1)}{2} - \frac{a\mu}{2\cos\frac{\alpha\pi}{2}} \left(1 + \binom{\alpha}{2}\right) \right) y_{j+1}^{(n)}$$

$$- \frac{1}{1 - k\tau^{\beta-1}} \frac{a\mu}{2\cos\frac{\alpha\pi}{2}} \sum_{s\geq 3} (-1)^s \binom{\alpha}{s} \left(y_{j+1-s}^{(n)} + y_{j-1+s}^{(n)} \right) . \quad (2.9)$$

For ease of writing, we use the same parameters b_n and c_m defining in [19]. Define the scaling relation as

$$\mu = \frac{\tau^{\beta}}{h^{\alpha}} . \quad (2.10)$$

For easing the computation, we rewrite Eq. (2.9) in the matrix form as

$$y^{(n+1)} = \frac{b_n}{1 - k\tau^{\beta-1}} y^{(0)} + \frac{1}{1 - k\tau^{\beta-1}} \sum_{m=2}^{n} c_m y^{(n+1-m)} + Q^T y^{(n)} . \quad (2.11)$$

Since $(-1)^s \binom{\alpha}{s} \to 0$ as $s \to \infty$, then we can ignore the terms which are corresponding to large values of s. Define $Q = \{q_{ij}\}$ to be the diagonally matrix whose elements q_{ij} are defined as

$$q_{ij} = \begin{cases} q_{ij}^{(1)} = \frac{a\mu}{2\cos\frac{\alpha\pi}{2}(1-k\tau^{\beta-1})} (-1)^{j-i+1} \binom{\alpha}{|j-i+1|} & j = i + M, i = -R, \cdots, R - M \\ q_{ij}^{(2)} = \frac{1}{1-k\tau^{\beta-1}} \left(\frac{b\tau^{\beta}(j+1)}{2} - \frac{a\mu}{2\cos\frac{\alpha\pi}{2}} \left(1 + \binom{\alpha}{2}\right) \right) & j = i + 1, i = -R, \cdots, R - 1 \\ q_{ij}^{(3)} = \frac{\beta - k\tau^{\beta-1}}{1-k\tau^{\beta-1}} + \frac{a\mu}{\cos\frac{\alpha\pi}{2}(1-k\tau^{\beta-1})} \binom{\alpha}{1} & j = i, i = -R, \cdots, R \\ q_{ij}^{(4)} = \frac{-1}{1-k\tau^{\beta-1}} \left(\frac{b\tau^{\beta}(j-1)}{2} + \frac{a\mu}{2\cos\frac{\alpha\pi}{2}} \left(1 + \binom{\alpha}{2}\right) \right) & j = i - 1, i = -R + 1, \cdots, R \\ q_{ij}^{(5)} = \frac{a\mu}{2\cos\frac{\alpha\pi}{2}(1-k\tau^{\beta-1})} (-1)^{j-i+1} \binom{\alpha}{|i-j+1|} & j = i - M, i = -R + M, \cdots, R , \end{cases} \quad (2.12)$$

where $2 \leq M \leq R$, to cover all the elements of the matrix. Now, we discuss the case $0 < \alpha < 1$ and $1 < \beta < 2$. Joining the discretization (2.8) with the discretization (2.4) with the common finite difference rules for finding the approximate solution of Eq. (2.1) and solving for $y_j^{(n+1)}$ after using the scaling relation (2.10), to get

$$y_j^{(n+1)} = \frac{b_n}{1 - k\tau^{\beta-1}} y_j^{(0)} + \frac{1}{1 - k\tau^{\beta-1}} \sum_{m=2}^{n} c_m y_j^{(n+1-m)}$$

$$+ \left(\frac{\beta - k\tau^{\beta-1}}{1 - k\tau^{\beta-1}} - \frac{a\mu}{\cos\frac{\alpha\pi}{2}(1 - k\tau^{\beta-1})} \right) y_j^{(n)}$$

$$+ \frac{1}{1 - k\tau^{\beta-1}} \left(\frac{a\mu}{2\cos\frac{\alpha\pi}{2}} \binom{\alpha}{1} - \frac{b\tau^{\beta}(j-1)}{2} \right) y_{j-1}^{(n)}$$

$$+ \frac{1}{1 - k\tau^{\beta-1}} \left(\frac{a\mu}{2\cos\frac{\alpha\pi}{2}} \binom{\alpha}{1} + \frac{b\tau^{\beta}(j+1)}{2} \right) y_{j+1}^{(n)}$$

$$+ \frac{1}{1 - k\tau^{\beta-1}} \frac{a\mu}{2\cos\frac{\alpha\pi}{2}} \sum_{s\geq 2} (-1)^{s+1} \binom{\alpha}{s} \left(y_{j-s}^{(n)} + y_{j+s}^{(n)} \right) . \quad (2.13)$$

Equation (2.13) can be written in the same matrix form (2.11), where the diagonal elements of the matrix Q are defined in this case as

$$q_{ij} = \begin{cases} q_{ij}^{(1)} = \frac{a\mu}{2\cos\frac{\alpha\pi}{2}(1-k\tau^{\beta-1})} (-1)^{j-i+1} \binom{\alpha}{|j-i|} & j = i + M, i = -R, \cdots, R - M \\ q_{ij}^{(2)} = \frac{1}{1-k\tau^{\beta-1}} \left(\frac{a\mu}{2\cos\frac{\alpha\pi}{2}} \binom{\alpha}{1} + \frac{b\tau^{\beta}(j+1)}{2} \right) & j = i + 1, i = -R, \cdots, R - 1 \\ q_{ij}^{(3)} = \frac{\beta-k\tau^{\beta-1}}{1-k\tau^{\beta-1}} - \frac{a\mu}{\cos\frac{\alpha\pi}{2}(1-k\tau^{\beta-1})} & j = i, i = -R, \cdots, R \\ q_{ij}^{(4)} = \frac{1}{1-k\tau^{\beta-1}} \left(\frac{a\mu}{2\cos\frac{\alpha\pi}{2}} \binom{\alpha}{1} - \frac{b\tau^{\beta}(j-1)}{2} \right) & j = i - 1, i = -R + 1, \cdots, R \\ q_{ij}^{(5)} = \frac{a\mu}{2\cos\frac{\alpha\pi}{2}(1-k\tau^{\beta-1})} (-1)^{j-i+1} \binom{\alpha}{|i-j|} & j = i - M, i = -R + M, \cdots, R , \end{cases}$$
$$\quad (2.14)$$

where $2 \leq M \leq R$. Finally, we discuss the singular case $\alpha = 1$ and $1 < \beta < 2$. In this case, Gorenflo and Mainardi [25] deduced the discretization of D_0^1 from the Cauchy density $p_1(x, 0) = \frac{1}{\pi}\frac{1}{1+x^2}$. They replaced the factor $(-1)^s \binom{\alpha}{s}$, $s \in \mathbb{Z}$, in equation (2.7) and in Eq. (2.8) by $\frac{-2}{\pi}$ for $s = 0$, and $\frac{1}{\pi|s|(|s|+1)}$ for $s \neq 0$, $s \in \mathbb{Z}$. Substituting with these factors with the scaling relation (2.10) for $\alpha = 1$, and finally with common finite difference rules in Eq. (2.1), then solving for $y_j^{(n+1)}$, to get

$$y_j^{(n+1)} = \frac{b_n}{1 - k\tau^{\beta-1}} y_j^{(0)} + \frac{1}{1 - k\tau^{\beta-1}} \sum_{m=2}^{n} c_m y_j^{(n+1-m)} + \left(\frac{\beta - k\tau^{\beta-1}}{1 - k\tau^{\beta-1}} - \frac{2a\mu}{\pi(1 - k\tau^{\beta-1})} \right) y_j^{(n)}$$

$$+ \frac{1}{1 - k\tau^{\beta-1}} \left(\frac{a\mu}{2\pi} - \frac{b\tau^{\beta}(j-1)}{2} \right) y_{j-1}^{(n)} + \frac{1}{1 - k\tau^{\beta-1}} \left(\frac{a\mu}{2\pi} + \frac{b\tau^{\beta}(j+1)}{2} \right) y_{j+1}^{(n)}$$

$$+ \frac{a\mu}{\pi(1 - k\tau^{\beta-1})} \sum_{s\geq 2} \frac{1}{s(s+1)} \left(y_{j+s}^{(n)} + y_{j-s}^{(n)} \right) . \quad (2.15)$$

This equation can also be written in the same matrix form (2.11). Where Q is a matrix whose diagonal elements are defined as

$$q_{ij} = \begin{cases} q_{ij}^{(1)} = \frac{a\mu}{\pi(1-k\tau^{\beta-1})} \frac{1}{(j-i)(j-i+1)} & j = i+M, i = -R, \cdots, R-M \\ q_{ij}^{(2)} = \frac{1}{1-k\tau^{\beta-1}} \left(\frac{a\mu}{2\pi} + \frac{b\tau^{\beta}(j+1)}{2} \right) & j = i+1, i = -R, \cdots, R-1 \\ q_{ij}^{(3)} = \frac{\beta-k\tau^{\beta-1}}{1-k\tau^{\beta-1}} - \frac{2a\mu}{\pi(1-k\tau^{\beta-1})} & j = i, i = -R, \cdots, R \\ q_{ij}^{(4)} = \frac{1}{1-k\tau^{\beta-1}} \left(\frac{a\mu}{2\pi} - \frac{b\tau^{\beta}(j-1)}{2} \right) & j = i-1, i = -R+1, \cdots, R \\ q_{ij}^{(5)} = \frac{a\mu}{\pi(1-k\tau^{\beta-1})} \frac{1}{(i-j)(i-j+1)} & j = i-M, i = -R+M, \cdots, R-1 , \end{cases}$$

$$(2.16)$$

where $2 \leq M \leq R$. In what follows, we give the stability of the above studied difference schemes.

3 The Proof of the Stability

In this section, we give the necessary conditions of stability, for the three previous cases separately, according to the *Gerschorgins* theorem. First, we use this theorem to find the eigenvalues λ_i, where $-R \leq i \leq R$, of the discrete scheme (2.9), of case a, from the inequality

$$|\lambda_i - q_{ii}| \leq \sum_{j \neq i} q_{ij} = \rho , \quad -R \leq j \leq R . \tag{3.1}$$

The eigenvalues are contained in circles centered at

$$\left\{ \frac{\beta - k\tau^{\beta-1}}{1 - k\tau^{\beta-1}} + \frac{a\mu}{\cos \frac{\alpha\pi}{2}(1 - k\tau^{\beta-1})} \binom{\alpha}{1} \right\} ,$$

with radius

$$\frac{1}{1-k\tau^{\beta-1}} \left\{ \frac{b\tau^{\beta}(j+1) - b\tau^{\beta}(j-1)}{2} - \frac{a\mu}{\cos \frac{\alpha\pi}{2}}\left(1 + \binom{\alpha}{2}\right) - \frac{a\mu}{\cos \frac{\alpha\pi}{2}} \sum_{s=3}^{\infty} (-1)^s \binom{\alpha}{s} \right\} ,$$

such that

$$\left| \lambda_i - \left\{ \frac{\beta - k\tau^{\beta-1}}{1 - k\tau^{\beta-1}} + \frac{a\mu}{\cos \frac{\alpha\pi}{2}(1 - k\tau^{\beta-1})} \binom{\alpha}{1} \right\} \right| \leq \frac{1}{1 - k\tau^{\beta-1}}$$

$$\left\{ \frac{b\tau^{\beta}(j+1) - b\tau^{\beta}(j-1)}{2} - \frac{a\mu}{\cos \frac{\alpha\pi}{2}}\left(1 + \binom{\alpha}{2}\right) - \frac{a\mu}{\cos \frac{\alpha\pi}{2}} \sum_{s=3}^{\infty} (-1)^s \binom{\alpha}{s} \right\} = \rho ,$$

$$(3.2)$$

and as a result

$$\frac{1}{1-k\tau^{\beta-1}}\left\{(\beta-b\tau^{\beta}-k\tau^{\beta-1}+\frac{2a\mu\alpha}{\cos\frac{\alpha\pi}{2}})+\frac{a\mu}{\cos\frac{\alpha\pi}{2}}\sum_{s=0}^{\infty}(-1)^{s+1}\binom{\alpha}{s}\right\}\leq\lambda_i\leq$$

$$\frac{1}{1-k\tau^{\beta-1}}\left\{(b\tau^{\beta}-k\tau^{\beta-1}+\beta)+\frac{a\mu}{\cos\frac{\alpha\pi}{2}}\sum_{s=0}^{\infty}(-1)^{s+1}\binom{\alpha}{s}\right\}. \tag{3.3}$$

But the term $\sum_{s=\infty}^{\infty}(-1)^{s+1}\binom{\alpha}{s}=0$, then Eq. (3.3) reduces to

$$\frac{1}{1-k\tau^{\beta-1}}\left(\beta-b\tau^{\beta}-k\tau^{\beta-1}+\frac{2a\mu\alpha}{\cos\frac{\alpha\pi}{2}}\right)\leq\lambda_i\leq\frac{1}{1-k\tau^{\beta-1}}\left(b\tau^{\beta}-k\tau^{\beta-1}+\beta\right).$$
$$\tag{3.4}$$

So far, as $0<k<1$ and by taking the limit as $\tau\to 0$, one gets

$$\frac{1}{1-k\tau^{\beta-1}}\left(b\tau^{\beta}-k\tau^{\beta-1}+\beta\right)\approx\beta,$$

and consequently $|\lambda_1|\leq\beta$. For λ_2 to satisfy the condition $|\lambda_2|\leq\beta$, the following condition must be satisfied:

$$\left|\frac{1}{1-k\tau^{\beta-1}}\left(\beta-b\tau^{\beta}-k\tau^{\beta-1}+\frac{2a\mu\alpha}{\cos\frac{\alpha\pi}{2}}\right)\right|\leq\frac{1}{1-k\tau^{\beta-1}}\left(b\tau^{\beta}-k\tau^{\beta-1}+\beta\right).$$

Solving this inequality, one gets the condition of the stability of this model as

$$0\leq\mu\leq\left(\frac{-\beta+k\tau^{\beta-1}}{a\alpha}\right)\cos\frac{\alpha\pi}{2}. \tag{3.5}$$

Second, we use also *Gerschorgins'* theorem to find the eigenvalues λ_i of the discrete scheme (2.13) from the inequality (3.1). In this case, the eigenvalues are contained in circles centered at

$$\left\{\frac{\beta-k\tau^{\beta-1}}{1-k\tau^{\beta-1}}-\frac{a\mu}{\cos\frac{\alpha\pi}{2}(1-k\tau^{\beta-1})}\right\}$$

with radius

$$\frac{1}{1-k\tau^{\beta-1}}\left\{b\tau^{\beta}+\frac{a\mu}{\cos\frac{\alpha\pi}{2}}\binom{\alpha}{1}+\frac{a\mu}{\cos\frac{\alpha\pi}{2}}\sum_{s=2}^{\infty}(-1)^{s+1}\binom{\alpha}{s}\right\},$$

such that

$$\left| \lambda_i - \left\{ \frac{\beta - k\tau^{\beta-1}}{1 - k\tau^{\beta-1}} - \frac{a\mu}{\cos\frac{\alpha\pi}{2}(1 - k\tau^{\beta-1})} \right\} \right| \le \frac{1}{1 - k\tau^{\beta-1}}$$

$$\left\{ b\tau^\beta + \frac{a\mu}{\cos\frac{\alpha\pi}{2}} \binom{\alpha}{1} + \frac{a\mu}{\cos\frac{\alpha\pi}{2}} \sum_{s=2}^{\infty} (-1)^{s+1} \binom{\alpha}{s} \right\} = \rho, \quad (3.6)$$

and as a result

$$\frac{1}{1 - k\tau^{\beta-1}} \left\{ (\beta - b\tau^\beta - k\tau^{\beta-1} - \frac{2a\mu}{\cos\frac{\alpha\pi}{2}}) - \frac{a\mu}{\cos\frac{\alpha\pi}{2}} \sum_{s=0}^{\infty} (-1)^{s+1} \binom{\alpha}{s} \right\} \le \lambda_i \le$$

$$\frac{1}{1 - k\tau^{\beta-1}} \left\{ (b\tau^\beta - k\tau^{\beta-1} + \beta) + \frac{a\mu}{\cos\frac{\alpha\pi}{2}} \sum_{s=0}^{\infty} (-1)^{s+1} \binom{\alpha}{s} \right\}. \quad (3.7)$$

But the term $\sum_{s=0}^{\infty} (-1)^{s+1} \binom{\alpha}{s} = 0$, then Eq. (3.7) reduces to

$$\frac{1}{1 - k\tau^{\beta-1}} \left(\beta - b\tau^\beta - k\tau^{\beta-1} - \frac{2a\mu}{\cos\frac{\alpha\pi}{2}} \right) \le \lambda_i \le \frac{1}{1 - k\tau^{\beta-1}} \left(b\tau^\beta - k\tau^{\beta-1} + \beta \right). \quad (3.8)$$

So far, as $0 < k < 1$ and by taking the limit as $\tau \to 0$, one gets

$$\frac{1}{1 - k\tau^{\beta-1}} \left(b\tau^\beta - k\tau^{\beta-1} + \beta \right) \approx \beta,$$

and consequently $|\lambda_1| \le \beta$. For λ_2 to satisfy the condition $|\lambda_2| \le \beta$, the following condition must be satisfied:

$$\left| \frac{1}{1 - k\tau^{\beta-1}} \left(\beta - b\tau^\beta - k\tau^{\beta-1} - \frac{2a\mu}{\cos\frac{\alpha\pi}{2}} \right) \right| \le \frac{1}{1 - k\tau^{\beta-1}} \left(b\tau^\beta - k\tau^{\beta-1} + \beta \right).$$

By solving this inequality, one gets the condition of the stability of this model as

$$0 \le \mu \le \left(\frac{\beta - k\tau^{\beta-1}}{a} \right) \cos\frac{\alpha\pi}{2}. \quad (3.9)$$

Finally, we use the same theorem to find the eigenvalues of the discrete scheme (2.15) from the inequality (3.1). In this case the eigenvalues are contained in circles centered at

$$\left\{ \frac{\beta - k\tau^{\beta-1}}{1 - k\tau^{\beta-1}} - \frac{2a\mu}{\pi(1 - k\tau^{\beta-1})} \right\}$$

with radius

$$\frac{1}{1 - k\tau^{\beta-1}} \left\{ b\tau^{\beta} + \frac{a\mu}{\pi} + \frac{2a\mu}{\pi} \sum_{s=2}^{\infty} \frac{1}{s(s+1)} \right\} \ ,$$

such that

$$\left| \lambda_i - \left\{ \frac{\beta - k\tau^{\beta-1}}{1 - k\tau^{\beta-1}} - \frac{2a\mu}{\pi(1 - k\tau^{\beta-1})} \right\} \right| \leq \frac{1}{1 - k\tau^{\beta-1}} \left\{ b\tau^{\beta} + \frac{a\mu}{\pi} + \frac{2a\mu}{\pi} \sum_{s=2}^{\infty} \frac{1}{s(s+1)} \right\} = \rho \ ,$$

(3.10)

and as a result

$$\frac{1}{1 - k\tau^{\beta-1}} \left\{ (\beta - b\tau^{\beta} - k\tau^{\beta-1} - \frac{4a\mu}{\pi}) + \frac{2a\mu}{\pi} - \frac{2a\mu}{\pi} \sum_{s=1}^{\infty} \frac{1}{s(s+1)} \right\} \leq \lambda_i \leq$$

$$\frac{1}{1 - k\tau^{\beta-1}} \left\{ (b\tau^{\beta} - k\tau^{\beta-1} + \beta) + \frac{-2a\mu}{\pi} + \frac{2a\mu}{\pi} \sum_{s=1}^{\infty} \frac{1}{s(s+1)} \right\} \ . \quad (3.11)$$

But the term $\sum_{s=1}^{\infty} \frac{1}{s(s+1)} = 1$, then Eq. (3.11) reduces to

$$\frac{1}{1 - k\tau^{\beta-1}} \left(\beta - b\tau^{\beta} - k\tau^{\beta-1} - \frac{4a\mu}{\pi} \right) \leq \lambda_i \leq \frac{1}{1 - k\tau^{\beta-1}} \left(b\tau^{\beta} - k\tau^{\beta-1} + \beta \right) \ .$$

(3.12)

Again, as $k < 1$ and by taking the limit as $\tau \to 0$, one gets

$$\frac{1}{1 - k\tau^{\beta-1}} \left(b\tau^{\beta} - k\tau^{\beta-1} + \beta \right) \approx \beta \ ,$$

and consequently $|\lambda_1| \leq \beta$. For λ_2 to satisfy the condition $|\lambda_2| \leq \beta$, the following condition must be satisfied:

$$\left| \frac{1}{1 - k\tau^{\beta-1}} \left(\beta - b\tau^{\beta} - k\tau^{\beta-1} - \frac{4a\mu}{\pi} \right) \right| \leq \frac{1}{1 - k\tau^{\beta-1}} \left(b\tau^{\beta} - k\tau^{\beta-1} + \beta \right) \ .$$

By solving this inequality, one gets the condition of the stability of this model is

$$0 < \mu \leq \frac{\pi(\beta - k\tau^{\beta-1})}{2a} \ . \quad (3.13)$$

In what follows, we prove the stability of these difference schemes by using the *von Neumann* stability condition but we have to put in mind that, the time-fractional means that the solution depends on all the history of the approximate solutions, i.e., $y^{(n+1)}$ depends on $y^{(n)}$, $y^{(n-1)}$, $y^{(n-2)}$, $\cdots$, and back to $y^{(0)}$. In other words, the wave propagation has a memory. Von Neumann method has the Fourier image

$$y_j^{(n)} = \zeta^{(n)} e^{i\kappa x_j} \ , \quad (3.14)$$

where $\zeta = \zeta(\kappa)$ is a complex number and this method does not depend on the boundary conditions. The approximate solution $y^{(n)}$ is stable if the amplification factor $|\zeta|^2 \leq 1$. Now, we substitute Eq. (3.14) into Eq. (2.9). To prove the stability, one has to do it on steps. First, we ignore the coefficients of $y^{(n-1)}$, $y^{(n-2)}$, $\cdots$, $y^{(0)}$, and substitute Eq. (3.14) on the rest of Eq. (2.9), to get after some mathematical manipulations

$$\zeta = \frac{1}{1 - k\tau^{\beta-1}}\{ \beta - k\tau^{\beta-1} + b\tau^{\beta}\cos(\kappa h) + ijb\tau^{\beta}\sin(\kappa h)+$$

$$\frac{a\mu}{\cos\frac{\alpha\pi}{2}}\sum_{s=0}^{\infty}(-1)^{s+1}\binom{\alpha}{s}\cos((1-s)\kappa h) \} . \qquad (3.15)$$

Taking the limit as $h \to 0$ and then putting the term $\sum_{s=0}^{\infty}(-1)^{s+1}\binom{\alpha}{s} = 0$, Eq. (3.15) reduces to

$$\zeta = \frac{1}{1 - k\tau^{\beta-1}}\left(\beta - k\tau^{\beta-1} + b\tau^{\beta}\right) . \qquad (3.16)$$

Calculate $|\zeta|^2$ as

$$|\zeta|^2 = \frac{1}{(1 - k\tau^{\beta-1})^2}\left(\beta^2 - 2\beta k\tau^{\beta-1} + k^2\tau^{2\beta-2} - b^2\tau^{2\beta}\right) . \qquad (3.17)$$

So far, as $0 < k < 1$ and by taking the limit as $\tau \to 0$, one gets $|\zeta|^2 = \beta^2 > 1$. Second, we take into consideration the dependence of $y^{(n+1)}$ on $y^{(n)}$ and $y^{(n-1)}$, only, to get after some calculations

$$\zeta^2 - \frac{1}{1 - k\tau^{\beta-1}}\{ \beta - k\tau^{\beta-1} + b\tau^{\beta}\cos(\kappa h) + ijb\tau^{\beta}\sin(\kappa h)+$$

$$\frac{a\mu}{\cos\frac{\alpha\pi}{2}}\sum_{s=0}^{\infty}(-1)^{s+1}\binom{\alpha}{s}\cos((1-s)\kappa h) \}\zeta + \frac{1}{1 - k\tau^{\beta-1}}\binom{\beta}{2} = 0 . \qquad (3.18)$$

After taking the limit as $h \to 0$ and putting $\sum_{s=0}^{\infty}(-1)^{s+1}\binom{\alpha}{s} = 0$, then the above equation reduces to

$$\zeta^2 - \frac{1}{1 - k\tau^{\beta-1}}\{ \beta - k\tau^{\beta-1} + b\tau^{\beta} \}\zeta + \frac{1}{1 - k\tau^{\beta-1}}\binom{\beta}{2} = 0 . \qquad (3.19)$$

The roots of the above equation are

$$\zeta_{1,2} = \frac{1}{4(\tau - k\tau^\beta)}\{ 2\beta\tau - 2k\tau^\beta + 2b\tau^{1+\beta}\mp$$

$$\sqrt{-8(\beta^2\tau - \beta\tau)(\tau - 2k\tau^\beta) + (2k\tau^\beta - 2\beta\tau - 2b\tau^{1+\beta})^2} \} . \quad (3.20)$$

After some mathematical manipulations and taking the limit as $\tau \to 0$, one gets the roots of the above equation $|\zeta_1|^2 = |\zeta_2|^2 \approx |\frac{\beta(\beta-1)}{2}| \le 1$ as $\beta \le 2$. The space fractional order α has no effect on the stability condition. This proves the stability condition. The third step is by assuming that $y^{(n+1)}$ depends only on $y^{(n)}$, $y^{(n-1)}$, and $y^{(n-2)}$. At this step, one gets $|\zeta_i|^2$, $i = 1, 2, 3$ that are approximately less than one. In the next step, we add the dependence on $y^{(n-3)}$, and so on till reaching $y^{(0)}$. At each step, one has to solve the resulted equation and use the previous limits to get $|\zeta_i|^2 < 1$, $i \ge 1$. So far, the scheme (2.9) is stable for the space–time-fractional order α and β.

Now, we prove also the stability of the scheme (2.13) by using the von Neumann method. We substitute equation (3.14) into Eq. (2.13). First, we ignore the coefficients of $y^{(n-1)}$, $y^{(n-2)}$, $\cdots$, $y^{(0)}$, and substitute Eq. (3.14) on the rest of equation (2.13), to get after manipulations

$$\zeta = \frac{1}{1 - k\tau^{\beta-1}}\{ \beta - k\tau^{\beta-1} + b\tau^\beta \cos(\kappa h) + ijb\tau^\beta \sin(\kappa h)+$$

$$\frac{a\mu}{\cos\frac{\alpha\pi}{2}} \sum_{s=0}^{\infty}(-1)^{s+1}\binom{\alpha}{s}\cos(\kappa sh) \} . \quad (3.21)$$

After taking the limit as $h \to 0$ and putting $\sum_{s=0}^{\infty}(-1)^{s+1}\binom{\alpha}{s} = 0$, Eq. (3.21) reduces to the same Eq. (3.16) with the same condition $|\zeta|^2 = \beta^2 > 1$. Second, we take into consideration the dependence of $y^{(n+1)}$ on $y^{(n)}$ and $y^{(n-1)}$, only, to get after calculations

$$\zeta^2 - \frac{1}{1 - k\tau^{\beta-1}}\{ \beta - k\tau^{\beta-1} + b\tau^\beta \cos(\kappa h) + ijb\tau^\beta \sin(\kappa h)+$$

$$\frac{a\mu}{\cos\frac{\alpha\pi}{2}} \sum_{s=0}^{\infty}(-1)^{s+1}\binom{\alpha}{s}\cos(\kappa sh) \}\zeta + \frac{1}{1 - k\tau^{\beta-1}}\binom{\beta}{2} = 0 . \quad (3.22)$$

After taking the limit as $h \to 0$ and putting $\sum_{s=0}^{\infty}(-1)^{s+1}\binom{\alpha}{s} = 0$, Equation (3.22) reduces to the same equation as (3.19). Then after some mathematical manipulations, one gets the roots of the above equation $|\zeta_1|^2 = |\zeta_2|^2 \approx |\frac{\beta(\beta-1)}{2}| \le 1$ as $\beta \le 2$. As before, **the space-fractional order α has no effect on the stability condition** The third step is to assume that $y^{(n+1)}$ depends only on $y^{(n)}$, $y^{(n-1)}$, and $y^{(n-2)}$. At this step, one gets $|\zeta_i|^2$, $i = 1, 2, 3$ are approximately less than one. In the next step, we add the dependence on $y^{(n-3)}$, and so on till reaching $y^{(0)}$. At each step, one has to solve the resulted equation and use the previous limits, to get $|\zeta_i|^2 < 1$, $i \geq 1$. So far, the scheme (2.13) is stable for the space–time-fractional order α and β.

Finally, we apply von Neumann method to prove the stability of the scheme (2.15). First, we ignore the coefficients of $y^{(n-1)}$, $y^{(n-2)}$, $\cdots$, $y^{(0)}$, and substitute Eq. (3.14) on the rest of Eq. (2.15), to get after manipulating

$$\zeta = \frac{1}{1 - k\tau^{\beta-1}}\{ \beta - k\tau^{\beta-1} + b\tau^{\beta}\cos(\kappa h) - \frac{2a\mu}{\pi} + ijb\tau^{\beta}\sin(\kappa h) +$$

$$\frac{2a\mu}{\pi} \sum_{s=1}^{\infty} \frac{1}{s(s+1)}\cos(\kappa s h) \} . \quad (3.23)$$

After taking the limit as $h \to 0$ and putting the term $\sum_{s=1}^{\infty} \frac{1}{s(s+1)} = 1$, the above equation reduces to the same Eq. (3.16). That means one gets the condition $|\zeta|^2 = \beta^2 > 1$. Second, we take into consideration the dependence of $y^{(n+1)}$ on $y^{(n)}$ and $y^{(n-1)}$, only, to get after some calculations

$$\zeta^2 - \frac{1}{1 - k\tau^{\beta-1}}\{ \beta - k\tau^{\beta-1} + b\tau^{\beta}\cos(\kappa h) - \frac{2a\mu}{\pi} + ijb\tau^{\beta}\sin(\kappa h) +$$

$$\frac{2a\mu}{\pi} \sum_{s=1}^{\infty} \frac{1}{s(s+1)}\cos(\kappa s h) \}\zeta + \frac{1}{1 - k\tau^{\beta-1}}\binom{\beta}{2} = 0 . \quad (3.24)$$

After taking the limit as $h \to 0$ and putting $\sum_{s=1}^{\infty} \frac{1}{s(s+1)} = 1$, Eq. (3.24) reduces to the same equation as (3.19). Then after some mathematical manipulations, one gets $|\zeta_1|^2 = |\zeta_2|^2 \approx |\frac{\beta(\beta-1)}{2}| \leq 1$ as $\beta \leq 2$. Again, the space-fractional order α has no effect on the stability condition. The third step is to assume that $y^{(n+1)}$ depends only

Table 1 The values of parameters which are used in the calculations

Case	Figures	$f(x)$	α	β	μ	t_d	t_f
$1 < \alpha < 2$	Figures 1, 2, 3, 4, 5 and 6	$\sin(\frac{\pi x}{2R+1})$	2	2	0.9	$t = 6$	$t = 50$
	Figures 1, 2, 3, 4, 5 and 6	$\sin(\frac{\pi x}{2R+1})$	1.8	1.9	0.97	$t = 5$	$t = 38$
	Figures 7, 8, 9 and 10	$\sin(\frac{\pi x}{2R+1})$	1.7	1.8	0.9	$t = 5$	$t = 38$
	Figures 11, 12, 13 and 14	$\sin(\frac{\pi x}{2R+1})$	1.7	2	1	$t = 29$	$t = 48$
$0 < \alpha < 1$	Figures 15 and 16	$\sin(\frac{\pi x}{2R+1})$	0.9	2	0.3	$t = 30$	$t = 30$
	Figures 17 and 18	$\delta(x)$	0.8	1.7	0.4	$t = 50$	$t = 50$
$\alpha = 1$	Figures 19 and 20	$\delta(x)$	1	2	0.92	–	$t = 30$
	Figures 21 and 22	$\sin(\frac{\pi x}{2R+1})$	1	1.7	0.6	–	$t = 20$

on $y^{(n)}$, $y^{(n-1)}$, and $y^{(n-2)}$. At this step, one gets $|\zeta_i|^2$, $i = 1, 2, 3$ are approximately less than one. In the next step, we add the dependence on $y^{(n-3)}$, and so on till reaching $y^{(0)}$. At each step, one has to solve the resulted equation and use the previous limits, to get $|\zeta_i|^2 < 1$, $i \geq 1$. So far, the scheme (2.15) is stable for the space–time-fractional order α and β (Table 1).

Fig. 1 $t = 1$

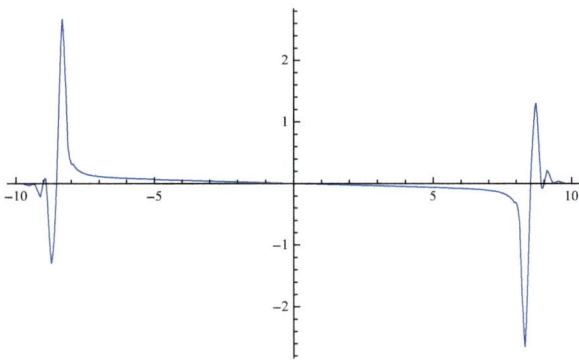

Fig. 2 $t = 2$

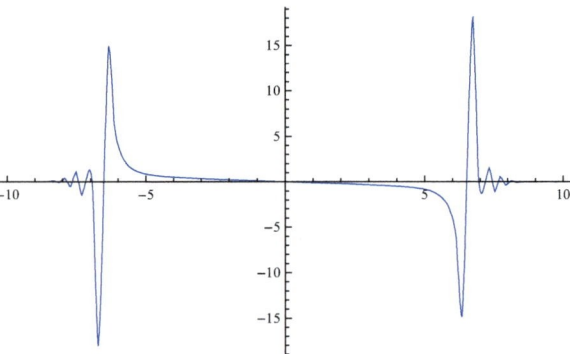

Fig. 3 $t = 5$

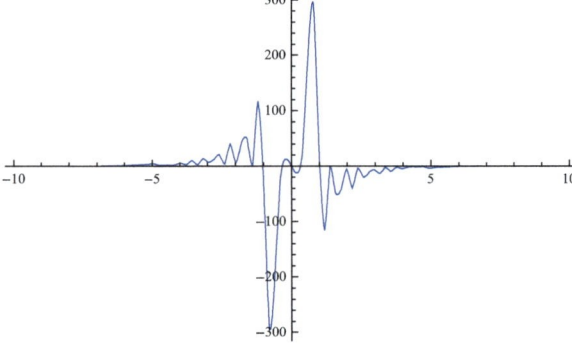

Fig. 4 t = 10

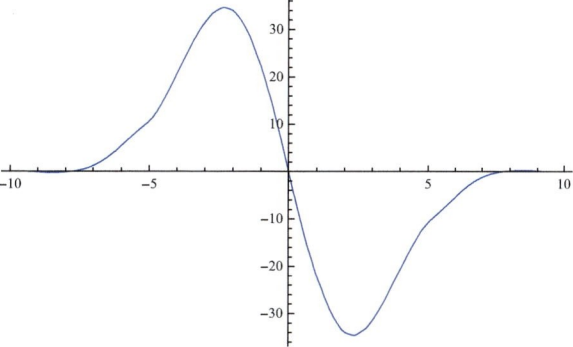

Fig. 5 t = 35

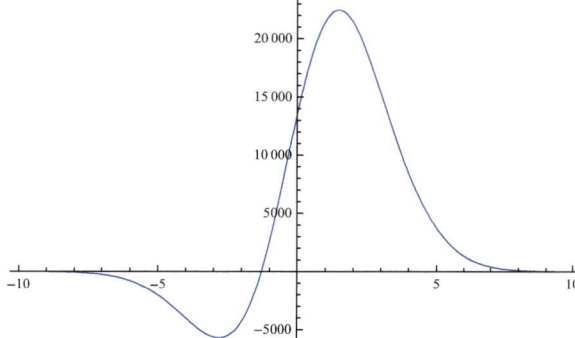

Fig. 6 t = 38

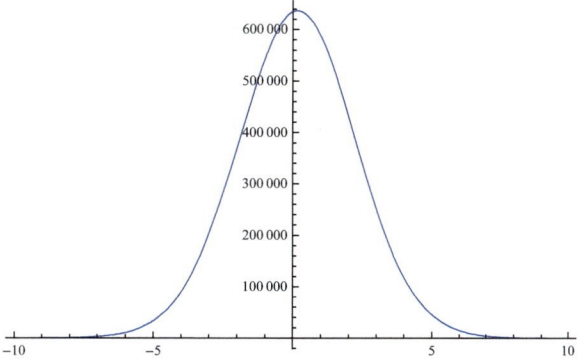

Fig. 7 t = 5

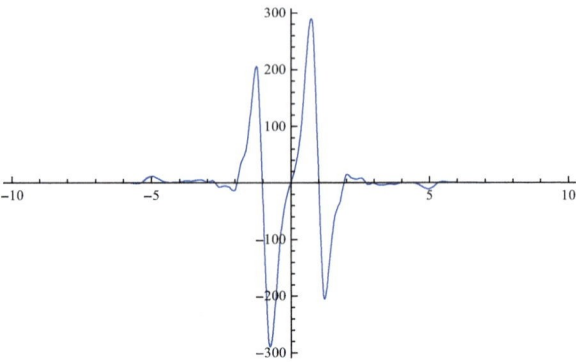

Fig. 8 t = 8

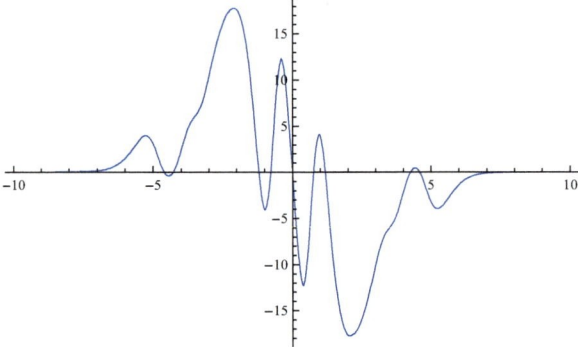

Fig. 9 t = 20

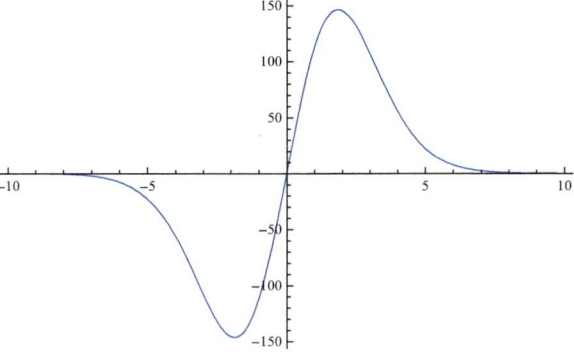

Fig. 10 t = 38

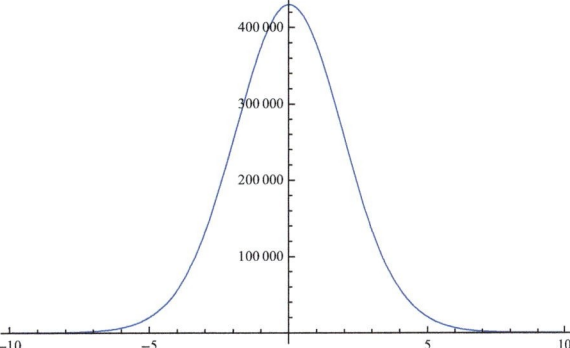

Fig. 11 t = 5

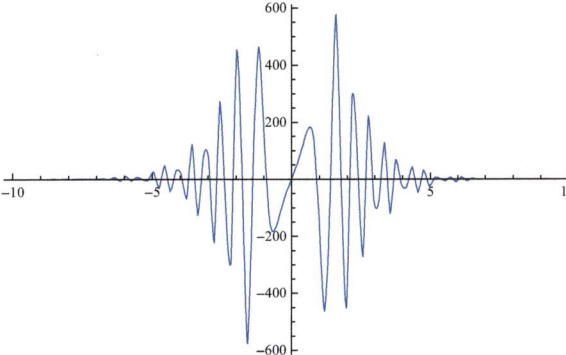

Fig. 12 t = 29

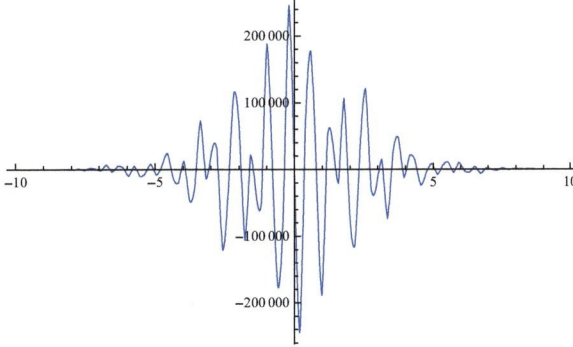

Fig. 13 t = 44

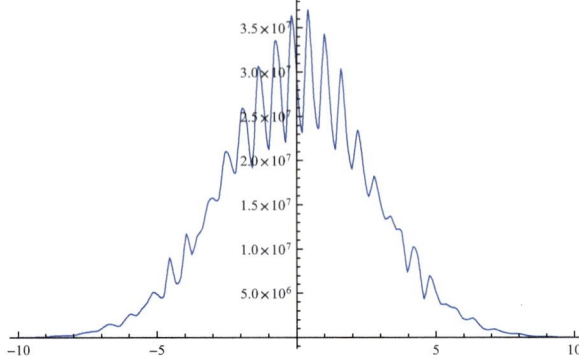

Fig. 14 t = 48

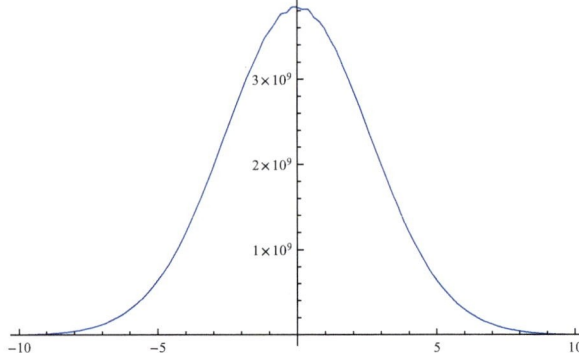

Fig. 15 t = 10

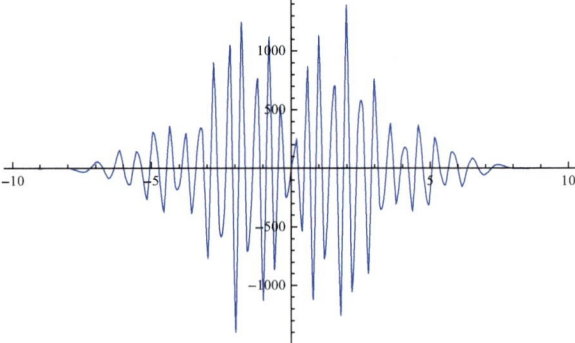

Fig. 16 t = 30

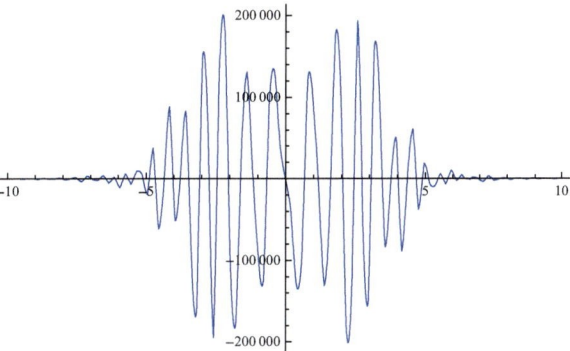

Fig. 17 t = 5

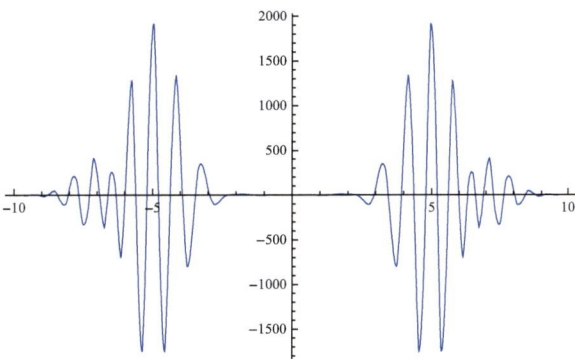

Fig. 18 t = 50

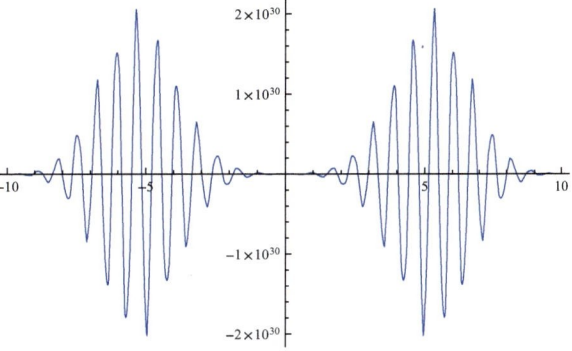

Fig. 19 t = 20

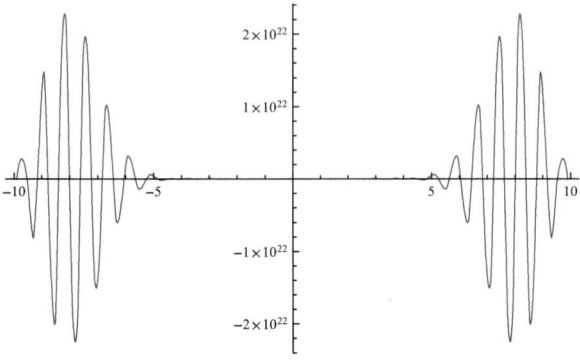

Fig. 20 t = 30

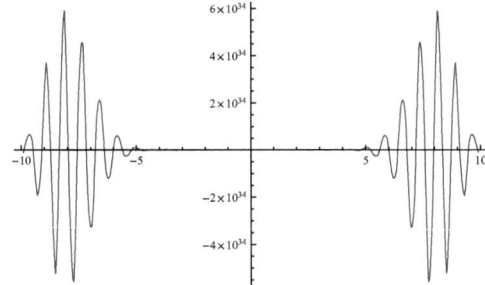

Fig. 21 t = 10

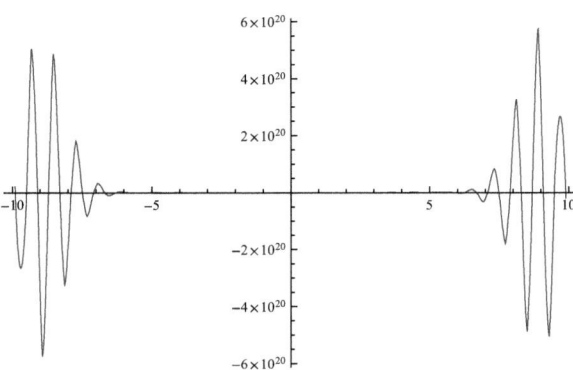

Fig. 22 t = 20

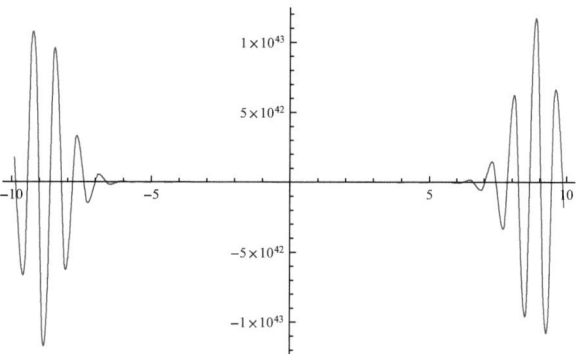

4 Convergence to the Stationary Approximate Solutions

We seek to find the stationary approximate solution which does not depend on the time, i.e., the solution of the space–time-fractional differential equation as $t \to \infty$, for the above-discussed cases. To do so, we omit the terms depending on the time in the matrix Eq. (2.11), where Q is defined for each cases in Eqs. (2.12), (2.14), and (2.16). For these cases, we get the same matrix equation of the form $z.H = 0$, i.e., $H^T.y = 0$. The matrix H is obtained from the matrices defined in (2.12), (2.14), and (2.16) after omitting all the terms depending on t and β. The sum of the rows of the resulted matrix H is zero. The matrix H^T has an eigenvector y^* of eigenvalue zero. Our stationary approximation solution $\bar{y} = vy^*$ with $v = 1/\sum_{j=-R}^{R} y_j^*$ is a vector, whose elements sum to 1. We simulate the stationary approximate solutions at Figs. 23 and 24 for the classical case and for $0 < \alpha < 1$.

To study the convergence of the approximate solution, we constitute the sequence $d = \{d(t_1), d(t_2), \cdots\}$, where $t_1 < t_2 < \cdots \to \infty$. The numbers $d(t_i), i : 1 \to 16$ is defined as

$$d(t_i) = \sum_{j=-R}^{R} |y_j(t_i) - \bar{y}_j| , \quad i = 1, 2, \cdots .$$

The convergent approximate solutions are simulated at Figs. 25, 26 and 27. In these figures, we plot $Log_{10}d$ against the number of points. We compare Figs. 25, 26 and 27 with Fig. 28 to prove that the convergent approximate solutions of the discussed cases are related to e^{-t} by the relation $d(t) = c_1 e^{-c_2 t}$, where c_1 and c_2 are positive constants related to α and β for each case.

Fig. 23 $\alpha = \beta = 2$

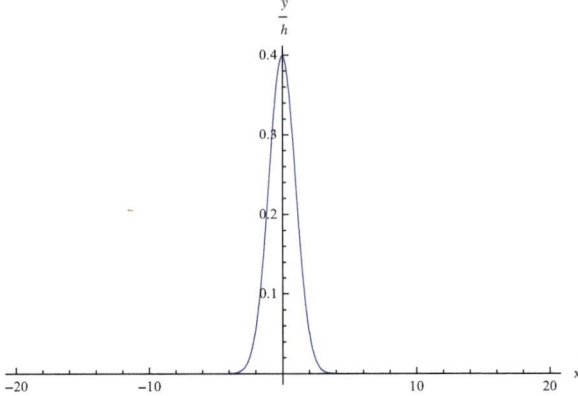

Fig. 24 $0 < \alpha < 1, \beta = 1.7$

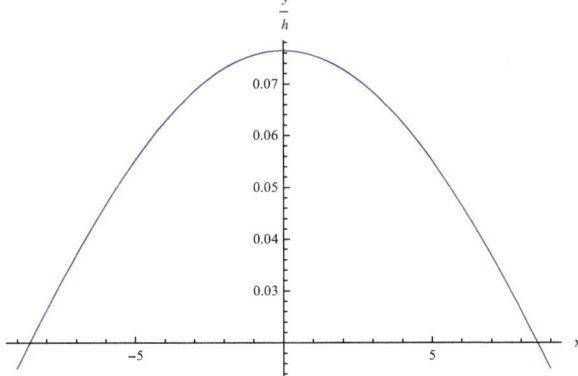

Fig. 25 Convergence

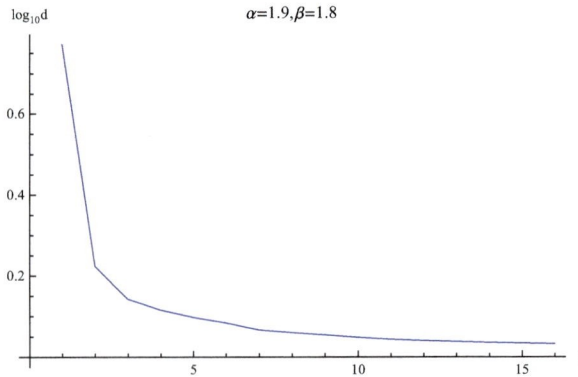

Fig. 26 Convergence

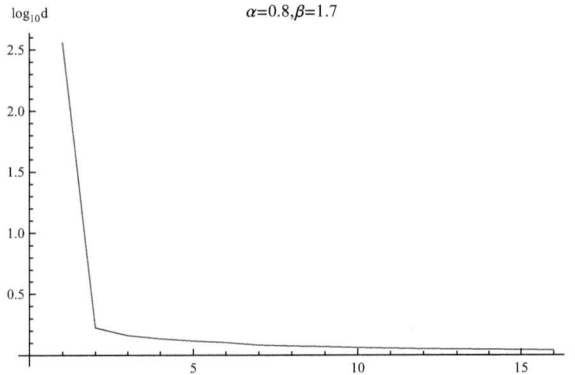

Fig. 27 Convergence

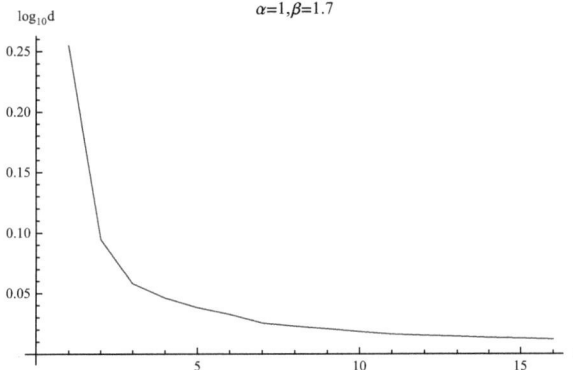

Fig. 28 Convergence

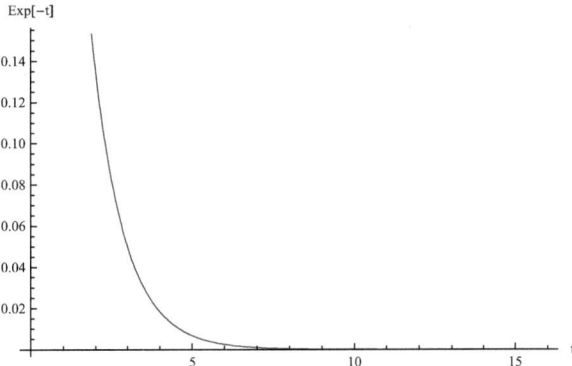

5 Numerical Results and Discussions

In this section, we give the numerical approximate solution for Eq. (2.1). We give the evolution of $y^{(n)} = y(t_n)$ with different values of t_n and with different values of the space-fractional order α and different values of the time-fractional order β. In these simulations, we have enlarged the $x-$axis as $-10 \leq x \leq 10$ with $h = 0.2$. We fix the value of the attenuation coefficient $k = 0.5$. The value of μ must satisfy the required conditions depending on the values of α and β. Since $\{t_0, t_1, t_2, \cdots\} = \{0, 1, 2, \cdots\}$, then the iteration index $n = \frac{t_n}{\tau}$ while τ is calculated from the scaling parameter of the specified model. We calculate all the numerical results for $a = b = 1$. In the following table, we summarize the values of $f(x)$, μ, β, α, t_d, t_f being used in the numerical results. Here t_d is the time when the wave is starting to damp and t_f is the time when the wave reaches to the get its stationary solution. For all the values of the fractional orders α and β, the studied approximate solutions have the same start for $t = 1$ till $t = 3$, i.e., have the same start . For $1 < \alpha < 2$, the results show that the effect of the damping force is much bigger than the external force $F(x)$. Also, we observe that for the fractional values α and β, the propagating waves reach its stationary solutions faster than as in the classical case, $\alpha = 2$ and $\beta = 2$, see [9, 18].

The convergence of the approximate solutions (the first norm) are simulated at Figs. 25, 26, and 27. Figure 28 represents the plot of the rapid convergent function e^{-t}. Then by comparing the last four figures, one can deduce that the approximate solution of the studied model is convergent for all the values of the fractional orders α and β.

References

1. Szabo, T.L.: Causal theories and data for acoustic attenuation obeying a frequency power-law. J. Acoust. Soc. Am. **97**, 14–24 (1995)
2. Mandelbrot, B.B.: The Fractal Geometry of Nature. Freeman, New York (1982)
3. Family, F., Masters, B.R., Platt, D.E.: Fractal pattern formation in the human retinal vessels. Phys. D: Nonlinear Phenom. J. **38**, 98–103 (1989)
4. Lemehaute, A.: Fractal Geometries Theory and Applications. CRC Press, Boca Raton (1991)
5. Daxer, A.A.: Fractals and Retinal vessels. Lancet J., 339–618 (1992)
6. Cross, S.S.: Fractals in pathology. Pathol. J. **182**, 1–8 (1997)
7. Gazit, Y., Baish, J.W., Safabakhsh, N., Leunig, M., Baxter, L.T., Jain, R.K.: Fractal Characteristics of tumor vascular architecture during tumor growth and regression. Microcircul. J., 395–402 (1997)
8. Baish, J.W., Jain, R.K.: Fractals and cancer. Cancer Res., 3683–3688 (2000)
9. Abdel-Rehim, E.A., El-Sayed, A.M.A., Hashem, A.S.: Simulation of the approximate solution of the time fractional multi-term wave equations. J. Comput.Math. Appl. **73**, 1134–1154 (2017)
10. Liebler, M., Ginter, S., Dreyer, T., Riedlinge, R.E.: Full wave modeling of therapeutic ultrasound: efficient time-domain, implemetation of the frequency power-law attenuation. J. Acoust. Soc. Am. **116**, 2742–2750 (2004)
11. Tarasov, V.E.: Fractional hydrodynamic equations for fractal media. Ann. Phys. J. **318**, 286–307 (2005)

12. Gorenflo, R., Mainardi, F.: Fractional calculus: integral and differential equations of fractional order. In: Carpinteri, A., Mainardi, F. (eds.) Fractals and Fractional Calculus in Continuum Mechanics. Springer, Wien and New York (1997)
13. Saichev, A.I., Zaslavsky, G.M.: Fractional kinetic equations, solutions and applications. Chaos **7**, 743–764 (1997)
14. Caputo, M.: Linear models of dissipation whose Q is almost independent II. Geophys. J. **13**, 529–539 (1967)
15. Treeby, B.E., Cox, B.T.: Modeling power law absorption and dispersion for acoustic propagation using the fractional Laplacian. J. Acoust. Soc. Am., 2741–2748 (2010)
16. Fellah, M.M., Fellah, Z.E.A., Depollier, C.: Transient wave propagation in inhomogeneous porous materials: application of fractional derivatives. Signal Process. J. **86**, 2658–2667 (2006)
17. Casasanta, G., Garra, R.: Fractional calculus approach to the acoustic wave propagation with space-dependent sound speed. Signal Image Video Process. J. **6**, 389–392 (2016)
18. El-Sayed, A.M.A., Abdel-Rehim, E.A., Hashem, A.S.: Time evolution of the approximate and stationary solutions of the time-fractional forced-damped wave equation. Tbillisi Math. J. **10**, 127–144 (2017)
19. Gorenflo, R., Mainardi, F., Moretti, D., Paradisi, P.: Time-Fractional diffusion: a discrete random walk approach. J. Nonlinear Dyn. **29**, 129–143 (2002)
20. Abdel-Rehim, E.A.: Modelling and simulating of classical and non-classical diffusion processes by random walks. Mensch& Buch Verlag (2004)
21. Feller, W.: On a generalization of marcel Riesz potentials and the semi-groups generated by them, In: Meddelanden Lunds Universitetes Matematiska Seminar-ium (Comm. Sm. Mathm. Universit de Lund), Tome Suppl. ddi a M. Riesz, Lund, pp. 73–81 (1952)
22. Oldham, K.B.: Spanier Journal of the Fractional Calculus. Academic Press, New York (1974)
23. Miller, K.S., Ross, B.: An Introduction to the Fractional Calculus and Fractional Differential Equations. Wily (1993)
24. Samko, S.G., Kilbas, A.A., Marichevm, O.I.: Fractional Integrals and Derivatives. Theory and Applications), OPA, Amsterdam (1993)
25. Gorenflo, R., Mainardi, F.: Approximation of Lévy-Feller diffusion by random walk. J. Anal. Appl. (ZAA) **18**, 231–246 (1999)
26. Abdel-Rehim, E.A.: Explicit approximation solutions and proof of convergence of the space-time fractional advection dispersion equations. J. Appl. Math. **4**, 1427–1440 (2013)

Certain Properties of Konhauser Polynomial via Generalized Mittag-Leffler Function

J. C. Prajapati, N. K. Ajudia, Shilpi Jain, Anjali Goswami
and Praveen Agarwal

Abstract The principal aim of this paper is to establish several new properties of generalized Mittag-Leffler function via Konhauser polynomials. Properties like mixed recurrence relations, Differential equations, pure recurrence relations, finite summation formulae, and Laplace transform have been obtained.

Keywords Konhauser polynomials · Generalized Mittag-Leffler function · Laguerre polynomials · Laplace transform

AMS(2010) Subject Classification: 33E12, 33C45, 33C47.

J. C. Prajapati
Department of Mathematics, Sardar Patel University, Vallabh Vidyanagar 388120, Gujarat, India
e-mail: jyotindra18@rediffmail.com

N. K. Ajudia
H & H B Kotak Institute of Science, Saurashtra University, Rajkot, Gujarat, India
e-mail: nka121@gmail.com

S. Jain
Department of Mathematics, Poornima College of Engineering, Jaipur 302022, India
e-mail: shilpijain1310@gmail.com

A. Goswami
College of Science and Theoretical Studies, Main Branch Riyadh, Female, Saudi Electronic University, Abu Bakr Street, PO. Box: 93499, Riyadh KSA, Saudi Arabia
e-mail: dranjaligoswami09@gmail.com

P. Agarwal (✉)
Department of Mathematics, Anand International College of Engineering, Jaipur 303012, India
e-mail: goyal.praveen2011@gmail.com

Department of Mathematics, Harish Chandra Research Institute, Chhatnag Road, Jhunsi 211019, Allahabad, India

International Center for Basic and Applied Sciences, Jaipur 302029, India

© Springer Nature Singapore Pte Ltd. 2019
P. Agarwal et al. (eds.), *Fractional Calculus*, Springer Proceedings
in Mathematics & Statistics 303, https://doi.org/10.1007/978-981-15-0430-3_11

1 Introduction

Konhauser polynomial has drawn the attention of several researchers. Recently, Prajapati et al. [7] introduced a class of polynomials

$$L_{\left[\frac{m}{q}\right]}^{(\alpha,\beta)}(z) = \frac{\Gamma(\alpha m + \beta + 1)}{m!} \sum_{n=0}^{\left[\frac{m}{q}\right]} \frac{(-m)_{qn}}{\Gamma(\alpha n + \beta + 1)} \frac{z^n}{n!}, \tag{1.1}$$

where $\alpha, \beta \in \mathbb{C}$; $m, q \in \mathbb{N}$, $\left[\frac{m}{q}\right]$ denotes integral part of $\frac{m}{q}$, $Re(\beta) > -1$.
   This is generalized form of Konhauser polynomials (Konhauser [5]),

$$Z_m^\mu(x; k) = \frac{\Gamma(km + \mu + 1)}{m!} \sum_{j=0}^{n} (-1)^j \binom{m}{j} \frac{x^{kj}}{\Gamma(kj + \mu + 1)}, \tag{1.2}$$

where $\mu > -1$.
   Note that

$$L_m^{(k,\mu)}(z^k) = Z_m^\mu(z; k). \tag{1.3}$$

The Laguerre polynomials (Rainville [8]) defined as

$$L_m^\mu(x) = \frac{\Gamma(m + \mu + 1)}{m!} \sum_{j=0}^{n} (-1)^j \binom{m}{j} \frac{x^j}{\Gamma(j + \mu + 1)}, \tag{1.4}$$

where $\mu > -1$.
   This is special case of (1.2) as

$$Z_m^\mu(x; 1) = L_m^\mu(x). \tag{1.5}$$

In 1970, Prabhakar [6] defined the generalized Mittag-Leffler function as

$$E_{\alpha,\beta}^\gamma(z) = \sum_{n=0}^{\infty} \frac{(\gamma)_n \, z^n}{\Gamma(\alpha n + \beta) \, n!}, \quad Re(\alpha) > 0. \tag{1.6}$$

This is an entire function of order $(Re(\alpha))^{-1}$.
   Kilbas et al. [4] studied the relation between (1.2) and (1.6) as

$$E_{k,\mu+1}^{-m}(z^k) = \frac{\Gamma(m + 1)}{\Gamma(km + \mu + 1)} Z_m^\mu(z; k), \tag{1.7}$$

where $m, k \in \mathbb{N}$; $\mu \in \mathbb{C}$ with $Re(\mu) > -1$. If $k = 1$ then (1.7) reduces to

$$E_{1,\mu+1}^{-m}(z) = \frac{\Gamma(m+1)}{\Gamma(m+\mu+1)}L_m^{\mu}(z), \qquad m \in \mathbb{N}; \mu \in \mathbb{C}. \tag{1.8}$$

In 2007, Shukla and Prajapati [9] introduced generalized Mittag-Leffler function as

$$E_{\alpha,\,\beta}^{\gamma,\,q}(z) = \sum_{n=0}^{\infty} \frac{(\gamma)_{qn}}{\Gamma(\alpha n + \beta)}\frac{z^n}{n!}, \tag{1.9}$$

where $\alpha, \beta, \gamma \in \mathbb{C}$; $\mathrm{Re}(\alpha) > 0, \mathrm{Re}(\beta) > 0, \mathrm{Re}(\gamma) > 0$ and $q \in (0, 1) \cup \mathbb{N}$

In 2014, Prajapati et al. [7] established relation between (1.9) and (1.1) as

$$E_{k,\mu+1}^{-m,q}(z^k) = \frac{\Gamma(m+1)}{\Gamma(km+\mu+1)}L_{\left[\frac{m}{q}\right]}^{(k,\mu)}(z^k). \tag{1.10}$$

Furthermore, some useful results are obtained in (1.11)–(1.15) (Prajapati et al. [7]).

$$L_{\left[\frac{m}{q}\right]}^{(\alpha,\beta)}(t) = (\alpha m + \beta)\int_0^1 z^{\beta-1}L_{\left[\frac{m}{q}\right]}^{(\alpha,\beta-1)}(tz^\alpha)dz \tag{1.11}$$

$$(z-t)^\beta L_{\left[\frac{m}{q}\right]}^{(\alpha,\beta)}\left((z-t)^\alpha\right) = \frac{\Gamma(\alpha m + \beta + 1)}{\Gamma(\alpha m + \beta - \gamma + 1)\Gamma(\gamma)} \times$$
$$\int_t^z (z-u)^{\gamma-1}(u-t)^{\beta-\gamma}L_{\left[\frac{m}{q}\right]}^{(\alpha,\beta-\gamma)}\left((u-t)^\alpha\right)du, \tag{1.12}$$

where $\alpha, \beta, \gamma \in \mathbb{C}$ with $\mathrm{Re}(\beta) > \mathrm{Re}(\gamma) > -1$,

$$\sum_{m=0}^{\infty}\frac{(\gamma)_m L_{\left[\frac{m}{q}\right]}^{(\alpha,\beta)}(z)t^m}{\Gamma(\alpha m + \beta + 1)} = (1-t)^{-\gamma}E_{\alpha,\beta+1}^{\gamma,q}\left(\frac{z(-t)^q}{(1-t)^q}\right), \quad |t| < 1 \tag{1.13}$$

where $\alpha, \beta, \gamma \in \mathbb{C}$ with $Re(\beta) > -1$ and $q \in \mathbb{N}$,

$$\sum_{m=0}^{\infty}\frac{L_{\left[\frac{m}{q}\right]}^{(\alpha,\beta)}(z^\alpha)t^m}{\Gamma(\alpha m + \beta + 1)} = e^t W(\alpha; \beta + 1; z^\alpha(-t)^q), \tag{1.14}$$

where $\alpha, \beta \in \mathbb{C}$ with $Re(\beta) > -1$ and $q \in \mathbb{N}$, and

$$L_{\left[\frac{m}{q}\right]}^{(k,\beta)}(z^k) = \left(\frac{z}{y}\right)^{km}\sum_{r=0}^m \frac{(\beta+1)_{km}}{(\beta+1)_{km-kr}}\frac{\left[\left(\frac{y}{z}\right)^k - 1\right]^r}{r!}L_{\left[\frac{m-r}{q}\right]}^{(k,\beta)}(y^k), \tag{1.15}$$

where $k \in \mathbb{N}$ and $\beta \in \mathbb{C}$ with $\mathrm{Re}(\beta) > -1$.

In 2010, Maged Gumaan Bin-Saad [1] investigated Hermite–Konhauser polynomial as

$$
{}_k H_m^\mu(x, y; z) = m! \sum_{n=0}^m \sum_{r=0}^{\left[\frac{m-n}{2}\right]} \frac{(-1)^{n+r} x^r y^{kn+\mu} z^{m-n-2r}}{n! r! (m-n-2r)! \Gamma(kn+\mu+1)}, \qquad (1.16)
$$

and the relation between Konhauser polynomial and Hermite–Konhauser polynomial as

$$
Z_m^\mu(x; k) = \frac{x^{-\mu} \Gamma(km + \mu + 1)}{m!} {}_k H_m^\mu(0, x; 1). \qquad (1.17)
$$

Konhauser [5] obtained mixed recurrence relation, differential equation, and pure recurrence relation of (1.2) as (1.18)–(1.20)

$$
x D Z_m^\mu(x; k) = mk Z_m^\mu(x; k) - k(km - k + \mu + 1) {}_k Z_{m-1}^\mu(x; k), \qquad (1.18)
$$

$$
D^k[x^{\mu+1} D Z_m^\mu(x; k)] = x^{\mu+1} D Z_m^\mu(x; k) - mk x^\mu Z_m^\mu(x; k), \qquad (1.19)
$$

$$
\sum_{i=0}^k \binom{k}{i} [D^{k-i} x^{\mu+1}][D^{i+1} Z_m^\mu(x; k)] = -kx^\mu(km - k + c + 1) {}_k Z_{m-1}^\mu(x; k). \qquad (1.20)
$$

Srivastava [10] gives another form of $Z_m^\mu(y; k)$ as

$$
Z_m^\mu(x; k) = \left(\frac{x}{y}\right)^{km} \sum_{r=0}^m \binom{\mu + km}{kr} \frac{(kr)!}{r!} \left[\left(\frac{y}{x}\right)^k - 1\right]^r Z_{m-r}^\mu(y; k). \quad (1.21)
$$

The recurrence relations, differential equations, pure recurrence relations, finite summation formulae, and Laplace transforms of (1.2) studied by Srivastava [12] as (1.22) to (1.30)

$$
D Z_m^\mu(x; k) = -kx^{k-1} Z_{m-1}^{\mu+k}(x; k), \qquad (1.22)
$$

$$
(x^{1-k} D)^n Z_m^\mu(x; k) = (-k)^n Z_{m-n}^{\mu+kn}(x; k), \quad m \ge n \ge 0, \qquad (1.23)
$$

$$
x D Z_m^\mu(x; k) = (mk + \mu) Z_m^{\mu-1}(x; k) - \mu Z_m^\mu(x; k), \qquad (1.24)
$$

$$
Z_m^\mu(x; k) - Z_m^{\mu-1}(x; k) = \frac{k\Gamma(km + \mu)}{\Gamma(k(m-1) + \mu + 1)} Z_{m-1}^\mu(x; k), \qquad (1.25)
$$

$$
x^k Z_m^{\mu+k}(x; k) = (km + \mu + 1) {}_k Z_m^\mu(x; k) - (m+1) Z_{m+1}^\mu(x; k), \qquad (1.26)
$$

$$
kx^k Z_m^{\mu+k}(x; k) = \mu Z_{m+1}^\mu(x; k) - (km + \mu + k) Z_{m+1}^{\mu-1}(x; k), \qquad (1.27)
$$

$$Z_m^\mu(\delta x; k) = \sum_{j=0}^{m} \binom{\mu + km}{kj} \delta^{k(m-j)} (1 - \delta^k)^j Z_{m-j}^\mu(x; k), \qquad (1.28)$$

$$L\{t^\nu Z_m^\mu(xt; k); s\} = \frac{(\mu + 1)_{km} \Gamma(\nu + 1)}{s^{\nu+1} m!} {}_{k+1}F_k \left[\begin{array}{c} -m, \frac{\nu+1}{k}, \frac{\nu+2}{k}, ..., \frac{\nu+k}{k}; \\ \frac{\mu+1}{k}, \frac{\mu+2}{k}, ..., \frac{\mu+k}{k}; \end{array} \left(\frac{x}{s} \right)^k \right], \quad (1.29)$$

$Re(s) > 0$, $Re(\nu) > -1$;
if $\nu = \mu$ then (1.29) reduces to

$$L\{t^\mu Z_m^\mu(xt; k); s\} = \frac{\Gamma(km + \mu + 1)}{s^{km+\mu+1}} (s^k - x^k)^m, \qquad (1.30)$$

where $Re(s) > 0$, $Re(\mu) > -1$.

In 1970, Prabhakar [6] gives relations as follows:

$$D^n \left[x^{\mu+n} Z_m^{\mu+n}(x; k) \right] = \frac{\Gamma(km + \mu + n + 1)}{\Gamma(km + \mu + 1)} x^\mu Z_m^\mu(x; k), \quad \mu > -1, \qquad (1.31)$$

$$I^\nu [z^\mu Z_m^\mu(z; k)] = \frac{\Gamma(km + \mu + 1)}{\Gamma(km + \mu + \nu + 1)} z^{\mu+\nu} Z_m^{\mu+\nu}(z; k), \qquad (1.32)$$

$Re(\mu) > -1$, $Re(\nu) > -Re(\mu + 1)$, where for suitable f and complex ν, $I^\nu f(x)$ denotes the νth-order fractional integral (or fractional derivative) of $f(x)$. He also obtained

$$x^{k(\gamma-1)} Z_m^\mu(x; k) = \frac{\Gamma(km + \mu + 1)}{(\gamma)_m 2\pi i} \int_C \frac{t^{m+\gamma-1} E_{k, \mu+1}^\gamma(x^k - t)}{(t - x^k)^{m+1}} dt, \qquad (1.33)$$

where C is a circle: $|t - x^k| = \epsilon$, for small radius ϵ.

In 1976, Karande and Thakare [3], studied the relation

$$\sigma \left[\frac{Z_m^\mu(z; k)}{(1 + \mu)_{km}} \right] = -\frac{Z_{m-1}^\mu(z; k)}{k^k (1 + \mu)_{k(m-1)}}, \qquad (1.34)$$

where $\sigma = \frac{z^{-k+1}}{k} D \prod_{i=1}^{k} \left(\frac{\theta}{k} + \frac{\mu + i}{k} - 1 \right), \theta = zD = z\frac{d}{dz}$.

Orthogonal property of Konhauser polynomials is given by Konhauser [5] as

$$\int_0^\infty e^{-z} z^\mu Z_m^\mu(z; k) Y_n^\mu(z; k) dz = \frac{\Gamma(km + \mu + 1)}{\Gamma(m + 1)} \delta_{mn}, \qquad \forall m, n \in \{0, 1, 2, ...\} \quad (1.35)$$

where δ_{mn} is Kronecker's delta.

Srivastava [10] derived summation formulae as

$$\sum_{m=0}^{\infty} Z_m^{\mu}(x; k) \frac{\left(\frac{y}{k}\right)^{km} t^m}{(1+\mu)_{km}} = exp\left(\left(\frac{y}{k}\right)^k t\right) {}_0F_k\left[-; \frac{\mu+1}{k}, \frac{\mu+2}{k}, ..., \frac{\mu+k}{k}; -\left(\frac{xy}{k^2}\right)^k t\right], \quad (1.36)$$

and

$$\sum_{m=0}^{\infty} Z_m^{\mu}(x; k) \frac{\left(\frac{y}{k}\right)^{km} t^m}{(1+\mu)_{km}} = exp\left(\left\{\left(\frac{y}{k}\right)^k - \left(\frac{x}{k}\right)^k\right\} t\right) \sum_{m=0}^{\infty} Z_m^{\mu}(y; k) \frac{\left(\frac{x}{k}\right)^{km} t^m}{(1+\mu)_{km}} \quad (1.37)$$

Srivastava [11] proved bilateral generating function as

$$\sum_{m=0}^{\infty} \frac{m! Z_m^{\mu}(x; k) Y_m^{\alpha-lm}(y; l) t^m}{\Gamma(\mu+km+1)} = (1+t)^{-1+\frac{\alpha+1}{l}} e^{(x[1-(1+t)^{\frac{1}{l}}])} H\left[x(1+t)^{\frac{1}{l}}, \frac{-y^k t}{(1+t)}\right], \quad (1.38)$$

where $H[x, t] = \sum_{m=0}^{\infty} \frac{Y_m^{\alpha-lm}(x; l) t^m}{\Gamma(\mu+km+1)}$.

Srivastava [12] studied generating function for Konhauser polynomials as

$$\sum_{m=0}^{\infty} Z_m^{\mu}(z; k) \frac{(\gamma)_m t^m}{(1+\mu)_{km}} = (1-t)^{-\gamma} {}_1F_k\left[\gamma; \frac{\mu+1}{k}, \frac{\mu+2}{k}, ..., \frac{\mu+k}{k}; \left(\frac{x}{k}\right)^k \frac{t}{t-1}\right] \quad (1.39)$$

and a summation formula as

$$\sum_{m=0}^{\infty} \binom{m+n}{m} \frac{Z_{m+n}^{\mu}(z; k) t^m}{(1+\mu)_{k(m+n)}} = \sum_{m=n}^{\infty} \binom{m}{n} \frac{t^{m-n}}{m!} \frac{(-z^k)^m}{(1+\mu)_{km}} {}_1F_1[m+1; m-n+1; t] \quad (1.40)$$

In 1981, Karande and Patil [2] obtained double integral in the form of orthogonal property

$$\int_0^{\infty} \int_0^{\infty} e^{-(x+y)} x^{\mu} y^{\nu} Z_m^{\mu+\nu+1}(x+y; k) Y_n^{\mu+\nu+1}(x+y; k) dx dy,$$

$$= \begin{cases} \dfrac{\Gamma(1+\mu)\Gamma(1+\nu)(\mu+\nu+2)_{km}}{m!} & \text{if } n = m \\ 0 & \text{if } m \neq n. \end{cases} \quad (1.41)$$

2 Mixed Recurrence Relations

Consider (1.18)

$$x D Z_m^{\mu}(z; k) = mk Z_m^{\mu}(z; k) - k(km - k + \mu + 1)_k Z_{m-1}^{\mu}(z; k),$$

this can be written as

$$x D \frac{\Gamma(km + \mu + 1)}{\Gamma(m + 1)} E_{k,\mu+1}^{-m}(z^k) = mk \frac{\Gamma(km + \mu + 1)}{\Gamma(m + 1)} E_{k,\mu+1}^{-m}(z^k)$$
$$- k \frac{\Gamma(km - k + \mu + 1 + k)}{\Gamma(km - k + \mu + 1)} \frac{\Gamma(km - k + \mu + 1)}{\Gamma(m)} E_{k,\mu+1}^{-m+1}(z^k),$$

i.e., $\quad x D E_{k,\mu+1}^{-m}(z^k) = mk[E_{k,\mu+1}^{-m}(z^k) - E_{k,\mu+1}^{-m+1}(z^k)].$ \hfill (2.1)

From (1.20), we have

$$\sum_{i=0}^{k} \binom{k}{i} [D^{k-i} z^{\mu+1}][D^{i+1} Z_m^{\mu}(z; k)] = -kz^{\mu}(km - k + \mu + 1)_k Z_{m-1}^{\mu}(z; k),$$

this gives

$$\sum_{i=0}^{k} \binom{k}{i} [D^{k-i} z^{\mu+1}][D^{i+1} \frac{\Gamma(km + \mu + 1)}{\Gamma(m + 1)} E_{k,\mu+1}^{-m}(z^k)]$$
$$= -kz^{\mu} \frac{\Gamma(km - k + \mu + 1 + k)}{\Gamma(km - k + \mu + 1)} \frac{\Gamma(km - k + \mu + 1)}{\Gamma(m)} E_{k,\mu+1}^{-m+1}(z^k),$$

i.e., $\quad \displaystyle\sum_{i=0}^{k} \binom{k}{i} [D^{k-i} z^{\mu+1}][D^{i+1} E_{k,\mu+1}^{-m}(z^k)] = -kmz^{\mu} E_{k,\mu+1}^{-m+1}(z^k).$ \hfill (2.2)

Now, consider (1.22),

$$D Z_m^{\mu}(z; k) = -kz^{k-1} Z_{m-1}^{\mu+k}(z; k),$$

this reduces to

$$D \frac{\Gamma(km + \mu + 1)}{\Gamma(m + 1)} E_{k,\mu+1}^{-m}(z^k) = -kz^{k-1} \frac{\Gamma(km - k + \mu + k + 1)}{\Gamma(m)} E_{k,\mu+k+1}^{-m+1}(z^k),$$

the simplification gives

$$D E_{k,\mu+1}^{-m}(z^k) = -kmz^{k-1} E_{k,\mu+k+1}^{-m+1}(z^k).$$ \hfill (2.3)

Consider (1.23),

$$(z^{1-k} D)^p Z_m^{\mu}(z; k) = (-k)^p Z_{m-p}^{\mu+kp}(z; k),$$

this leads to

$$(z^{1-k} D)^p E_{k,\mu+1}^{-m}(z^k) = (-k)^p \frac{\Gamma(m+1)}{\Gamma(m-p+1)} E_{k,\mu+kp+1}^{-m+p}(z^k). \qquad (2.4)$$

Consider (1.24)

$$zDZ_m^\mu(z;k) = (mk+\mu)Z_m^{\mu-1}(z;k) - \mu Z_m^\mu(z;k),$$

this gives

$$zDE_{k,\mu+1}^{-m}(z^k) = E_{k,\mu}^{-m}(z^k) - \mu E_{k,\mu+1}^{-m}(z^k). \qquad (2.5)$$

Consider (1.31)

$$D^n \left[x^{\mu+n} Z_m^{\mu+n}(x;k) \right] = \frac{\Gamma(km+\mu+n+1)}{\Gamma(km+\mu+1)} x^\mu Z_m^\mu(x;k)$$

with $Re(\mu) > -1$.
   Using (1.7), this reduces to

$$D^n \left[x^{\mu+n} E_{k,\mu+n+1}^{-m}(z^k) \right] = x^\mu E_{k,\mu+1}^{-m}(z^k). \qquad (2.6)$$

Consider (1.32)

$$I^\nu[z^\mu Z_m^\mu(z;k)] = \frac{\Gamma(km+\mu+1)}{\Gamma(km+\mu+\nu+1)} z^{\mu+\nu} Z_m^{\mu+\nu}(z;k).$$

Using (1.7), this can be written as

$$I^\nu[z^\mu E_{k,\mu+1}^{-m}(z^k)] = z^{\mu+\nu} E_{k,\mu+\nu}^{-m}(z^k). \qquad (2.7)$$

From (1.34), we have

$$\sigma \left[\frac{Z_m^\mu(z;k)}{(1+\mu)_{km}} \right] = -\frac{Z_{m-1}^\mu(z;k)}{k^k(1+\mu)_{k(m-1)}}.$$

this gives

$$\sigma \left[\frac{E_{k,\mu+1}^{-m}(z^k)\Gamma(\mu+1)}{\Gamma(m+1)} \right] = -\frac{\Gamma(km-k+\mu+1)E_{k,\mu+1}^{-m+1}(z^k)}{\Gamma(m)k^k(1+\mu)_{k(m-1)}},$$

this reduces to

$$\sigma \left[E_{k,\mu+1}^{-m}(z^k) \right] = -\frac{m}{k^k} E_{k,\mu+1}^{-m+1}(z^k). \qquad (2.8)$$

3 Pure Recurrence Relations

In this section, the authors obtained some recurrence relations.

Consider (1.25)

$$Z_m^\mu(z;k) - Z_m^{\mu-1}(z;k) = \frac{k\Gamma(km+\mu)}{\Gamma(k(m-1)+\mu+1)} Z_{m-1}^\mu(z;k),$$

this follows

$$(km+\mu)E_{k,\mu+1}^{-m}(z^k) - E_{k,\mu}^{-m}(z^k) = km E_{k,\mu+1}^{-m+1}(z^k). \tag{3.1}$$

Consider (1.26)

$$z^k Z_m^{\mu+k}(z;k) = (km+\mu+1)_k Z_m^\mu(z;k) - (m+1)Z_{m+1}^\mu(z;k),$$

this leads to

$$z^k E_{k,\mu+k+1}^{-m}(z^k) = E_{k,\mu+1}^{-m}(z^k) - E_{k,\mu+1}^{-m-1}(z^k). \tag{3.2}$$

Consider (1.27)

$$kz^k Z_m^{\mu+k}(z;k) = \mu Z_{m+1}^\mu(z;k) - (km+\mu+k)Z_{m+1}^{\mu-1}(z;k),$$

this gives

$$(m+1)kz^k E_{k,\mu+k+1}^{-m}(z^k) = \mu E_{k,\mu+1}^{-m-1}(z^k) - E_{k,\mu}^{-m-1}(z^k). \tag{3.3}$$

4 Differential Equation

Consider (1.19)

$$D^k[z^{\mu+1} DZ_m^\mu(z;k)] = z^{\mu+1} DZ_m^\mu(z;k) - kmz^\mu Z_m^\mu(z;k).$$

Using (1.7), this follows

$$D^k[z^{\mu+1} DE_{k,\mu+1}^{-m}(z^k)] = z^{\mu+1} DE_{k,\mu+1}^{-m}(z^k) - mkz^\mu E_{k,\mu+1}^{-m}(z^k). \tag{4.1}$$

5   Finite Summation Formulae

From (1.10) and (1.15), we have

$$
E_{k,\beta+1}^{-m,q}(z^k) = \left(\frac{z}{y}\right)^{km} \sum_{r=0}^{m} \binom{m}{r} \left[\left(\frac{y}{z}\right)^k - 1\right]^r E_{k,\beta+1}^{-m+r,q}(y^k), \qquad (5.1)
$$

where $k \in \mathbb{N}$ and $\beta \in \mathbb{C}$ with $\mathrm{Re}(\beta) > -1$.

Now, consider (1.21)

$$
Z_m^{\mu}(z; k) = \left(\frac{z}{y}\right)^{km} \sum_{r=0}^{m} \binom{\mu+km}{kr} \frac{(kr)!}{r!} \left[\left(\frac{y}{z}\right)^k - 1\right]^r Z_{m-r}^{\mu}(y; k).
$$

Using (1.7), this can be written as

$$
E_{k,\mu+1}^{-m}(z^k) = \left(\frac{z}{y}\right)^{km} \frac{m!}{(km+\mu)!} \sum_{r=0}^{m} \frac{(\mu+km)!}{(kr)!(\mu+km-kr)!} \frac{(kr)!}{r!}
$$
$$
\left[\left(\frac{y}{z}\right)^k - 1\right]^r \frac{\Gamma(km-kr+\mu+1)}{\Gamma(m-r+1)} E_{k,\mu+1}^{-m+r}(y^k).
$$

Finally, we arrived at

$$
E_{k,\mu+1}^{-m}(z^k) = \left(\frac{z}{y}\right)^{km} \sum_{r=0}^{m} \binom{m}{r} \left[\left(\frac{y}{z}\right)^k - 1\right]^r E_{k,\mu+1}^{-m+r}(y^k). \qquad (5.2)
$$

Consider (1.28)

$$
Z_m^{\mu}(\delta z; k) = \sum_{j=0}^{m} \binom{\mu+km}{kj} \frac{(kj)!}{j!} \delta^{k(m-j)} (1-\delta^k)^j Z_{m-j}^{\mu}(z; k).
$$

By using (1.7), this leads to

$$
E_{k,\,\mu+1}^{-m}((\delta z)^k) = \sum_{j=0}^{m} \frac{\Gamma(m+1)}{j!\Gamma(m-j+1)} \delta^{k(m-j)} (1-\delta^k)^j E_{k,\mu+1}^{-m+j}(z^k),
$$

this follows:

$$
E_{k,\,\mu+1}^{-m}((\delta z)^k) = \sum_{j=0}^{m} \binom{m}{j} \delta^{k(m-j)} (1-\delta^k)^j E_{k,\mu+1}^{-m+j}(z^k). \qquad (5.3)
$$

6 Integral Representation and Orthogonal Property

Keeping $\alpha = k \in \mathbb{N}$ and replacing t by t^k in (1.11), we have

$$L_{\left[\frac{m}{q}\right]}^{(k,\beta)}(t^k) = (km + \beta) \int_0^1 z^{\beta-1} L_{\left[\frac{m}{q}\right]}^{(k,\beta-1)}((tz)^k) dz.$$

By using (1.10), this reduces to

$$E_{k,\beta+1}^{-m,q}(t^k) = \int_0^1 t^{\beta-1} E_{k,\beta}^{-m,q}((tz)^k) dz. \tag{6.1}$$

Consider (1.12)

$$(z-t)^\beta L_{\left[\frac{m}{q}\right]}^{(\alpha,\beta)}\left((z-t)^\alpha\right) = \frac{\Gamma(\alpha m + \beta + 1)}{\Gamma(\alpha m + \beta - \gamma + 1)\Gamma(\gamma)} \times$$

$$\int_t^z (z-u)^{\gamma-1}(u-t)^{\beta-\gamma} L_{\left[\frac{m}{q}\right]}^{(\alpha,\beta-\gamma)}\left((u-t)^\alpha\right) du, \tag{6.2}$$

where $\beta, \gamma \in \mathbb{C}$ with $\text{Re}(\beta) > \text{Re}(\gamma) > -1$ and $\alpha \in \mathbb{N}$.
Using (1.10), this gives

$$(z-t)^\beta E_{k,\beta+1}^{-m,q}\left((z-t)^\alpha\right) = \frac{1}{\Gamma(\gamma)} \int_t^z (z-u)^{\gamma-1}(u-t)^{\beta-\gamma} E_{k,\beta-\gamma+1}^{-m,q}\left((u-t)^\alpha\right) du. \tag{6.3}$$

Let C is a circle $|\, t - x^k| = \epsilon$ for small radius ϵ, consider (1.33)

$$x^{k(\gamma-1)} Z_m^\mu(x; k) = \frac{\Gamma(km + \mu + 1)}{(\gamma)_m 2\pi i} \int_C \frac{t^{m+\gamma-1} E_{k,\mu+1}^\gamma(x^k - t)}{(t - x^k)^{m+1}} dt.$$

Using (1.7), the above equation can be written in the form

$$x^{k(\gamma-1)} E_{k,\mu+1}^{-m}(z^k) = \frac{\Gamma(m + 1)}{(\gamma)_m 2\pi i} \int_C \frac{t^{m+\gamma-1} E_{k,\mu+1}^\gamma(x^k - t)}{(t - x^k)^{m+1}} dt. \tag{6.4}$$

From (1.35) and (1.7), we have

$$\int_0^\infty e^{-z} z^\mu E_{k,\mu+1}^{-m}(z^k) Y_n^\mu(z;k) dz = \delta_{mn}, \quad \forall m, n \in \{0, 1, 2, ...\}. \tag{6.5}$$

Consider (1.41)

$$\int_0^\infty \int_0^\infty e^{-(x+y)} x^\mu y^\nu Z_m^{\mu+\nu+1}(x+y;k) Y_n^{\mu+\nu+1}(x+y;k) dxdy$$

$$= \begin{cases} \dfrac{\Gamma(1+\mu)\Gamma(1+\nu)(\mu+\nu+2)_{km}}{m!} & \text{if } n = m \\ 0 & \text{if } m \neq n \end{cases}$$

By using (1.7), this immediately follows:

$$\int_0^\infty \int_0^\infty e^{-(x+y)} x^\mu y^\nu E_{k,\mu+\nu+2}^{-m}((x+y)^k) Y_n^{\mu+\nu+1}(x+y;k) dxdy$$

$$= \begin{cases} \mathfrak{B}(1+\mu, 1+\nu) & \text{if } n = m \\ 0 & \text{if } m \neq n \end{cases}$$

where $\mathfrak{B}(m, n)$ is the usual beta function.

7 Laplace Transform

From equation (1.29), we have

$$L\{t^\nu Z_m^\mu(zt;k); s\} = \frac{(\mu+1)_{km}\Gamma(\nu+1)}{s^{\nu+1}m!} {}_{k+1}F_k \left[\begin{matrix} -m, \frac{\nu+1}{k}, \frac{\nu+2}{k}, ..., \frac{\nu+k}{k}; \\ \frac{\mu+1}{k}, \frac{\mu+2}{k}, ..., \frac{\mu+k}{k}; \end{matrix} \left(\frac{z}{s}\right)^k \right].$$

This immediately leads to

$$L\left\{t^\nu E_{k,\mu+1}^{-m}((zt)^k); s\right\} = \frac{\Gamma(\nu+1)}{\Gamma(\mu+1)s^{\nu+1}} {}_{k+1}F_k \left[\begin{matrix} -m, \frac{\nu+1}{k}, \frac{\nu+2}{k}, ..., \frac{\nu+k}{k}; \\ \frac{\mu+1}{k}, \frac{\mu+2}{k}, ..., \frac{\mu+k}{k}; \end{matrix} \left(\frac{z}{s}\right)^k \right].$$

If $\mu = \nu$, then the above equation follows:

$$L\left\{t^\nu E_{k,\nu+1}^{-m}((zt)^k); s\right\} = \frac{(s^k - z^k)^n}{s^{km+\nu+1}}.$$

If $k = 1$, then the above results reduces in the form of Laguerre polynomials.

8 Generating Functions

Keeping $\alpha = k \in \mathbb{N}$ and replacing z by z^k in (1.13) gives

$$\sum_{m=0}^{\infty} \frac{(\gamma)_m L_{\left[\frac{m}{q}\right]}^{(k,\beta)}(z^k) t^m}{\Gamma(km + \beta + 1)} = (1-t)^{-\gamma} E_{k,\beta+1}^{\gamma,q}\left(\frac{z^k(-t)^q}{(1-t)^q}\right), \quad |t| < 1. \tag{8.1}$$

By using (1.10), this gives

$$\sum_{m=0}^{\infty} (\gamma)_m E_{k,\beta+1}^{-m,q}(z^k) \frac{t^m}{m!} = (1-t)^{-\gamma} E_{k,\beta+1}^{\gamma,q}\left(\frac{z^k(-t)^q}{(1-t)^q}\right), \quad |t| < 1 \tag{8.2}$$

where $\beta, \gamma \in \mathbb{C}$ with $Re(\beta) > -1$ and $k, q \in \mathbb{N}$.
Keeping $\alpha = k \in \mathbb{N}$ then (1.14) yields

$$\sum_{m=0}^{\infty} L_{\left[\frac{m}{q}\right]}^{(k,\beta)}(z^k) \frac{t^m}{\Gamma(km + \beta + 1)} = e^t W(k; \beta + 1; z^k(-t)^q), \tag{8.3}$$

where $\beta \in \mathbb{C}$ with $Re(\beta) > -1$ and $k, q \in \mathbb{N}$,
Using (1.10), (8.3) can be written in the form

$$\sum_{m=0}^{\infty} E_{k,\beta+1}^{-m,q}(z^k) \frac{t^m}{m!} = e^t W(k; \beta + 1; z^k(-t)^q). \tag{8.4}$$

Again from (1.38) and (1.7), we have

$$\sum_{m=0}^{\infty} E_{k,\mu+1}^{-m}(z^k) Y_m^{\alpha-lm}(y; l) t^m = (1+t)^{-1+\frac{\alpha+1}{l}} e^{(x[1-(1+t)^{\frac{1}{l}}])} H\left[x(1+t)^{\frac{1}{l}}, \frac{-y^k t}{(1+t)}\right],$$

where $H[x, t] = \sum_{m=0}^{\infty} \frac{Y_m^{\alpha-lm}(x; l) t^m}{\Gamma(\mu + km + 1)}.$
From (1.39) and (1.7), we get

$$\sum_{m=0}^{\infty} E_{k,\mu+1}^{-m}(z^k)(\gamma)_m t^m = \frac{(1-t)^{-\gamma}}{\Gamma(1+\mu)} {}_1F_k\left[\gamma; \frac{\mu+1}{k}, \frac{\mu+2}{k}, ..., \frac{\mu+k}{k}; \left(\frac{x}{k}\right)^k \frac{t}{t-1}\right]. \tag{8.5}$$

9 Miscellaneous

Equations (1.17) and (1.7) gives

$$E_{k,\mu+1}^{-m}(z^k) = x^{-\mu}{}_k H_m^{\mu}(0, x; 1).$$

Consider (1.36)

$$\sum_{m=0}^{\infty} Z_m^{\mu}(x; k) \frac{\left(\frac{y}{k}\right)^{km} t^m}{(1+\mu)_{km}} = exp\left(\left(\frac{y}{k}\right)^k t\right) {}_0F_k\left[-; \frac{\mu+1}{k}, \frac{\mu+2}{k}, ..., \frac{\mu+k}{k}; -\left(\frac{xy}{k^2}\right)^k t\right].$$

By using (1.7), the above equation can be written as

$$\sum_{m=0}^{\infty} E_{k,\mu+1}^{-m}(z^k)\left(\frac{y}{k}\right)^{km} \frac{\Gamma(\mu+1)t^m}{m!} = exp\left(\left(\frac{y}{k}\right)^k t\right) {}_0F_k\left[-; \frac{\mu+1}{k}, \frac{\mu+2}{k}, ..., \frac{\mu+k}{k}; -\left(\frac{xy}{k^2}\right)^k t\right].$$

From (1.37) and (1.7), we have

$$\sum_{m=0}^{\infty} E_{k,\mu+1}^{-m}(x^k)\left(\frac{y}{k}\right)^{km} \frac{t^m}{m!} = exp\left(\left\{\left(\frac{y}{k}\right)^k - \left(\frac{x}{k}\right)^k\right\} t\right) \sum_{m=0}^{\infty} E_{k,\mu+1}^{-m}(y^k)\left(\frac{x}{k}\right)^{km} \frac{t^m}{m!}. \quad (9.1)$$

Consider (1.40)

$$\sum_{m=0}^{\infty} \binom{m+n}{m} \frac{Z_{m+n}^{\mu}(z; k)t^m}{(1+\mu)_{k(m+n)}} = \sum_{m=n}^{\infty} \binom{m}{n} \frac{t^{m-n}}{m!} \frac{(-z^k)^m}{(1+\mu)_{km}} {}_1F_1[m+1; m-n+1; t].$$

Using (1.7), the above equation yields

$$\sum_{m=0}^{\infty} E_{k,\mu+1}^{-(m+n)}(z^k) \frac{t^m}{m!} = \sum_{m=n}^{\infty} \frac{t^{m-n}}{(m-n)!} \frac{(-z^k)^m}{\Gamma(km+\mu+1)} {}_1F_1[m+1; m-n+1; t].$$

The simplification gives

$$\sum_{m=0}^{\infty} E_{k,\mu+1}^{-(m+n)}(z^k) \frac{t^m}{m!} = \sum_{j=0}^{\infty} {}_1F_1[j+n+1; j+1; t] \frac{(-z^k)^{j+n}}{\Gamma(k(j+n)+\mu+1)} \frac{t^j}{j!}. \quad (9.2)$$

Acknowledgements This work was supported to third author [S Jain] by the Science & Engineering Research Board (SERB), India (No: MTR/2017/000194) and fifth author's [P Agarwal] research grant is supported by the Department of Science & Technology(DST), India (No: INT/RUS/RFBR/P-308) and Science & Engineering Research Board (SERB), India (No: TAR/2018/000001).

References

1. Bin-Saad, M.G., New, A.: Class of Hermite-Konhauser polynomials together with differential equations. Kyungpook Math. J. **50**, 237–253 (2010)
2. Karande, B.K., Patil, K.R.: Note on Konhauser Biorthogonal polynomials. Indian J. pure appl. Math. **12**(2), 222–225 (1981)
3. Karande, B.K., Thakare, N.K.: Some Results for Konhauser-Biorthogonal polynomials and dual series equations. Indian J. Pure Appl. Math. **7**(6), 635–646 (1976)
4. Kilbas, A.A., Saigo, M., Saxsena, R.K.: Generalized Mittag-Leffler function and generalized fractional calculus operators. Integral Transforms Spec. Funct. **15**(1), 31–49 (2004)
5. Konhauser, J.D.E.: Biorthogomal polynomials suggested by the laguerre polynomials. Pacific J. Math. **21**(2), 303–314 (1967)
6. Prabhakar, T.R.: On a set of polynomials suggested by laguerre polynomials. Pacific J. Math. **35**(1), 213–219 (1970)
7. Prajapati, J.C., Ajudia, N.K., Agarwal, P.: Some results due to Konhauser polynomials of first kind and laguerre polynomials. Appl. Math. Comput. **247**, 639–650 (2014)
8. Rainville, E.D.: Special Functions. The Macmillan Company, New York (1960)
9. Shukla, A.K., Prajapati, J.C.: On a generalization of Mittag-Leffler function and its properties. J. Math. Anal. Appl. **336**(2), 779–811 (2007)
10. Srivastava, H.M.: On the Konhauser sets of Biorthogonal polynomials suggested by the laguerre polynomials. Pacific J. Math. **49**(2), 489–492 (1973)
11. Srivastava, H.M.: A Note on the Konhauser Sets of Biorthogonal polynomials suggested by the laguerre polynomials. Pacific J. Math. **90**(1), 197–200 (1980)
12. Srivastava, H.M.: Some Biorthogonal polynomials suggested by the Laguerre polynomials. Pacific J. Math. **98**(1), 235–250 (1982)

An Effective Numerical Technique Based on the Tau Method for the Eigenvalue Problems

Maryam Attary and Praveen Agarwal

Abstract We consider the (presumably new) effective numerical scheme based on the Legendre polynomials for an approximate solution of eigenvalue problems. First, a new operational matrix, which can be represented by a sparse matrix defined by using the Tau method and orthogonal functions. Sparse data is by nature more compressed and thus requires significantly less storage. A comparison of the results for some examples reveals that the presented method is convenient and effective, also we consider the problem of column buckling to show the validity of the proposed method.

Keywords Eigenvalue problems · Legendre polynomials · Numerical treatment

Mathematics Subject Classifications 65L15, 65L05, 65L10, 65N35.

1 Introduction

A special class of boundary-value problems are eigenvalue problems. They are used in a wide variety of engineering contexts beyond boundary-value problems and play a very important role in many scientific fields such as vibrations, elasticity, and other oscillating systems. A simple context to illustrate how eigenvalues occur in physical problems is the mass–spring system. Detailed description and application of these problems may be found in [3] and references therein.

M. Attary
Department of Mathematics, Karaj Branch, Islamic Azad University, Karaj, Iran

P. Agarwal (✉)
Department of Mathematics, ANAND International College of Engineering, Jaipur 303012, India
e-mail: goyal.praveen2011@gmail.com

Department of Mathematics, Harish Chandra Research Institute, Chhatnag Road, Jhunsi 211019, Allahabad, India

International Center for Basic and Applied Sciences, Jaipur 302029, India

© Springer Nature Singapore Pte Ltd. 2019
P. Agarwal et al. (eds.), *Fractional Calculus*, Springer Proceedings in Mathematics & Statistics 303, https://doi.org/10.1007/978-981-15-0430-3_12

The numerical solvability of eigenvalue problems and other related equations has been considered by several authors. In [5], a software package has been introduced and discussed, which deals with the computation of the eigenvalue of Strum–Liouville problems. Gamel et al. [4] were concerned with the Chebyshev method for solving eigenvalue problems of fourth order of ODEs.

Due to the good approximation properties of spectral methods, these methods have been discussed intensively in recent years. A special case of them is the Tau method, which has been applied for the numerical solution of many operator equations.

In this paper, we intend to introduce a new Tau approach by using Legendre polynomials to solve the following eigenvalue problems:

$$u^{(k)}(x) + \lambda^2 u(x) = 0, \qquad k = 2 \ or \ 4, \tag{1.1}$$

$$\sum_{r=0}^{k-1} \alpha_{j,r} u^{(r)}(c_r) = 0, \qquad j = 1, ..., k. \tag{1.2}$$

s.t.

$$c_r = \begin{cases} -1, & r = 0, \\ 1, & o.w. \end{cases}$$

where $\alpha_{j,r} \in \mathbb{R}$ are constants, $(r = 0, \cdots, k-1)$ and λ is a solution to be determined.

The rest of the article is organized as follows: In Sect. 2, we describe some preliminaries of the Legendre polynomials and their properties. Section 3 explains the new scheme and introduces matrix representation of the method for problem (1.1). To clarify the efficiency of the method, the proposed algorithm is applied to some numerical experiments and also the obtained results are compared with some existing methods in the literature.

2 Basic Definitions of the Legendre Polynomials

We recall some definitions, which are required for the present study (see, for example, [1]).

Definition 2.1 The starting point of Legendre polynomials is Rodrigues formula, which is introduced as

$$P_n(x) = \frac{1}{2^n n!} \frac{d^n}{dx^n} (x^2 - 1)^n,$$

the orthogonality of Legendre polynomials in $[-1, 1]$ with $r(x) = 1$ can be shown.

Definition 2.2 The Legendre polynomials $P_n(x)$ satisfy the recurrence relation:

$$P_{n+1}(x) = \frac{1}{n+1} [(2n+1)x P_n(x) - n P_{n-1}(x)], \qquad n = 1, 2, \cdots. \tag{2.1}$$

Also, Legendre polynomials $P_n(x)$ can be represented in the following recurrence form:

$$P_0(x) = P_1'(x),$$

$$P_n(x) = \frac{1}{2n+1}[P_{n+1}'(x) - P_{n-1}'(x)], \qquad n = 1, 2, \cdots. \qquad (2.2)$$

Using Rodrigues' formula, the orthogonality of Legendre polynomials can be obtained as follow

$$\int_{-1}^{1} P_m(x)P_n(x)dx = \begin{cases} 0, & m \neq n, \\ \frac{2}{2n+1} & m = n. \end{cases}$$

Since the Legendre polynomials are defined on the interval $[-1, 1]$, for using these polynomials on the interval $[a, b]$, we convert it to $[-1, 1]$ by introducing $x = \frac{b-a}{2}t + \frac{b+a}{2}$.

3 Numerical Treatment of the Problem

In this section, we replace the differential part of Eq. (1.1) by an operational matrix. Our main result is asserted by Theorem 1.

Theorem 1 *Let $P_i(x)$ be a Legendre polynomials in $[-1, 1]$. Suppose that functions $u(x)$ and $u'(x)$ can be expressed as*

$$u(x) = \sum_{i=0}^{\infty} a_i P_i(x) = \mathbf{a}\mathbf{P}_x, \qquad (3.1)$$

$$u'(x) = \sum_{i=0}^{\infty} b_{1,i} P_i(x) = \mathbf{b}_1\mathbf{P}_x, \qquad (3.2)$$

where $\mathbf{a} = [a_0, a_1, a_2, \ldots]^T$, $\mathbf{b}_1 = [b_{1,0}, b_{1,1}, b_{1,2}, \ldots]^T$ and $\mathbf{P}_x = [P_0, P_1, P_2, \cdots]$. Then we have:

$$u'(x) = \mathbf{Map}_x, \qquad (3.3)$$

where

$$M = \begin{bmatrix} 0 & 1 & 0 & 1 & 0 & 1 & 0 & \cdots \\ 0 & 0 & 3 & 0 & 3 & 0 & 3 & \cdots \\ 0 & 0 & 0 & 5 & 0 & 5 & 0 & \cdots \\ 0 & 0 & 0 & 0 & 7 & 0 & 7 & \cdots \\ \vdots & \vdots & \vdots & \vdots & \vdots & \vdots & \vdots & \ddots \end{bmatrix}.$$

Proof Taking the derivative of (3.1) and due to (3.2), we can write

$$\sum_{i=1}^{\infty} a_i P_i'(x) = \sum_{i=0}^{\infty} b_{1,i} P_i(x) = b_{1,0} P_0(x) + \sum_{i=1}^{\infty} b_{1,i} P_i(x). \tag{3.4}$$

Using (2.2), we rewrite the above relation as

$$\sum_{i=1}^{\infty} a_i P_i'(x) = b_{1,0} P_1'(x) + \sum_{i=1}^{\infty} \frac{b_{1,i}}{2i+1}[P_{i+1}'(x) - P_{i-1}'(x)]. \tag{3.5}$$

Therefore

$$a_1 = b_{1,0} - \frac{b_{1,2}}{5}, \quad a_2 = \frac{b_{1,1}}{3} - \frac{b_{1,3}}{7}, \ldots, \quad a_n = \frac{b_{1,n-1}}{2n-1} - \frac{b_{1,n+1}}{2n+3}, \ldots, \tag{3.6}$$

and so

$$b_{1,n} = \sum_{i=n+1}^{\infty} (2n+1)\lambda_{n,i} a_i, \quad s.t. \ \lambda_{n,i} = \begin{cases} 1, & i+n \ odd, \\ 0, & o.w. \end{cases} \tag{3.7}$$

(3.7) can be transformed to the following matrix form:

$$\mathbf{b}_1 = \mathbf{Ma}, \tag{3.8}$$

where

$$\mathbf{M} = \begin{bmatrix} 0 & 1 & 0 & 1 & 0 & 1 & 0 & \cdots \\ 0 & 0 & 3 & 0 & 3 & 0 & 3 & \cdots \\ 0 & 0 & 0 & 5 & 0 & 5 & 0 & \cdots \\ 0 & 0 & 0 & 0 & 7 & 0 & 7 & \cdots \\ \vdots & \vdots & \vdots & \vdots & \vdots & \vdots & \vdots & \ddots \end{bmatrix}.$$

Due to the last equation, (3.2) can be written as

$$u'(x) = \mathbf{b}_1 \mathbf{P}_x = \mathbf{Ma}\mathbf{P}_x. \tag{3.9}$$

□

Lemma 1 *Let* $u^{(n)}(x) = \sum_{i=0}^{\infty} b_{n,i} P_i(x) = \mathbf{b}_n \mathbf{P}_x$, *be a Legendre polynomiyal with* $\mathbf{b}_n = [b_{n,0}, b_{n,1}, b_{n,2}, \ldots]^T$, *and* $\mathbf{M}$ *is a matrix, which is defined in Theorem1. Then we have*

$$u^{(n)}(x) = \mathbf{M}^n \mathbf{a}\mathbf{P}_x. \tag{3.10}$$

Proof According to Theorem 1, the validity of Lemma 1 for $n = 1$ is obvious. From (2.2), we can write

$$p'_i(x) = \frac{1}{2i+1}[P_{i+1}''(x) - P_{i-1}''(x)].$$

Given the assumption, it follows that

$$u^{(2)}(x) = \sum_{i=0}^{\infty} b_{2,i} P_i(x) = \mathbf{b}_2 \mathbf{P}_x. \tag{3.11}$$

Using the given scheme in Theorem 1, we conclude

$$\mathbf{b}_2 = \mathbf{M}\mathbf{b}_1, \tag{3.12}$$

Due to (3.8) and the last equation, we get $\mathbf{b}_2 = \mathbf{M}^2\mathbf{a}$. Therefore, by repeating this scheme, it follows that $\mathbf{b}_n = \mathbf{M}^n\mathbf{a}$.
  Finally,

$$u^{(n)}(x) = \mathbf{M}^n\mathbf{a}\mathbf{P}_x. \tag{3.13}$$

$\square$

  We are now ready to obtain the algebraic form of the eigenvalue problems (1.1) based on the operational matrix of the Legendre polynomials. We define $u_m(x)$ as an approximation function of the exact solution $u(x)$ as follows:

$$u_m(x) = \sum_{i=0}^{m} a_i P_i(x) = \mathbf{a}_m \mathbf{P}_{x,m}. \tag{3.14}$$

First, we consider the following form of (1.1):

$$u^{(2)}(x) + \lambda^2 u(x) = 0. \tag{3.15}$$

We define $u_m^{(2)}(x)$ as an approximation function of the exact solution $u^{(2)}(x)$ as follows:

$$u_m^{(2)}(x) = \mathbf{M}_m^2 \mathbf{a}_m \mathbf{P}_{x,m}, \tag{3.16}$$

where $\mathbf{M}_m$ and $\mathbf{M}_m^2$ are finite forms of $\mathbf{M}$ and $\mathbf{M}^2$, respectively.
  Also boundary conditions of (3.15) can be written as

$$\begin{cases} \alpha_{1,0} \sum_{i=0}^{m} a_i P_i(-1) + \alpha_{1,1} \sum_{i=0}^{m} a_i P_i'(1) = 0, \\ \alpha_{2,0} \sum_{i=0}^{m} a_i P_i(-1) + \alpha_{2,1} \sum_{i=0}^{m} a_i P_i'(1) = 0, \end{cases} \tag{3.17}$$

or equivalently

$$\begin{cases} \sum_{i=0}^{m} a_i \underbrace{[\alpha_{1,0} P_i(-1) + \alpha_{1,1} P_i'(1)]}_{d_{i,1}} = 0, \\ \sum_{i=0}^{m} a_i \underbrace{[\alpha_{2,0} P_i(-1) + \alpha_{2,1} P_i'(1)]}_{d_{i,2}} = 0. \end{cases} \tag{3.18}$$

Due to (3.14) and (3.16), Eq. (3.15) rewritten as

$$\mathbf{M}_m^2 \mathbf{a}_m \mathbf{P}_{x,m} + \lambda^2 \mathbf{I}_m \mathbf{a}_m \mathbf{P}_{x,m} = 0. \tag{3.19}$$

Since (3.15) has two boundary conditions, we need to remove the last two equations from (3.19) and replace boundary conditions (3.18) instead of them.

Due to orthogonality of $\{P_i(x)\}_{i=0}^{\infty}$ and using simple computations, we derive

$$\begin{cases} \overline{\mathbf{M}}_m^2 \mathbf{a}_m = -\lambda^2 \overline{\mathbf{I}}_m \mathbf{a}_m, \\ d_{i,1} \mathbf{a}_m = 0, \\ d_{i,2} \mathbf{a}_m = 0, \end{cases} \tag{3.20}$$

where $\overline{\mathbf{M}}_m^2$ and $\overline{\mathbf{I}}_m$ are obtained by removing the last two rows of $\mathbf{M}_m^2$ and $\mathbf{I}_m$, respectively. Also, $\mathbf{I}_m$ is a m+1-dimensionl identity matrix.

(3.20) can be symbolically expressed as

$$\mathbf{H}\mathbf{a}_m = \lambda^2 \mathbf{G}\mathbf{a}_m, \tag{3.21}$$

where $\mathbf{H}$ and $\mathbf{G}$ are defined as

$$\mathbf{H} = \begin{bmatrix} \overline{\mathbf{M}}_m^2 \\ d_{i,1} \\ d_{i,2} \end{bmatrix}, \qquad \mathbf{G} = \begin{bmatrix} -\overline{\mathbf{I}}_m \\ \overline{\mathbf{0}} \\ \overline{\mathbf{0}} \end{bmatrix}. \tag{3.22}$$

In the following, we consider another form of (1.1)

$$u^{(4)}(x) + \lambda^2 u(x) = 0. \tag{3.23}$$

In a similar manner, let $u_m^{(4)}(x)$ be an approximation function of the exact solution $u^{(4)}(x)$ as follows:

$$u_m^{(4)}(x) = \mathbf{M}_m^4 \mathbf{a}_m \mathbf{P}_{x,m}, \tag{3.24}$$

where $\mathbf{M}_m^4$ be a finite form of $\mathbf{M}^4$. by substituting (3.14) and (3.24) in (3.23), we have

$$\mathbf{M}_m^4 \mathbf{a}_m \mathbf{P}_{x,m} + \lambda^2 \mathbf{I}_m \mathbf{a}_m \mathbf{P}_{x,m} = 0. \tag{3.25}$$

Also boundary conditions of (3.23) can be written as

$$
\begin{cases}
\sum_{i=0}^{m} a_i \underbrace{[\alpha_{1,0} P_i(-1) + \alpha_{1,1} P_i'(1) + \alpha_{1,2} P_i''(1) + \alpha_{1,3} P_i'''(1)]}_{e_{i,1}} = 0, \\[2mm]
\sum_{i=0}^{m} a_i \underbrace{[\alpha_{2,0} P_i(-1) + \alpha_{2,1} P_i'(1) + \alpha_{2,2} P_i''(1) + \alpha_{2,3} P_i'''(1)]}_{e_{i,2}} = 0, \\[2mm]
\sum_{i=0}^{m} a_i \underbrace{[\alpha_{3,0} P_i(-1) + \alpha_{3,1} P_i'(1) + \alpha_{3,2} P_i''(1) + \alpha_{3,3} P_i'''(1)]}_{e_{i,3}} = 0, \\[2mm]
\sum_{i=0}^{m} a_i \underbrace{[\alpha_{4,0} P_i(-1) + \alpha_{4,1} P_i'(1) + \alpha_{4,2} P_i''(1) + \alpha_{4,3} P_i'''(1)]}_{e_{i,4}} = 0.
\end{cases}
\tag{3.26}
$$

These boundary conditions should be applied in Eq. (3.25), so we conclude

$$
\begin{cases}
\overline{\mathbf{M}}_m^4 \mathbf{a}_m = -\lambda^2 \overline{\mathbf{I}}_m \mathbf{a}_m, \\
e_{i,1} \mathbf{a}_m = 0, \\
e_{i,2} \mathbf{a}_m = 0, \\
e_{i,3} \mathbf{a}_m = 0, \\
e_{i,4} \mathbf{a}_m = 0,
\end{cases}
\tag{3.27}
$$

or equivalently

$$
\Pi \mathbf{a}_m = \lambda^2 \Theta \mathbf{a}_m,
\tag{3.28}
$$

where Π and Θ are defined as

$$
\Pi = \begin{bmatrix} \overline{\mathbf{M}}_m^4 \\ e_{i,1} \\ e_{i,2} \\ e_{i,3} \\ e_{i,4} \end{bmatrix}, \qquad
\Theta = \begin{bmatrix} -\overline{\mathbf{I}}_m \\ \mathbf{0} \\ \mathbf{0} \\ \mathbf{0} \\ \mathbf{0} \end{bmatrix}.
\tag{3.29}
$$

The same as before, $\overline{\mathbf{M}}_m^4$ and $\overline{\mathbf{I}}_m$ are obtained by removing the last four rows of $\mathbf{M}_m^4$ and $\mathbf{I}_m$, respectively. Finally, due to (3.21) and (3.28), the values of λ can be computed.

4 Numerical Results

In this section, we present some examples to show the accuracy of the proposed method. These examples are solved by Legendre polynomials. Numerical results are compared with some existing numerical methods. Obtained results are reported in Tables 1, 2, 3, 4, and 5.

Example 1 Consider the following second-order eigenvalue problem:

$$
u''(x) + \lambda^2 u(x) = 0,
\tag{4.1}
$$

with the conditions:

$$u(0) = 0,$$

$$u'(1) + u(1) = 0,$$

and the exact solution for λ is $\lambda = -\tan \lambda$.

Here, we consider computational details of the presented method for Example 1. As we pointed out, the Legendre polynomials are defined in the interval $[-1, 1]$. So (4.1), which is stated on the interval $[0, 1]$, will be converted to the interval $[-1, 1]$ by choosing $x = \frac{1}{2}(t + 1)$ or $t = 2x - 1$.

Therefore, above example can be written as

$$\begin{cases} (\frac{dt}{dx})^2 \frac{d^2u}{dt^2} + \lambda^2 u = 0, \\ u(-1) = 0, \\ ((\frac{dt}{dx})\dfrac{du}{dt})(1) + u(1) = 0, \end{cases}$$

or

$$\begin{cases} 4\frac{d^2u}{dt^2} + \lambda^2 u = 0, \\ u(-1) = 0, \\ 2\dfrac{du}{dt}(1) + u(1) = 0 \end{cases}$$

by choosing $m = 5$, for numerical implimentation of the proposed method, we will obtain the following matrices :

$$\mathbf{M}_5 = \begin{bmatrix} 0 & 1 & 0 & 1 & 0 & 1 \\ 0 & 0 & 3 & 0 & 3 & 0 \\ 0 & 0 & 0 & 5 & 0 & 5 \\ 0 & 0 & 0 & 0 & 7 & 0 \\ 0 & 0 & 0 & 0 & 0 & 9 \\ 0 & 0 & 0 & 0 & 0 & 0 \end{bmatrix}, \quad \mathbf{M}_5^2 = \begin{bmatrix} 0 & 0 & 3 & 0 & 10 & 0 \\ 0 & 0 & 0 & 15 & 0 & 42 \\ 0 & 0 & 0 & 0 & 35 & 0 \\ 0 & 0 & 0 & 0 & 0 & 63 \\ 0 & 0 & 0 & 0 & 0 & 0 \\ 0 & 0 & 0 & 0 & 0 & 0 \end{bmatrix},$$

$$\mathbf{H} = \begin{bmatrix} 0 & 0 & 3 & 0 & 10 & 0 \\ 0 & 0 & 0 & 15 & 0 & 42 \\ 0 & 0 & 0 & 0 & 35 & 0 \\ 0 & 0 & 0 & 0 & 0 & 63 \\ 1 & -1 & 1 & -1 & -1 & 1 \\ 1 & 3 & 7 & 13 & 21 & 31 \end{bmatrix}, \quad \mathbf{G} = \begin{bmatrix} -1 & 0 & 0 & 0 & 0 & 0 \\ 0 & -1 & 0 & 0 & 0 & 0 \\ 0 & 0 & -1 & 0 & 0 & 0 \\ 0 & 0 & 0 & -1 & 0 & 0 \\ 0 & 0 & 0 & 0 & 0 & 0 \\ 0 & 0 & 0 & 0 & 0 & 0 \end{bmatrix}.$$

According to (3.21), we obtain the following values of λ:

$$\lambda = [-, -, 16.3648, 8.54625, 4.92652, 2.02877].$$

Table 1 Numerical results of Example 1, using proposed method

λ_i	m = 5	m = 6	m = 7	m = 8	m = 9	Exact sol.
λ_1	2.02877	2.02876	2.02876	2.02876	2.02876	2.02876
λ_2	4.92652	4.9145	4.9132	4.91318	4.91318	4.91318
λ_3	8.54625	8.06465	7.9931	7.97981	7.97877	7.97867
λ_4	16.3648	12.391	11.3681	11.1517	11.0946	11.0855

Table 2 Numerical results of Example 2, using the proposed method

λ_i	m = 14	m = 15	m = 17	Exact sol.
λ_1	237.72106753	237.72106753	237.72106753	237.72106753
λ_2	2496.48743849	2496.48743786	2496.48743786	2496.48743786
λ_3	10867.583360842	10867.5824827024	10867.582217387	10867.58221698
λ_4	31782.7593574787	31780.1535527804	31780.096714447	31780.09645408

The obtained numerical results for different values of m have been reported in Table 1. Our proposed method has produced highly numerical results and the reported results, show that we can obtain good numerical results for $m \geq 9$.

Example 2 Consider the following fourth-order eigenvalue problem:

$$u^{(4)}(x) - \lambda u(x) = 0 \tag{4.2}$$

with the conditions:

$$u(0) = u'(0) = 0,$$

$$u''(1) = u(1) = 0,$$

and the exact solution for λ is $\tan \sqrt{\lambda} - \tanh \sqrt{\lambda} = 0$.

We have reported the obtained results for $m = 14, 15, 17$, in Table 2. Also, as we expected the reported results show that high accuracy is obtained in comparison to the numerical results in [2] and [6]. Table 3 represents that we can achieve better results for a lower values of m.

Example 3 To show the validity of the proposed method, consider the problem of column buckling. A slender column which is subjected to a concentric axial compressive load, P, as it shown in Fig. 1, and is simply supported at its both ends. The equation representing the bending of the column is

$$u^{(2)}(x) = \frac{M}{EI}, \tag{4.3}$$

Table 3 Numerical results of Example 2, using the proposed method

| λ_i | Presented method | | m = 22 | |
	m = 18	m = 21	Method in [2]	Method in [6]
λ_1	237.72106753	237.72106753	237.72106753	237.72106753
λ_2	2496.48743786	2496.48743786	2496.48743784	2496.48743843
λ_3	10867.58221698	10867.58221698	10867.59367145	10867.58221699
λ_4	31780.09651687	31780.09645408	31475.48355038	31780.09650785

Table 4 Numerical results of Example 3, using proposed method

| λ_i | Presented method | | Method in [3] | | Analytical sol. |
	m = 6	m = 8	h = 3/4	h = 3/5	
λ_1	1.0472	1.0472	1.0205	1.0301	1.0472
λ_2	2.10198	2.09443	1.8856	1.9593	2.0944
λ_3	3.18373	3.14225	2.4637	2.6967	3.1416
λ_4	6.49987	4.31943	–	3.1702	4.1888
λ_5	9.06094	5.53544	–	–	5.2360

where $u^{(2)}(x)$ specifies the curvature, M is the bending moment, E is the modulus of elasticity, and I is the moment of inertia of the cross section about its neutral axis. Corresponding to Fig. 1b, it is clear that the total moment at the free side of the column is equal to $M = -Pu$, by substituting the moment into the equation (4.3), the following second-order differential equation for Euler buckling will be obtained:

$$u^{(2)}(x) + k^2 u(x) = 0, \tag{4.4}$$

where

$$k^2 = \frac{P}{EI},$$

with the conditions:

$$u(0) = 0, \quad u(L) = 0,$$

and $k = \dfrac{n\pi}{L}$ are the eigenvalues for the column.

For this example, we take $E = 10 \times 10^9$ pa, $I = 1.25 \times 10^{-5}$ m^4, and $L = 3$. The numerical results can be seen from Table 5. Table 4 shows our results in comparison with the results of [3]. By increasing m, additional eigenvalues are determined and the previously determined values become progressively more accurate.

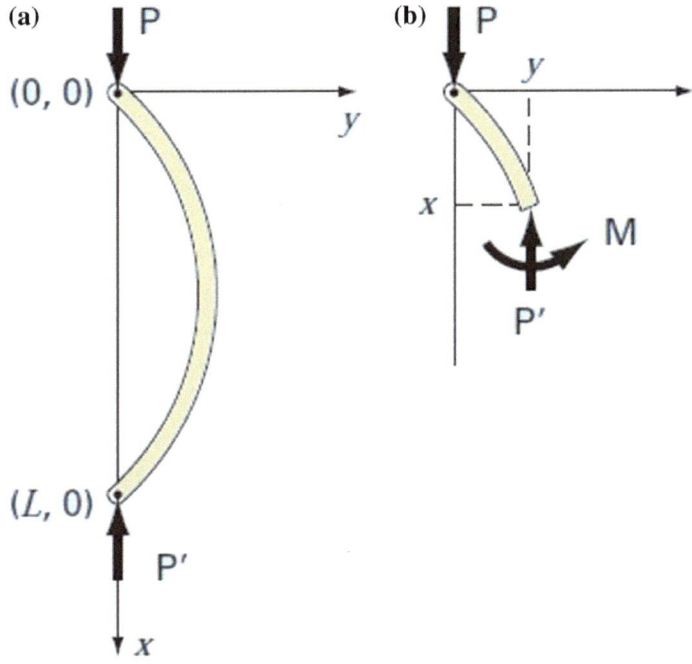

Fig. 1 **a** A slender rod. **b** A free- body diagram of a rod

Table 5 Numerical results of Example 3

λ_i	m = 10	m = 12	m = 14
λ_1	1.0472	1.0472	1.0472
λ_2	2.0944	2.0944	2.0944
λ_3	3.1416	3.1416	3.1416
λ_4	4.1933	4.1888	4.1888
λ_5	5.2540	5.2364	5.2360

5   Conclusion

In this research, a numerical technique based on the Tau method was presented for solving the eigenvalue problems. This method converts eigenvalue problems into a system of algebraic equations. The comparison of the obtained results with the other numerical methods and the exact solution indicates that the desired accuracy is obtained. By using some modifications, the proposed method can be applied to solve other ordinary differential equations.

Acknowledgements This work was supported to the second author [P Agarwal] by the research grant supported by the Department of Science & Technology(DST), India (No:INT/RUS/RFBR/P-308) and Science & Engineering Research Board (SERB), India (No:TAR/2018/000001).

References

1. Agarwal, R.P., Regan, D.O.: Ordinary and Partial Differential Equations. Springer (2009)
2. Attili, B., Lesnic, D.: An efficient method for computing eigenelements of Sturm-Liouville fourth-order boundary value problems. Appl. Math. Comput. **182**(2), 1247–1254 (2006)
3. Chapra, S.C., Canale, R.P.: Numerical Methods for Engineers. McGraw-Hill (2010)
4. EI-Gamel, M., Sameeh, M.: An efficient technique for Finding the Eigenvalues of fourth-order Sturm-Liouville problems. Appl. Math. **3**, 920–925 (2012)
5. Greenberg, L., Marletta, M.: Algorithm 775: The code SLEUTH for solving fourth-order Sturm-Liouville problems. ACM Trans. Math. Softw. **23**(4), 453–493 (1997)
6. Syam, M., Siyyam, H.: An efficient technique for finding the Eigenvalues of fourth-order Sturm-Liouville problems. Chaos Solitons Fractals **39**, 659–665 (2009)

On Hermite–Hadamard-Type Inequalities for Coordinated Convex Mappings Utilizing Generalized Fractional Integrals

Hüseyin Budak and Praveen Agarwal

Abstract In this chapter, we obtain the Hermite–Hadamard-type inequalities for coordinated convex function via generalized fractional integrals, which generalize some important fractional integrals such as the Riemann–Liouville fractional integrals, the Hadamard fractional integrals, and Katugampola fractional integrals. The results given in this chapter provide a generalization of several inequalities obtained in earlier studies.

Keywords Hermite–Hadamard's inequalities · Generalized fractional integral · Coordinated convex · Integral inequalities

2010 Mathematics Subject Classification. 26D07, 26D10, 26D15, 26B15, 26B25.

1 Introduction

The Hermite–Hadamard inequality discovered by Hermite and Hadamard see, e.g., [13, 28], p. 137) is one of the most well-established inequalities in the theory of convex functions with a geometrical interpretation and many applications. These inequalities state that if $f : I \to \mathbb{R}$ is a convex function on the interval I of real numbers and $a, b \in I$ with $a < b$, then

H. Budak (✉)
Department of Mathematics, Anand International College of Engineering,
Near Kanota, Agra Road, Jaipur 303012, Rajasthan, India
e-mail: hsyn.budak@gmail.com

P. Agarwal
Department of Mathematics, Faculty of Science and Arts, Düzce University, Düzce, Turkey
e-mail: goyal.praveen2011@gmail.com

© Springer Nature Singapore Pte Ltd. 2019
P. Agarwal et al. (eds.), *Fractional Calculus*, Springer Proceedings
in Mathematics & Statistics 303, https://doi.org/10.1007/978-981-15-0430-3_13

$$f\left(\frac{a+b}{2}\right) \leq \frac{1}{b-a} \int_a^b f(x)dx \leq \frac{f(a)+f(b)}{2}. \tag{1.1}$$

Both inequalities hold in the reverse direction if f is concave. We note that Hermite–Hadamard inequality may be regarded as a refinement of the concept of convexity and it follows easily from Jensen's inequality. Hermite–Hadamard inequality for convex functions has received renewed attention in recent years and a remarkable variety of refinements and generalizations have been studied (see, for example, [3, 14–16, 18, 20, 27, 34, 35, 40, 47, 48]).

A formal definition for coordinated convex function may be stated as follows:

Definition 1 A function $f : \Delta \to \mathbb{R}$ is called coordinated convex on Δ, for all $(x, u), (y, v) \in \Delta$ and $t, s \in [0, 1]$, if it satisfies the following inequality:

$$f(tx + (1-t)y, su + (1-s)v) \tag{1.2}$$
$$\leq ts\, f(x, u) + t(1-s)f(x, v) + s(1-t)f(y, u) + (1-t)(1-s)f(y, v).$$

The mapping f is a coordinated concave on Δ if the inequality (1.2) holds in reverse direction for all $t, s \in [0, 1]$ and $(x, u), (y, v) \in \Delta$.

In [12], Dragomir proved the following inequalities which is Hermite–Hadamard-type inequalities for coordinated convex functions on the rectangle from the plane $\mathbb{R}^2$.

Theorem 1 *Suppose that $f : \Delta \to \mathbb{R}$ is a coordinated convex, then we have the following inequalities:*

$$f\left(\frac{a+b}{2}, \frac{c+d}{2}\right) \leq \frac{1}{2}\left[\frac{1}{b-a}\int_a^b f\left(x, \frac{c+d}{2}\right)dx + \frac{1}{d-c}\int_c^d f\left(\frac{a+b}{2}, y\right)dy\right]$$

$$\leq \frac{1}{(b-a)(d-c)}\int_a^b \int_c^d f(x, y)\,dydx \tag{1.3}$$

$$\leq \frac{1}{4}\left[\frac{1}{b-a}\int_a^b f(x, c)dx + \frac{1}{b-a}\int_a^b f(x, d)dx\right.$$

$$\left. + \frac{1}{d-c}\int_c^d f(a, y)dy + \frac{1}{d-c}\int_c^d f(b, y)dy\right]$$

$$\leq \frac{f(a, c) + f(a, d) + f(b, c) + f(b, d)}{4}.$$

The above inequalities are sharp. The inequalities in (1.3) hold in reverse direction if the mapping f is a coordinated concave mapping.

For the other Hermite–Hadamard-type inequalities for coordinated convex functions, please refer to [2, 4, 24, 26, 41, 44].

In the following, we give the definition of Riemann–Liouville fractional integrals:

Definition 2 Let $f \in L_1[a, b]$. The Riemann–Liouville fractional integrals $J_{a+}^\alpha f$ and $J_{b-}^\alpha f$ of order $\alpha > 0$ with $a \geq 0$ are defined by

$$J_{a+}^\alpha f(x) = \frac{1}{\Gamma(\alpha)} \int_a^x (x - t)^{\alpha-1} f(t)dt, \quad x > a$$

and

$$J_{b-}^\alpha f(x) = \frac{1}{\Gamma(\alpha)} \int_x^b (t - x)^{\alpha-1} f(t)dt, \quad x < b$$

respectively. Here, $\Gamma(\alpha)$ is the Gamma function and $J_{a+}^0 f(x) = J_{b-}^0 f(x) = f(x)$.

More details on Riemann–Liouville fractional integrals, one can see [19, 23, 25, 30].

It is remarkable that Sarikaya et al. [32] first gave the following interesting integral inequalities of Hermite–Hadamard-type involving Riemann–Liouville fractional integrals.

Theorem 2 *Let $f : [a, b] \to \mathbb{R}$ be a positive function with $0 \leq a < b$ and $f \in L_1[a, b]$. If f is a convex function on $[a, b]$, then the following inequalities for fractional integrals hold:*

$$f\left(\frac{a+b}{2}\right) \leq \frac{\Gamma(\alpha+1)}{2(b-a)^\alpha} \left[J_{a+}^\alpha f(b) + J_{b-}^\alpha f(a)\right] \leq \frac{f(a) + f(b)}{2} \tag{1.4}$$

with $\alpha > 0$.

Moreover, Hermite–Hadamard-type inequality for coordinated convex functions utilizing Riemann–Liouville fractional integrals is obtained by Sarıkaya in [36]. One can find some recent Hermite–Hadamard inequalities for the function of one and two variables via Riemann–Liouville fractional integrals in [1, 5–11, 17, 22, 31, 33, 37–39, 45, 49, 50].

Hadamard fractional integrals are as follows:

Definition 3 Let $f \in L_1([a, b])$. The Hadamard fractional integrals $\mathbf{H}_{a+}^\alpha f$, and $\mathbf{H}_{b-}^\alpha f$ of order $\alpha > 0$ with $a \geq 0$ are defined by

$$\mathbf{H}_{a+}^\alpha f(x) := \frac{1}{\Gamma(\alpha)} \int_a^x \left(\ln \frac{x}{t}\right)^{\alpha-1} \frac{f(t)}{t} dt, \quad x > a,$$

and

$$\mathbf{H}_{b-}^{\alpha} f(x) := \frac{1}{\Gamma(\alpha)} \int_x^b \left(\ln \frac{t}{x} \right)^{\alpha-1} \frac{f(t)}{t} dt, \quad x < b,$$

respectively.

Recently, some papers are devoted to Hermite–Hadamard inequalities via Hadamard fractional integrals, see [29, 42, 43, 51, 52].

Now we give following generalized fractional integrals:

Definition 4 Let $g : [a, b] \to \mathbb{R}$ be an increasing and positive monotone function on $(a, b]$ having a continuous derivative $g'(x)$ on (a, b). The left-side $(I_{a+;g}^{\alpha} f(x))$ and right-side $(I_{b-;g}^{\alpha} f(x))$ fractional integral of f with respect to the function g on $[a, b]$ of order $\alpha < 0$ are defined by

$$I_{a+;g}^{\alpha} f(x) = \frac{1}{\Gamma(\alpha)} \int_a^x \frac{g'(t) f(t)}{[g(x) - g(t)]^{1-\alpha}} dt, \quad x > a$$

and

$$I_{b-;g}^{\alpha} f(x) = \frac{1}{\Gamma(\alpha)} \int_x^b \frac{g'(t) f(t)}{[g(t) - g(x)]^{1-\alpha}} dt, \quad x < b$$

respectively.

Jleli and Samet establish following Hermite–Hadamard inequalities:

Theorem 3 ([21]) *Let* $g : [a, b] \to \mathbb{R}$ *be an increasing and positive monotone function on* $(a, b]$*, having a continuous derivative* $g'(x)$ *on* (a, b) *and let* $\alpha > 0$*. If* f *is a convex function on* $[a, b]$*, then*

$$\varphi \left(\frac{a+b}{2} \right) \leq \frac{\Gamma(\alpha+1)}{4 [g(b) - g(a)]^{\alpha}} \left[I_{a+;g}^{\alpha} \Phi(b) + \Phi_{b-;g}^{\alpha} f(a) \right] \leq \frac{\varphi(a) + f(b)}{2} \tag{1.5}$$

where $\Phi(x) = \varphi(x) + \widetilde{\varphi}(x)$ *and* $\widetilde{\varphi}(x) = \varphi(a + b - x)$ *for* $x \in [a, b]$*.*

Hadamard fractional integrals of a function with two variables can be given as follows:

Definition 5 Let $f \in L_1 ([a, b] \times [c, d])$. The Hadamard fractional integrals $\mathbf{J}_{a+,c+}^{\alpha,\beta} f$, $\mathbf{J}_{a+,d-}^{\alpha,\beta} f$, $\mathbf{J}_{b-,c+}^{\alpha,\beta} f$ and $\mathbf{J}_{b-,d-}^{\alpha,\beta} f$ of order $\alpha, \beta > 0$ with $a, c \geq 0$ are defined by

$$\mathbf{J}_{a+,c+}^{\alpha,\beta} f(x, y) := \frac{1}{\Gamma(\alpha)\Gamma(\beta)} \int_a^x \int_c^y \left(\ln \frac{x}{t} \right)^{\alpha-1} \left(\ln \frac{y}{s} \right)^{\beta-1} \frac{f(t, s)}{ts} ds dt, \quad x > a, \ y > c,$$

$$J_{a+,d-}^{\alpha,\beta} f(x,y) := \frac{1}{\Gamma(\alpha)\Gamma(\beta)} \int_a^x \int_y^d \left(\ln \frac{x}{t}\right)^{\alpha-1} \left(\ln \frac{s}{y}\right)^{\beta-1} \frac{f(t,s)}{ts} ds dt, \quad x > a, \ y < d,$$

$$J_{b-,c+}^{\alpha,\beta} f(x,y) := \frac{1}{\Gamma(\alpha)\Gamma(\beta)} \int_x^b \int_c^y \left(\ln \frac{t}{x}\right)^{\alpha-1} \left(\ln \frac{y}{s}\right)^{\beta-1} \frac{f(t,s)}{ts} ds dt, \quad x < b, \ y > c,$$

and

$$J_{b-,d-}^{\alpha,\beta} f(x,y) := \frac{1}{\Gamma(\alpha)\Gamma(\beta)} \int_x^b \int_y^d \left(\ln \frac{t}{x}\right)^{\alpha-1} \left(\ln \frac{y}{s}\right)^{\beta-1} \frac{f(t,s)}{ts} ds dt, \quad x < b, \ y < d,$$

respectively.

Now, we give following generalized fractional integral operators:

Definition 6 Let $g : [a,b] \to \mathbb{R}$ be an increasing and positive monotone function on $(a,b]$, having a continuous derivative $g'(x)$ on (a,b) and let $w : [c,d] \to \mathbb{R}$ be an increasing and positive monotone function on $(c,d]$, having a continuous derivative $w'(y)$ on (c,d). Let $f \in L_1([a,b] \times [c,d])$. The generalized fractional integral operators for functions of two variables are defined by

$$\mathcal{J}_{a+,c+;g,w}^{\alpha,\beta} f(x,y) := \frac{1}{\Gamma(\alpha)\Gamma(\beta)} \int_a^x \int_c^y \frac{g'(t)}{[g(x)-g(t)]^{1-\alpha}} \frac{w'(s)}{[w(y)-w(s)]^{1-\beta}} f(t,s) ds dt, \quad x > a, \ y > c,$$

$$\mathcal{J}_{a+,d-;g,w}^{\alpha,\beta} f(x,y) := \frac{1}{\Gamma(\alpha)\Gamma(\beta)} \int_a^x \int_y^d \frac{g'(t)}{[g(x)-g(t)]^{1-\alpha}} \frac{w'(s)}{[w(s)-w(y)]^{1-\beta}} f(t,s) ds dt, \quad x > a, \ y < d,$$

$$\mathcal{J}_{b-,c+;g,w}^{\alpha,\beta} f(x,y) := \frac{1}{\Gamma(\alpha)\Gamma(\beta)} \int_x^b \int_c^y \frac{g'(t)}{[g(t)-g(x)]^{1-\alpha}} \frac{w'(s)}{[w(y)-w(s)]^{1-\beta}} f(t,s) ds dt, \quad x < b, \ y > c,$$

and

$$\mathcal{J}_{b-,d-;g,w}^{\alpha,\beta} f(x,y) := \frac{1}{\Gamma(\alpha)\Gamma(\beta)} \int_x^b \int_y^d \frac{g'(t)}{[g(t)-g(x)]^{1-\alpha}} \frac{w'(s)}{[w(s)-w(y)]^{1-\beta}} f(t,s) ds dt, \quad x < b, \ y < d.$$

Similar to the above definitions, we can give the following integrals:

$$\mathcal{J}_{a+;g}^{\alpha} f\left(x, \frac{c+d}{2}\right) := \frac{1}{\Gamma(\alpha)} \int_a^x \frac{g'(t)}{[g(x)-g(t)]^{1-\alpha}} f\left(t, \frac{c+d}{2}\right) dt, \quad x > a,$$

$$\mathcal{J}_{b-;g}^{\alpha} f\left(x, \frac{c+d}{2}\right) := \frac{1}{\Gamma(\alpha)} \int_{x}^{b} \frac{g'(t)}{[g(t) - g(x)]^{1-\alpha}} f\left(t, \frac{c+d}{2}\right) dt, \quad x < b,$$

$$\mathcal{J}_{c+;w}^{\beta} f\left(\frac{a+b}{2}, y\right) := \frac{1}{\Gamma(\beta)} \int_{c}^{y} \frac{w'(t)}{[w(y) - w(s)]^{1-\beta}} f\left(\frac{a+b}{2}, y\right) ds, \quad y > c,$$

and

$$\mathcal{J}_{d-;w}^{\beta} f\left(\frac{a+b}{2}, y\right) := \frac{1}{\Gamma(\beta)} \int_{c}^{y} \frac{w'(t)}{[w(s) - w(y)]^{1-\beta}} f\left(\frac{a+b}{2}, y\right) ds, \quad y < d,$$

If we choose $g(t) = \frac{t^{\rho}}{\rho}$ and $w(s) = \frac{s^{\sigma}}{\sigma}$ in Definition 6, then we have the following Katugampola fractional integrals for function with two variables similar to definitions given by Yaldiz in [46]:

Definition 7 Let $f \in L_1([a, b] \times [c, d])$. The Katugampola fractional integrals for function with two variables are defined by

$$^{\rho,\sigma} \mathcal{I}_{a+,c+}^{\alpha,\beta} f(x, y) := \frac{\rho^{1-\alpha} \sigma^{1-\beta}}{\Gamma(\alpha) \Gamma(\beta)} \int_{a}^{x} \int_{c}^{y} \frac{t^{\rho-1}}{[x^{\rho} - t^{\rho}]^{1-\alpha}} \frac{s^{\sigma-1}}{[y^{\sigma} - s^{\sigma}]^{1-\beta}} f(t, s) ds dt, \quad x > a, \ y > c,$$

$$^{\rho,\sigma} \mathcal{I}_{a+,d-}^{\alpha,\beta} f(x, y) := \frac{\rho^{1-\alpha} \sigma^{1-\beta}}{\Gamma(\alpha) \Gamma(\beta)} \int_{a}^{x} \int_{y}^{d} \frac{t^{\rho-1}}{[x^{\rho} - t^{\rho}]^{1-\alpha}} \frac{s^{\sigma-1}}{[s^{\sigma} - y^{\sigma}]^{1-\beta}} f(t, s) ds dt, \quad x > a, \ y < d,$$

$$^{\rho,\sigma} \mathcal{I}_{b-,c+}^{\alpha,\beta} f(x, y) := \frac{\rho^{1-\alpha} \sigma^{1-\beta}}{\Gamma(\alpha) \Gamma(\beta)} \int_{x}^{b} \int_{c}^{y} \frac{t^{\rho-1}}{[t^{\rho} - x^{\rho}]^{1-\alpha}} \frac{s^{\sigma-1})}{[y^{\sigma} - s^{\sigma}]^{1-\beta}} f(t, s) ds dt, \quad x < b, \ y > c,$$

and

$$^{\rho,\sigma} \mathcal{I}_{b-,d-}^{\alpha,\beta} f(x, y) := \frac{\rho^{1-\alpha} \sigma^{1-\beta}}{\Gamma(\alpha) \Gamma(\beta)} \int_{x}^{b} \int_{y}^{d} \frac{t^{\rho-1}}{[t^{\rho} - x^{\rho}]^{1-\alpha}} \frac{s^{\sigma-1}}{[s^{\sigma} - y^{\sigma}]^{1-\beta}} f(t, s) ds dt, \quad x < b, \ y < d.$$

The aim of this study is to establish Hermite–Hadamard-type integral inequalities for a coordinated convex function involving generalized fractional integrals. The results presented in this paper provide extensions of those given in earlier works.

2 Main Results

Let $f : \Delta = [a, b] \times [c, d] \to \mathbb{R}$. First, we define the following functions which will be used frequently:

$$
\begin{aligned}
\widetilde{f}_1(x, y) &= f(a + b - x, y), \\
\widetilde{f}_2(x, y) &= f(x, c + d - y), \\
\widetilde{f}_3(x, y) &= f(a + b - x, c + d - y), \\
G(x, y) &= f(x, y) + \widetilde{f}_2(x, y) \\
H(x, y) &= f(x, y) + \widetilde{f}_1(x, y) \\
K(x, y) &= \widetilde{f}_1(x, y) + \widetilde{f}_3(x, y) \\
L(x, y) &= \widetilde{f}_2(x, y) + \widetilde{f}_3(x, y) \\
F(x, y) &= \widetilde{f}_1(x, y) + \widetilde{f}_2(x, y) + \widetilde{f}_3(x, y) + f(x, y) \\
&= \frac{G(x, y) + H(x, y) + K(x, y) + L(x, y)}{2}
\end{aligned}
\tag{2.1}
$$

for $(x, y) \in [a, b] \times [c, d]$.

Theorem 4 *Let $g : [a, b] \to \mathbb{R}$ be an increasing and positive monotone function on $(a, b]$, having a continuous derivative $g'(x)$ on (a, b) and let $w : [c, d] \to \mathbb{R}$ be an increasing and positive monotone function on $(c, d]$, having a continuous derivative $w'(y)$ on (c, d). If $f : \Delta \to \mathbb{R}$ is a coordinated convex on Δ, then for $\alpha, \beta > 0$ the following Hermite–Hadamard-type inequality hold:*

$$
f\left(\frac{a + b}{2}, \frac{c + d}{2}\right)
\tag{2.2}
$$
$$
\leq \frac{\Gamma(\alpha + 1)\Gamma(\beta + 1)}{16\,[g(b) - g(a)]^\alpha\,[w(d) - w(c)]^\beta}
$$
$$
\times \left[\mathcal{J}^{\alpha,\beta}_{a+,c+;g,w} F(b, d) + \mathcal{J}^{\alpha,\beta}_{a+,d-;g,w} F(b, c) + \mathcal{J}^{\alpha,\beta}_{b-,c+;g,w} F(a, d) + \mathcal{J}^{\alpha,\beta}_{b-,d-;g,w} F(a, c) \right]
$$
$$
\leq \frac{f(a, c) + f(a, d) + f(b, c) + f(b, d)}{4},
$$

where the function F is defined as in (2.1).

Proof Since f is a coordinated convex mapping on Δ, we have

$$
f\left(\frac{u + v}{2}, \frac{p + q}{2}\right) \leq \frac{f(u, p) + f(u, q) + f(v, p) + f(v, q)}{4}
\tag{2.3}
$$

for $(u, p), (v, q) \in \Delta$. Now, for $t, s \in [0, 1]$, let $u = ta + (1 - t)b$, $v = (1 - t)a + tb$, $p = cs + (1 - s)d$ and $q = (1 - s)c + sd$. Then we have

$$f\left(\frac{a+b}{2}, \frac{c+d}{2}\right) \tag{2.4}$$

$$\leq \frac{1}{4} f(ta + (1-t)b, cs + (1-s)d) + \frac{1}{4} f(ta + (1-t)b, (1-s)c + sd)$$

$$+ \frac{1}{4} f((1-t)a + tb, cs + (1-s)d) + \frac{1}{4} f((1-t)a + tb, (1-s)c + sd).$$

Multiplying both sides of (2.4) by

$$\frac{(b-a)(d-c)}{\Gamma(\alpha)\Gamma(\beta)} \frac{g'((1-t)a+tb)}{[g(b) - g((1-t)a+tb)]^{1-\alpha}} \frac{w'((1-s)c+sd)}{[w(d) - w((1-s)c+sd)]^{1-\beta}},$$

and integrating the resulting inequality with respect to t, s over $(0, 1) \times (0, 1)$, we get

$$\frac{(b-a)(d-c)}{\Gamma(\alpha)\Gamma(\beta)} f\left(\frac{a+b}{2}, \frac{c+d}{2}\right) \int_0^1 \int_0^1 \frac{g'((1-t)a+tb)}{[g(b) - g((1-t)a+tb)]^{1-\alpha}} \frac{w'((1-s)c+sd)}{[w(d) - w((1-s)c+sd)]^{1-\beta}} ds dt$$

$$\leq \frac{(b-a)(d-c)}{4\Gamma(\alpha)\Gamma(\beta)}$$

$$\times \int_0^1 \int_0^1 \frac{g'((1-t)a+tb)}{[g(b) - g((1-t)a+tb)]^{1-\alpha}} \frac{w'((1-s)c+sd)}{[w(d) - w((1-s)c+sd)]^{1-\beta}} f(ta + (1-t)b, cs + (1-s)d) ds dt$$

$$+ \frac{(b-a)(d-c)}{4\Gamma(\alpha)\Gamma(\beta)}$$

$$\times \int_0^1 \int_0^1 \frac{g'((1-t)a+tb)}{[g(b) - g((1-t)a+tb)]^{1-\alpha}} \frac{w'((1-s)c+sd)}{[w(d) - w((1-s)c+sd)]^{1-\beta}} f(ta + (1-t)b, (1-s)c + sd) ds dt$$

$$+ \frac{(b-a)(d-c)}{4\Gamma(\alpha)\Gamma(\beta)}$$

$$\times \int_0^1 \int_0^1 \frac{g'((1-t)a+tb)}{[g(b) - g((1-t)a+tb)]^{1-\alpha}} \frac{w'((1-s)c+sd)}{[w(d) - w((1-s)c+sd)]^{1-\beta}} f((1-t)a + tb, cs + (1-s)d) ds dt$$

$$+ \frac{(b-a)(d-c)}{4\Gamma(\alpha)\Gamma(\beta)}$$

$$\times \int_0^1 \int_0^1 \frac{g'((1-t)a+tb)}{[g(b) - g((1-t)a+tb)]^{1-\alpha}} \frac{w'((1-s)c+sd)}{[w(d) - w((1-s)c+sd)]^{1-\beta}} f((1-t)a + tb, (1-s)c + sd) ds dt.$$

By a simple calculation, we have

$$\int_0^1 \int_0^1 \frac{g'((1-t)a+tb)}{[g(b) - g((1-t)a+tb)]^{1-\alpha}} \frac{w'((1-s)c+sd)}{[w(d) - w((1-s)c+sd)]^{1-\beta}} ds dt = \frac{[g(b) - g(a)]^\alpha [w(d) - w(c)]^\beta}{\alpha\beta(b-a)(d-c)}.$$

Using the change of variables $\tau = (1-t)a + tb$ and $\eta = (1-s)c + sd$, we obtain

$$\frac{[g(b) - g(a)]^\alpha \, [w(d) - w(c)]^\beta}{\Gamma(\alpha + 1)\Gamma(\beta + 1)} f\left(\frac{a + b}{2}, \frac{c + d}{2}\right)$$

$$\leq \frac{1}{4\Gamma(\alpha)\Gamma(\beta)} \int_0^1 \int_0^1 \frac{g'(\tau)}{[g(b) - g(\tau)]^{1-\alpha}} \frac{w'(\eta)}{[w(d) - w(\eta)]^{1-\beta}} f(a + b - \tau, c + d - \eta) d\eta d\tau$$

$$+ \frac{1}{4\Gamma(\alpha)\Gamma(\beta)} \int_0^1 \int_0^1 \frac{g'(\tau)}{[g(b) - g(\tau)]^{1-\alpha}} \frac{w'(\eta)}{[w(d) - w(\eta)]^{1-\beta}} f(a + b - \tau, \eta) d\eta d\tau$$

$$+ \frac{1}{4\Gamma(\alpha)\Gamma(\beta)} \int_0^1 \int_0^1 \frac{g'(\tau)}{[g(b) - g(\tau)]^{1-\alpha}} \frac{w'(\eta)}{[w(d) - w(\eta)]^{1-\beta}} f(\tau, c + d - \eta) d\eta d\tau$$

$$+ \frac{1}{4\Gamma(\alpha)\Gamma(\beta)} \int_0^1 \int_0^1 \frac{g'(\tau)}{[g(b) - g(\tau)]^{1-\alpha}} \frac{w'(\eta)}{[w(d) - w(\eta)]^{1-\beta}} f(\tau, \eta) d\eta d\tau$$

$$= \frac{1}{4} \left[\mathcal{J}^{\alpha,\beta}_{a+,c+;g,w} \tilde{f_3}(b, d) + \mathcal{J}^{\alpha,\beta}_{a+,c+;g,w} \tilde{f_1}(b, d) + \mathcal{J}^{\alpha,\beta}_{a+,c+;g,w} \tilde{f_2}(b, d) + \mathcal{J}^{\alpha,\beta}_{a+,c+;g,w} f(b, d) \right]$$

$$= \frac{1}{4} \mathcal{J}^{\alpha,\beta}_{a+,c+;g,w} F(b, d).$$

That is, we have

$$\frac{[g(b) - g(a)]^\alpha \, [w(d) - w(c)]^\beta}{\Gamma(\alpha + 1)\Gamma(\beta + 1)} f\left(\frac{a + b}{2}, \frac{c + d}{2}\right) \leq \frac{1}{4} \mathcal{J}^{\alpha,\beta}_{a+,c+;g,w} F(b, d). \quad (2.5)$$

Similarly, multiplying both sides of (2.4) by

$$\frac{(b - a)(d - c)}{\Gamma(\alpha)\Gamma(\beta)} \frac{g'((1 - t)a + tb)}{[g(b) - g((1 - t)a + tb)]^{1-\alpha}} \frac{w'((1 - s)c + sd)}{[w((1 - s)c + sd) - w(c)]^{1-\beta}}$$

and integrating the obtained inequality with respect to t, s over $(0, 1) \times (0, 1)$, we obtain

$$\frac{[g(b) - g(a)]^\alpha \, [w(d) - w(c)]^\beta}{\Gamma(\alpha + 1)\Gamma(\beta + 1)} f\left(\frac{a + b}{2}, \frac{c + d}{2}\right) \leq \frac{1}{4} \mathcal{J}^{\alpha,\beta}_{a+,d-;g,w} F(b, c). \quad (2.6)$$

Moreover, multiplying both sides of (2.4) by

$$\frac{(b - a)(d - c)}{\Gamma(\alpha)\Gamma(\beta)} \frac{g'((1 - t)a + tb)}{[g((1 - t)a + tb) - g(s)]^{1-\alpha}} \frac{w'((1 - s)c + sd)}{[w(d) - w((1 - s)c + sd)]^{1-\beta}}$$

and

$$\frac{(b - a)(d - c)}{\Gamma(\alpha)\Gamma(\beta)} \frac{g'((1 - t)a + tb)}{[g((1 - t)a + tb) - g(s)]^{1-\alpha}} \frac{w'((1 - s)c + sd)}{[w((1 - s)c + sd) - w(c)]^{1-\beta}}$$

then integrating the established inequalities with respect to t, s over $(0, 1) \times (0, 1)$, we have the following inequalities:

$$\frac{[g(b) - g(a)]^\alpha [w(d) - w(c)]^\beta}{\Gamma(\alpha + 1)\Gamma(\beta + 1)} f\left(\frac{a + b}{2}, \frac{c + d}{2}\right) \leq \frac{1}{4}\mathcal{J}^{\alpha,\beta}_{b-,c+;g,w} F(a, d) \quad (2.7)$$

and

$$\frac{[g(b) - g(a)]^\alpha [w(d) - w(c)]^\beta}{\Gamma(\alpha + 1)\Gamma(\beta + 1)} f\left(\frac{a + b}{2}, \frac{c + d}{2}\right) \leq \frac{1}{4}\mathcal{J}^{\alpha,\beta}_{b-,d-;g,w} F(a, c), \quad (2.8)$$

respectively.

Summing the inequalities (2.5)–(2.8), we get

$$f\left(\frac{a + b}{2}, \frac{c + d}{2}\right)$$
$$\leq \frac{\Gamma(\alpha + 1)\Gamma(\beta + 1)}{16\,[g(b) - g(a)]^\alpha [w(d) - w(c)]^\beta}$$
$$\times \left[\mathcal{J}^{\alpha,\beta}_{a+,c+;g,w} F(b, d) + \mathcal{J}^{\alpha,\beta}_{a+,d-;g,w} F(b, c) + \mathcal{J}^{\alpha,\beta}_{b-,c+;g,w} F(a, d) + \mathcal{J}^{\alpha,\beta}_{b-,d-;g,w} F(a, c)\right].$$

This completes the proof of first inequality in (2.2).

For the proof of the second inequality in (2.2), since f is a coordinated convex, we have

$$f(ta + (1 - t)b, cs + (1 - s)d) + f(ta + (1 - t)b, (1 - s)c + sd) \quad (2.9)$$
$$+ f((1 - t)a + tb, cs + (1 - s)d) + f((1 - t)a + tb, (1 - s)c + sd)$$
$$\leq f(a, c) + f(a, d) + f(b, c) + f(b, d).$$

Multiplying both sides of (2.9) by

$$\frac{(b - a)(d - c)}{\Gamma(\alpha)\Gamma(\beta)} \frac{g'((1 - t)a + tb)}{[g(b) - g((1 - t)a + tb)]^{1-\alpha}} \frac{w'((1 - s)c + sd)}{[w(d) - w((1 - s)c + sd)]^{1-\beta}},$$

and integrating the resulting inequality with respect to t, s over $(0, 1) \times (0, 1)$, we get

$$\frac{(b - a)(d - c)}{\Gamma(\alpha)\Gamma(\beta)}$$

$$\times \int_0^1 \int_0^1 \frac{g'((1 - t)a + tb)}{[g(b) - g((1 - t)a + tb)]^{1-\alpha}} \frac{w'((1 - s)c + sd)}{[w(d) - w((1 - s)c + sd)]^{1-\beta}} f(ta + (1 - t)b, cs + (1 - s)d)dsdt$$

$$+ \frac{(b - a)(d - c)}{\Gamma(\alpha)\Gamma(\beta)}$$

$$\times \int_0^1 \int_0^1 \frac{g'((1 - t)a + tb)}{[g(b) - g((1 - t)a + tb)]^{1-\alpha}} \frac{w'((1 - s)c + sd)}{[w(d) - w((1 - s)c + sd)]^{1-\beta}} f(ta + (1 - t)b, (1 - s)c + sd)dsdt$$

$$+\frac{(b-a)(d-c)}{\Gamma(\alpha)\Gamma(\beta)}$$

$$\times \int_0^1 \int_0^1 \frac{g'((1-t)a+tb)}{[g(b)-g((1-t)a+tb)]^{1-\alpha}} \frac{w'((1-s)c+sd)}{[w(d)-w((1-s)c+sd)]^{1-\beta}} f((1-t)a+tb, cs+(1-s)d)dsdt$$

$$+\frac{(b-a)(d-c)}{\Gamma(\alpha)\Gamma(\beta)}$$

$$\times \int_0^1 \int_0^1 \frac{g'((1-t)a+tb)}{[g(b)-g((1-t)a+tb)]^{1-\alpha}} \frac{w'((1-s)c+sd)}{[w(d)-w((1-s)c+sd)]^{1-\beta}} f((1-t)a+tb, (1-s)c+sd)dsdt$$

$$\leq \frac{(b-a)(d-c)}{\Gamma(\alpha)\Gamma(\beta)} [f(a,c)+f(a,d)+f(b,c)+f(b,d)]$$

$$\times \int_0^1 \int_0^1 \frac{g'((1-t)a+tb)}{[g(b)-g((1-t)a+tb)]^{1-\alpha}} \frac{w'((1-s)c+sd)}{[w(d)-w((1-s)c+sd)]^{1-\beta}} dsdt.$$

Then, we get

$$\mathcal{J}_{a+,c+;g,w}^{\alpha,\beta}\tilde{f_3}(b,d) + \mathcal{J}_{a+,c+;g,w}^{\alpha,\beta}\tilde{f_1}(b,d) + \mathcal{J}_{a+,c+;g,w}^{\alpha,\beta}\tilde{f_2}(b,d) + \mathcal{J}_{a+,c+;g,w}^{\alpha,\beta}f(b,d)$$

$$\leq [f(a,c)+f(a,d)+f(b,c)+f(b,d)] \frac{[g(b)-g(a)]^\alpha [w(d)-w(c)]^\beta}{\Gamma(\alpha+1)\Gamma(\beta+1)},$$

that is,

$$\frac{\Gamma(\alpha+1)\Gamma(\beta+1)}{[g(b)-g(a)]^\alpha [w(d)-w(c)]^\beta} \mathcal{J}_{a+,c+;g,w}^{\alpha,\beta} F(b,d) \leq f(a,c)+f(a,d)+f(b,c)+f(b,d).$$

$$(2.10)$$

Similarly, multiplying both sides of (2.9) by

$$\frac{(b-a)(d-c)}{\Gamma(\alpha)\Gamma(\beta)} \frac{g'((1-t)a+tb)}{[g(b)-g((1-t)a+tb)]^{1-\alpha}} \frac{w'((1-s)c+sd)}{[w((1-s)c+sd)-w(c)]^{1-\beta}},$$

$$\frac{(b-a)(d-c)}{\Gamma(\alpha)\Gamma(\beta)} \frac{g'((1-t)a+tb)}{[g((1-t)a+tb)-g(a)]^{1-\alpha}} \frac{w'((1-s)c+sd)}{[w(d)-w((1-s)c+sd)]^{1-\beta}}$$

and

$$\frac{(b-a)(d-c)}{\Gamma(\alpha)\Gamma(\beta)} \frac{g'((1-t)a+tb)}{[g((1-t)a+tb)-g(a)]^{1-\alpha}} \frac{w'((1-s)c+sd)}{[w((1-s)c+sd)-w(c)]^{1-\beta}}$$

integrating the resulting inequalities with respect to t, s over $(0, 1) \times (0, 1)$, we establish the following inequalities:

$$\frac{\Gamma(\alpha+1)\Gamma(\beta+1)}{[g(b)-g(a)]^\alpha [w(d)-w(c)]^\beta} \mathcal{J}_{a+,d-;g,w}^{\alpha,\beta} F(b,c) \leq f(a,c)+f(a,d)+f(b,c)+f(b,d),$$

$$(2.11)$$

$$\frac{\Gamma(\alpha+1)\Gamma(\beta+1)}{[g(b)-g(a)]^\alpha\,[w(d)-w(c)]^\beta}\mathcal{J}^{\alpha,\beta}_{b-,c+;g,w}F(a,d)\le f(a,c)+f(a,d)+f(b,c)+f(b,d),$$
$$(2.12)$$

and

$$\frac{\Gamma(\alpha+1)\Gamma(\beta+1)}{[g(b)-g(a)]^\alpha\,[w(d)-w(c)]^\beta}\mathcal{J}^{\alpha,\beta}_{b-,d-;g,w}F(a,c)\le f(a,c)+f(a,d)+f(b,c)+f(b,d),$$
$$(2.13)$$

respectively.

By adding the inequalities (2.10)–(2.13), we have the inequality:

$$\frac{\Gamma(\alpha+1)\Gamma(\beta+1)}{[g(b)-g(a)]^\alpha\,[w(d)-w(c)]^\beta} \tag{2.14}$$

$$\times\left[\mathcal{J}^{\alpha,\beta}_{a+,c+;g,w}F(b,d)+\mathcal{J}^{\alpha,\beta}_{a+,d-;g,w}F(b,c)+\mathcal{J}^{\alpha,\beta}_{b-,c+;g,w}F(a,d)+\mathcal{J}^{\alpha,\beta}_{b-,d-;g,w}F(a,c)\right]$$

$$\le 4\left[f(a,c)+f(a,d)+f(b,c)+f(b,d)\right]$$

If we divide the both sides of inequality (2.14) by 16, then we have the second inequality in (2.2).

This completes the proof. $\qquad\qquad\qquad\qquad\qquad\qquad\qquad\qquad\qquad\qquad\qquad\Box$

Remark 1 If we choose $g(t)=t$ and $w(s)=s$ in Theorem 4, then we have the following inequalities for Riemann–Liouville fractional integrals

$$f\left(\frac{a+b}{2},\frac{c+d}{2}\right)$$

$$\le\frac{\Gamma(\alpha+1)\Gamma(\beta+1)}{4\,(b-a)^\alpha\,(d-c)^\beta}\left[J^{\alpha,\beta}_{a+,c+}f(b,d)+J^{\alpha,\beta}_{a+,d-}f(b,c)+J^{\alpha,\beta}_{b-,c+}f(a,d)+J^{\alpha,\beta}_{b-,d-}f(a,c)\right]$$

$$\le\frac{f(a,c)+f(a,d)+f(b,c)+f(b,d)}{4}$$

which are proved by Sarikaya in [36].

Corollary 1 *Under assumption of Theorem 4 with $g(t)=\ln t$ and $w(s)=\ln s$, then we have the following inequalities for Hadamard fractional integrals:*

$$f\left(\frac{a+b}{2},\frac{c+d}{2}\right)$$

$$\le\frac{\Gamma(\alpha+1)\Gamma(\beta+1)}{16\left[\ln\frac{b}{a}\right]^\alpha\left[\ln\frac{d}{c}\right]^\beta}\left[\mathbf{J}^{\alpha,\beta}_{a+,c+}F(b,d)+\mathbf{J}^{\alpha,\beta}_{a+,d-}F(b,c)+\mathbf{J}^{\alpha,\beta}_{b-,c+}F(a,d)+\mathbf{J}^{\alpha,\beta}_{b-,d-}F(a,c)\right]$$

$$\le\frac{f(a,c)+f(a,d)+f(b,c)+f(b,d)}{4}.$$

Corollary 2 *Under assumption of Theorem 4 with $g(t)=\frac{t^\rho}{\rho}$ and $w(s)=\frac{s^\sigma}{\sigma}$, then we have the following inequalities for Katugampola fractional integrals:*

$$f\left(\frac{a+b}{2},\frac{c+d}{2}\right)$$

$$\leq \frac{\Gamma(\alpha+1)\Gamma(\beta+1)\rho^\alpha\sigma^\beta}{16\,[b^\rho-a^\rho]^\alpha\,[d^\sigma-c^\sigma]^\beta}\left[{}^{\rho,\sigma}\mathcal{I}^{\alpha,\beta}_{a+,c+}F(b,d)+{}^{\rho,\sigma}\mathcal{I}^{\alpha,\beta}_{a+,d-}F(b,c)+{}^{\rho,\sigma}\mathcal{I}^{\alpha,\beta}_{b-,c+}F(a,d)+{}^{\rho,\sigma}\mathcal{I}^{\alpha,\beta}_{b-,d-}F(a,c)\right]$$

$$\leq \frac{f(a,c)+f(a,d)+f(b,c)+f(b,d)}{4}.$$

Theorem 5 *Let $g:[a,b]\to\mathbb{R}$ be an increasing and positive monotone function on $(a,b]$, having a continuous derivative $g'(x)$ on (a,b) and let $w:[c,d]\to\mathbb{R}$ be an increasing and positive monotone function on $(c,d]$, having a continuous derivative $w'(y)$ on (c,d). If $f\,\Delta\to\mathbb{R}$ is a coordinated convex on Δ, then for $\alpha,\beta>0$ the following Hermite–Hadamard-type inequality holds:*

$$f\left(\frac{a+b}{2},\frac{c+d}{2}\right) \tag{2.15}$$

$$\leq \frac{\Gamma(\alpha+1)}{8\,[g(b)-g(a)]^\alpha}\left[\mathcal{J}^\alpha_{a+;g}H\left(b,\frac{c+d}{2}\right)+\mathcal{J}^\alpha_{b-;g}H\left(a,\frac{c+d}{2}\right)\right]$$

$$+\frac{\Gamma(\beta+1)}{8\,[w(d)-w(c)]^\beta}\left[\mathcal{J}^\beta_{c+;w}G\left(\frac{a+b}{2},d\right)+\mathcal{J}^\beta_{d-;w}G\left(\frac{a+b}{2},c\right)\right]$$

$$\leq \frac{\Gamma(\alpha+1)\Gamma(\beta+1)}{16\,[g(b)-g(a)]^\alpha\,[w(d)-w(c)]^\beta}$$

$$\times\left[\mathcal{J}^{\alpha,\beta}_{a+,c+;g,w}F(b,d)+\mathcal{J}^{\alpha,\beta}_{a+,d-;g,w}F(b,c)+\mathcal{J}^{\alpha,\beta}_{b-,c+;g,w}F(a,d)+\mathcal{J}^{\alpha,\beta}_{b-,d-;g,w}F(a,c)\right]$$

$$\leq \frac{\Gamma(\alpha+1)}{16\,[g(b)-g(a)]^\alpha}\left[\mathcal{J}^\alpha_{a+;g}H(b,c)+\mathcal{J}^\alpha_{a+;g}H(b,d)+\mathcal{J}^\alpha_{b-;g}H(a,c)+\mathcal{J}^\alpha_{b-;g}H(a,d)\right]$$

$$+\frac{\Gamma(\beta+1)}{16\,[w(d)-w(c)]^\beta}\left[\mathcal{J}^\beta_{c+;w}G(a,d)+\mathcal{J}^\beta_{c+;w}G(b,d)+\mathcal{J}^\beta_{d-;w}G(a,c)+\mathcal{J}^\beta_{d-;w}G(b,c)\right]$$

$$\leq \frac{f(a,c)+f(a,d)+f(b,c)+f(b,d)}{4}$$

where the function H, F, and G are defined as in Eq. 2.1.

Proof Since f is a coordinated convex on Δ, if we define the mapping $h^1_x:[c,d]\to\mathbb{R}$, $h^1_x(y)=f(x,y)$, then $h^1_x(y)$ is convex for all $x\in[a,b]$ and $H^1_x(y)=h^1_x(y)+\widetilde{h}^1_x(y)=f(x,y)+\widetilde{f}_2(x,y)=G(x,y)$. If we apply the inequalities (1.5) for the convex function $h^1_x(y)$, then we have

$$h^1_x\left(\frac{c+d}{2}\right)\leq \frac{\Gamma(\beta+1)}{4\,[w(d)-w(c)]^\beta}\left[\mathcal{J}^\beta_{c+;w}H^1_x(d)+\mathcal{J}^\beta_{d-;w}H^1_x(c)\right]\leq \frac{h^1_x(c)+h^1_x(d)}{2},$$

that is,

$$f\left(x, \frac{c+d}{2}\right) \tag{2.16}$$

$$\leq \frac{\beta}{4\left[w(d) - w(c)\right]^\beta} \left[\int_c^d \frac{w'(y)}{\left[w(d) - w(y)\right]^{1-\beta}} G(x, y) dy + \int_c^d \frac{w'(y)}{\left[w(y) - w(c)\right]^{1-\beta}} G(x, y) dy\right]$$

$$\leq \frac{f(x, c) + f(x, d)}{2}.$$

Multiplying the inequalities (2.16) by

$$\frac{\alpha}{\left[g(b) - g(a)\right]^\alpha} \frac{g'(x)}{\left[g(b) - g(x)\right]^{1-\alpha}},$$

and

$$\frac{\alpha}{\left[g(b) - g(a)\right]^\alpha} \frac{g'(x)}{\left[g(x) - g(a)\right]^{1-\alpha}},$$

then by integrating the obtained results with respect to x from a to b, we get

$$\frac{\Gamma(\alpha + 1)}{\left[g(b) - g(a)\right]^\alpha} \mathcal{J}_{a+;g}^\alpha f\left(b, \frac{c+d}{2}\right) \tag{2.17}$$

$$\leq \frac{\Gamma(\alpha + 1)\Gamma(\beta + 1)}{4\left[g(b) - g(a)\right]^\alpha \left[w(d) - w(c)\right]^\beta} \left[\mathcal{J}_{a+,c+;g,w}^{\alpha,\beta} G(b, d) + \mathcal{J}_{a+,d-;g,w}^{\alpha,\beta} G(b, c)\right]$$

$$\leq \frac{\Gamma(\alpha + 1)}{2\left[g(b) - g(a)\right]^\alpha} \left[\mathcal{J}_{a+;g}^\alpha f(b, c) + \mathcal{J}_{a+;g}^\alpha f(b, d)\right],$$

and

$$\frac{\Gamma(\alpha + 1)}{\left[g(b) - g(a)\right]^\alpha} \mathcal{J}_{b-;g}^\alpha f\left(a, \frac{c+d}{2}\right) \tag{2.18}$$

$$\leq \frac{\Gamma(\alpha + 1)\Gamma(\beta + 1)}{4\left[g(b) - g(a)\right]^\alpha \left[w(d) - w(c)\right]^\beta} \left[\mathcal{J}_{b-,c+;g,w}^{\alpha,\beta} G(a, d) + \mathcal{J}_{b-,d-;g,w}^{\alpha,\beta} G(a, c)\right]$$

$$\leq \frac{\Gamma(\alpha + 1)}{2\left[g(b) - g(a)\right]^\alpha} \left[\mathcal{J}_{b-;g}^\alpha f(a, c) + \mathcal{J}_{b-;g}^\alpha f(a, d)\right],$$

respectively.

On the other hand, since f is a coordinated convex on Δ, if we define the mapping $h_x^2 : [c, d] \to \mathbb{R}$, $h_x^2(y) = \widetilde{f}_1(x, y)$, then $h_x^2(y)$ is convex for all $x \in [a, b]$ and $H_x^2(y) = h_x^2(y) + \widetilde{h}_x^2(y) = \widetilde{f}_1(x, y) + \widetilde{f}_3(x, y) = K(x, y)$. If we apply the inequalities (1.5) for the convex function $h_x^2(y)$, then we have

$$h_x^2\left(\frac{c+d}{2}\right) \leq \frac{\Gamma(\beta + 1)}{4\left[w(d) - w(c)\right]^\beta} \left[\mathcal{J}_{c+;w}^\beta H_x^2(d) + \mathcal{J}_{d-;w}^\beta H_x^2(c)\right] \leq \frac{h_x^2(c) + h_x^2(d)}{2},$$

i.e.,

$$\tilde{f}_1\left(x, \frac{c+d}{2}\right) \tag{2.19}$$

$$\leq \frac{\beta}{4\left[w(d) - w(c)\right]^{\beta}} \left[\int_c^d \frac{w'(y)}{[w(d) - w(y)]^{1-\beta}} K(x, y)dy + \int_c^d \frac{w'(y)}{[w(y) - w(c)]^{1-\beta}} K(x, y)dy\right]$$

$$\leq \frac{\tilde{f}_1(x, c) + \tilde{f}_1(x, d)}{2}.$$

Similarly, multiplying the inequalities (2.19) by

$$\frac{\alpha}{[g(b) - g(a)]^{\alpha}} \frac{g'(x)}{[g(b) - g(x)]^{1-\alpha}},$$

and

$$\frac{\alpha}{[g(b) - g(a)]^{\alpha}} \frac{g'(x)}{[g(x) - g(a)]^{1-\alpha}},$$

then by integrating the obtained results with respect to x from a to b, we get

$$\frac{\Gamma(\alpha+1)}{[g(b) - g(a)]^{\alpha}} \mathcal{J}_{a+;g}^{\alpha} \tilde{f}_1\left(b, \frac{c+d}{2}\right) \tag{2.20}$$

$$\leq \frac{\Gamma(\alpha+1)\Gamma(\beta+1)}{4[g(b) - g(a)]^{\alpha}[w(d) - w(c)]^{\beta}} \left[\mathcal{J}_{a+,c+;g,w}^{\alpha,\beta} K(b, d) + \mathcal{J}_{a+,d-;g,w}^{\alpha,\beta} K(b, c)\right]$$

$$\leq \frac{\Gamma(\alpha+1)}{2[g(b) - g(a)]^{\alpha}} \left[\mathcal{J}_{a+;g}^{\alpha} \tilde{f}_1(b, c) + \mathcal{J}_{a+;g}^{\alpha} \tilde{f}_1(b, d)\right],$$

and

$$\frac{\Gamma(\alpha+1)}{[g(b) - g(a)]^{\alpha}} \mathcal{J}_{b-;g}^{\alpha} \tilde{f}_1\left(a, \frac{c+d}{2}\right) \tag{2.21}$$

$$\leq \frac{\Gamma(\alpha+1)\Gamma(\beta+1)}{4[g(b) - g(a)]^{\alpha}[w(d) - w(c)]^{\beta}} \left[\mathcal{J}_{b-,c+;g,w}^{\alpha,\beta} K(a, d) + \mathcal{J}_{b-,d-;g,w}^{\alpha,\beta} K(a, c)\right]$$

$$\leq \frac{\Gamma(\alpha+1)}{2[g(b) - g(a)]^{\alpha}} \left[\mathcal{J}_{b-;g}^{\alpha} \tilde{f}_1(a, c) + \mathcal{J}_{b-;g}^{\alpha} \tilde{f}_1(a, d)\right],$$

respectively.

Moreover, if we define the mapping $h_y^1 : [a, b] \to \mathbb{R}$, $h_y^1(x) = f(x, y)$, then $h_y^1(x)$ is convex for all $y \in [c, d]$ and $H_y^1(x) = h_y^1(x) + \tilde{h}_y^1(x) = f(x, y) + \tilde{f}_1(x, y) = G(x, y)$. Applying the inequalities (1.5) for the convex function $h_y^1(x)$, then we have

$$h_y^1\left(\frac{a+b}{2}\right) \le \frac{\Gamma(\alpha+1)}{4\left[g(b)-g(a)\right]^\alpha}\left[\mathcal{J}_{a+;g}^\alpha H_y^1(b) + \mathcal{J}_{b-;w}^\alpha H_y^1(a)\right] \le \frac{h_y^1(a)+h_y^1(b)}{2},$$

that is,

$$f\left(\frac{a+b}{2}, y\right) \tag{2.22}$$

$$\le \frac{\alpha}{4\left[g(b)-g(a)\right]^\alpha}\left[\int_a^b \frac{g'(x)}{\left[g(b)-g(x)\right]^{1-\alpha}} H(x, y)dx + \int_a^b \frac{g'(x)}{\left[g(x)-g(a)\right]^{1-\alpha}} H(x, y)dx\right]$$

$$\le \frac{f(a, y) + f(b, y)}{2}.$$

Multiplying the inequalities (2.22) by

$$\frac{\beta}{\left[w(d)-w(c)\right]^\beta}\frac{w'(y)}{\left[w(d)-w(y)\right]^{1-\beta}}$$

and

$$\frac{\beta}{\left[w(d)-w(c)\right]^\beta}\frac{w'(y)}{\left[w(y)-w(c)\right]^{1-\beta}}$$

then integrating the established results with respect to y from c to d, we obtain the following inequalities:

$$\frac{\Gamma(\beta+1)}{\left[w(d)-w(c)\right]^\beta}\mathcal{J}_{c+;w}^\beta f\left(\frac{a+b}{2}, d\right) \tag{2.23}$$

$$\le \frac{\Gamma(\alpha+1)\Gamma(\beta+1)}{4\left[g(b)-g(a)\right]^\alpha\left[w(d)-w(c)\right]^\beta}\left[\mathcal{J}_{a+,c+;g,w}^{\alpha,\beta}H(b, d) + \mathcal{J}_{b-,c+;g,w}^{\alpha,\beta}H(a, d)\right]$$

$$\le \frac{\Gamma(\beta+1)}{2\left[w(d)-w(c)\right]^\beta}\left[\mathcal{J}_{c+;w}^\beta f(a, d) + \mathcal{J}_{c+;w}^\beta f(b, d)\right],$$

and

$$\frac{\Gamma(\beta+1)}{\left[w(d)-w(c)\right]^\beta}\mathcal{J}_{d-;w}^\beta f\left(\frac{a+b}{2}, c\right) \tag{2.24}$$

$$\le \frac{\Gamma(\alpha+1)\Gamma(\beta+1)}{4\left[g(b)-g(a)\right]^\alpha\left[w(d)-w(c)\right]^\beta}\left[\mathcal{J}_{a+,d-;g,w}^{\alpha,\beta}H(b, c) + \mathcal{J}_{b-,d-;g,w}^{\alpha,\beta}H(a, c)\right]$$

$$\le \frac{\Gamma(\beta+1)}{2\left[w(d)-w(c)\right]^\beta}\left[\mathcal{J}_{d-;w}^\beta f(a, c) + \mathcal{J}_{d-;w}^\beta f(b, c)\right],$$

respectively.

Furthermore, if we define the mapping $h_y^2 : [a, b] \to \mathbb{R}$, $h_y^2(x) = \tilde{f}_2(x, y)$, then $h_y^2(x)$ is convex for all $y \in [c, d]$ and $H_y^2(x) = h_y^2(x) + \tilde{h}_y^2(x) = \tilde{f}_2(x, y) +$

$\widetilde{f}_3(x, y) = L(x, y)$. Applying the inequalities (1.5) for the convex function $h_y^2(x)$, then we have

$$h_y^2\left(\frac{a+b}{2}\right) \leq \frac{\Gamma(\alpha+1)}{4\,[g(b)-g(a)]^\alpha}\left[\mathcal{J}_{a+;g}^\alpha H_y^2(b) + \mathcal{J}_{b-;w}^\alpha H_y^2(a)\right] \leq \frac{h_y^2(a) + h_y^2(b)}{2},$$

i.e.,

$$\widetilde{f}_2\left(\frac{a+b}{2}, y\right) \tag{2.25}$$

$$\leq \frac{\alpha}{4\,[g(b)-g(a)]^\alpha}\left[\int_a^b \frac{g'(x)}{[g(b)-g(x)]^{1-\alpha}}L(x, y)dx + \int_a^b \frac{g'(x)}{[g(x)-g(a)]^{1-\alpha}}L(x, y)dx\right]$$

$$\leq \frac{\widetilde{f}_2(a, y) + \widetilde{f}_2(b, y)}{2}.$$

Similarly, multiplying the inequalities (2.25) by

$$\frac{\beta}{[w(d)-w(c)]^\beta}\frac{w'(y)}{[w(d)-w(y)]^{1-\beta}}$$

and

$$\frac{\beta}{[w(d)-w(c)]^\beta}\frac{w'(y)}{[w(y)-w(c)]^{1-\beta}},$$

then integrating the obtained results with respect to y from c to d, we obtain the following inequalities:

$$\frac{\Gamma(\beta+1)}{[w(d)-w(c)]^\beta}\mathcal{J}_{c+;w}^\beta \widetilde{f}_2\left(\frac{a+b}{2}, d\right) \tag{2.26}$$

$$\leq \frac{\Gamma(\alpha+1)\Gamma(\beta+1)}{4\,[g(b)-g(a)]^\alpha\,[w(d)-w(c)]^\beta}\left[\mathcal{J}_{a+,c+;g,w}^{\alpha,\beta}L(b, d) + \mathcal{J}_{b-,c+;g,w}^{\alpha,\beta}L(a, d)\right]$$

$$\leq \frac{\Gamma(\beta+1)}{2\,[w(d)-w(c)]^\beta}\left[\mathcal{J}_{c+;w}^\beta \widetilde{f}_2(a, d) + \mathcal{J}_{c+;w}^\beta \widetilde{f}_2(b, d)\right]$$

and

$$\frac{\Gamma(\beta+1)}{[w(d)-w(c)]^\beta}\mathcal{J}_{d-;w}^\beta \widetilde{f}_2\left(\frac{a+b}{2}, c\right) \tag{2.27}$$

$$\leq \frac{\Gamma(\alpha+1)\Gamma(\beta+1)}{4\,[g(b)-g(a)]^\alpha\,[w(d)-w(c)]^\beta}\left[\mathcal{J}_{a+,d-;g,w}^{\alpha,\beta}L(b, c) + \mathcal{J}_{b-,d-;g,w}^{\alpha,\beta}L(a, c)\right]$$

$$\leq \frac{\Gamma(\beta+1)}{2\,[w(d)-w(c)]^\beta}\left[\mathcal{J}_{d-;w}^\beta \widetilde{f}_2(a, c) + \mathcal{J}_{d-;w}^\beta \widetilde{f}_2(b, c)\right],$$

respectively.

Summing the inequalities (2.17), (2.18), (2.20), (2.21), (2.23), (2.24), (2.26) and (2.27), we have the following inequalities:

$$
\frac{\Gamma(\alpha+1)}{[g(b)-g(a)]^\alpha}\left[\mathcal{J}_{a+;g}^\alpha f\left(b,\frac{c+d}{2}\right)+\mathcal{J}_{b-;g}^\alpha f\left(a,\frac{c+d}{2}\right)\right.
$$
$$
\left.+\mathcal{J}_{a+;g}^\alpha \tilde{f}_1\left(b,\frac{c+d}{2}\right)+\mathcal{J}_{b-;g}^\alpha \tilde{f}_1\left(a,\frac{c+d}{2}\right)\right]
$$
$$
+\frac{\Gamma(\beta+1)}{[w(d)-w(c)]^\beta}\left[\mathcal{J}_{c+;w}^\beta f\left(\frac{a+b}{2},d\right)+\mathcal{J}_{d-;w}^\beta f\left(\frac{a+b}{2},c\right)\right.
$$
$$
\left.+\mathcal{J}_{c+;w}^\beta \tilde{f}_2\left(\frac{a+b}{2},d\right)+\mathcal{J}_{d-;w}^\beta \tilde{f}_2\left(\frac{a+b}{2},c\right)\right]
$$
$$
\leq\frac{\Gamma(\alpha+1)\Gamma(\beta+1)}{4\left[g(b)-g(a)\right]^\alpha\left[w(d)-w(c)\right]^\beta}
$$
$$
\times\left[\mathcal{J}_{a+,c+;g,w}^{\alpha,\beta}G(b,d)+\mathcal{J}_{a+,d-;g,w}^{\alpha,\beta}G(b,c)+\mathcal{J}_{b-,c+;g,w}^{\alpha,\beta}G(a,d)+\mathcal{J}_{b-,d-;g,w}^{\alpha,\beta}G(a,c)\right.
$$
$$
+\mathcal{J}_{a+,c+;g,w}^{\alpha,\beta}K(b,d)+\mathcal{J}_{a+,d-;g,w}^{\alpha,\beta}K(b,c)+\mathcal{J}_{b-,c+;g,w}^{\alpha,\beta}K(a,d)+\mathcal{J}_{b-,d-;g,w}^{\alpha,\beta}K(a,c)
$$
$$
+\mathcal{J}_{a+,c+;g,w}^{\alpha,\beta}H(b,d)+\mathcal{J}_{b-,c+;g,w}^{\alpha,\beta}H(a,d)+\mathcal{J}_{a+,d-;g,w}^{\alpha,\beta}H(b,c)+\mathcal{J}_{b-,d-;g,w}^{\alpha,\beta}H(a,c)
$$
$$
\left.+\mathcal{J}_{a+,c+;g,w}^{\alpha,\beta}L(b,d)+\mathcal{J}_{b-,c+;g,w}^{\alpha,\beta}L(a,d)+\mathcal{J}_{a+,d-;g,w}^{\alpha,\beta}L(b,c)+\mathcal{J}_{b-,d-;g,w}^{\alpha,\beta}L(a,c)\right]
$$
$$
\leq\frac{\Gamma(\alpha+1)}{2\left[g(b)-g(a)\right]^\alpha}\left[\mathcal{J}_{a+;g}^\alpha f(b,c)+\mathcal{J}_{a+;g}^\alpha f(b,d)+\mathcal{J}_{b-;g}^\alpha f(a,c)+\mathcal{J}_{b-;g}^\alpha f(a,d)\right.
$$
$$
\left.+\mathcal{J}_{a+;g}^\alpha \tilde{f}_1(b,c)+\mathcal{J}_{a+;g}^\alpha \tilde{f}_1(b,d)+\mathcal{J}_{b-;g}^\alpha \tilde{f}_1(a,c)+\mathcal{J}_{b-;g}^\alpha \tilde{f}_1(a,d)\right]
$$
$$
+\frac{\Gamma(\beta+1)}{[w(d)-w(c)]^\beta}\left[\mathcal{J}_{c+;w}^\beta f(a,d)+\mathcal{J}_{c+;w}^\beta f(b,d)+\mathcal{J}_{d-;w}^\beta f(a,c)+\mathcal{J}_{d-;w}^\beta f(b,c)\right.
$$
$$
\left.+\mathcal{J}_{c+;w}^\beta \tilde{f}_2(a,d)+\mathcal{J}_{c+;w}^\beta \tilde{f}_2(b,d)+\mathcal{J}_{d-;w}^\beta \tilde{f}_2(a,c)+\mathcal{J}_{d-;w}^\beta \tilde{f}_2(b,c)\right].
$$

That is, we have

$$
\frac{\Gamma(\alpha+1)}{[g(b)-g(a)]^\alpha}\left[\mathcal{J}_{a+;g}^\alpha H\left(b,\frac{c+d}{2}\right)+\mathcal{J}_{b-;g}^\alpha H\left(a,\frac{c+d}{2}\right)\right]
$$
$$
+\frac{\Gamma(\beta+1)}{[w(d)-w(c)]^\beta}\left[\mathcal{J}_{c+;w}^\beta G\left(\frac{a+b}{2},d\right)+\mathcal{J}_{d-;w}^\beta G\left(\frac{a+b}{2},c\right)\right]
$$
$$
\leq\frac{\Gamma(\alpha+1)\Gamma(\beta+1)}{2\left[g(b)-g(a)\right]^\alpha\left[w(d)-w(c)\right]^\beta}
$$
$$
\times\left[\mathcal{J}_{a+,c+;g,w}^{\alpha,\beta}F(b,d)+\mathcal{J}_{a+,d-;g,w}^{\alpha,\beta}F(b,c)+\mathcal{J}_{b-,c+;g,w}^{\alpha,\beta}F(a,d)+\mathcal{J}_{b-,d-;g,w}^{\alpha,\beta}F(a,c)\right]
$$
$$
\leq\frac{\Gamma(\alpha+1)}{2\left[g(b)-g(a)\right]^\alpha}\left[\mathcal{J}_{a+;g}^\alpha H(b,c)+\mathcal{J}_{a+;g}^\alpha H(b,d)+\mathcal{J}_{b-;g}^\alpha H(a,c)+\mathcal{J}_{b-;g}^\alpha H(a,d)\right]
$$
$$
+\frac{\Gamma(\beta+1)}{2\left[w(d)-w(c)\right]^\beta}\left[\mathcal{J}_{c+;w}^\beta G(a,d)+\mathcal{J}_{c+;w}^\beta G(b,d)+\mathcal{J}_{d-;w}^\beta G(a,c)+\mathcal{J}_{d-;w}^\beta G(b,c)\right]
$$

which completes the proof of the second and third inequalities in (2.15).

On the other hand, from the first inequality in (1.5), we have

$$\varphi\left(\frac{a+b}{2}\right) \tag{2.28}$$

$$\leq \frac{\alpha}{4\,[g(b)-g(a)]^\alpha}\left[\int_a^b \frac{g'(x)}{[g(b)-g(x)]^\alpha}\,[\varphi(x)+\varphi(a+b-x)]\,dx\right.$$

$$\left.+\int_a^b \frac{g'(x)}{[g(x)-g(a)]^\alpha}\,[\varphi(x)+\varphi(a+b-x)]\,dx\right].$$

Since f is a coordinated convex on Δ, by using the inequality (2.28), we obtain

$$f\left(\frac{a+b}{2},\frac{c+d}{2}\right) \tag{2.29}$$

$$\leq \frac{\alpha}{4\,[g(b)-g(a)]^\alpha}\left[\int_a^b \frac{g'(x)}{[g(b)-g(x)]^\alpha}\left[f\left(x,\frac{c+d}{2}\right)+f\left(a+b-x,\frac{c+d}{2}\right)\right]dx\right.$$

$$\left.+\int_a^b \frac{g'(x)}{[g(x)-g(a)]^\alpha}\left[f\left(x,\frac{c+d}{2}\right)+f\left(a+b-x,\frac{c+d}{2}\right)\right]dx\right]$$

$$= \frac{\Gamma(\alpha+1)}{[g(b)-g(a)]^\alpha}\left[\mathcal{J}^\alpha_{a+;g}H\left(b,\frac{c+d}{2}\right)+\mathcal{J}^\alpha_{b-;g}H\left(a,\frac{c+d}{2}\right)\right],$$

and similarly we have

$$f\left(\frac{a+b}{2},\frac{c+d}{2}\right) \tag{2.30}$$

$$\leq \frac{\beta}{4\,[w(d)-w(c)]^\beta}\left[\int_c^d \frac{w'(y)}{[w(d)-w(y)]^\alpha}\left[f\left(\frac{a+b}{2},y\right)+f\left(\frac{a+b}{2},c+d-y\right)\right]dy\right.$$

$$\left.+\int_c^d \frac{w'(y)}{[w(y)-w(c)]^\alpha}\left[f\left(\frac{a+b}{2},y\right)+f\left(\frac{a+b}{2},c+d-y\right)\right]dy\right]$$

$$= \frac{\Gamma(\beta+1)}{[w(d)-w(c)]^\beta}\left[\mathcal{J}^\beta_{c+;w}G\left(\frac{a+b}{2},d\right)+\mathcal{J}^\beta_{d-;w}G\left(\frac{a+b}{2},c\right)\right].$$

Combining the inequalities (2.29) and (2.30), we obtain the first inequality in (2.15). From the second inequality in (1.5), we have

$$\frac{\alpha}{4\left[g(b)-g(a)\right]^{\alpha}}\left[\int_{a}^{b}\frac{g'(x)}{\left[g(b)-g(x)\right]^{\alpha}}\left[\varphi(x)+\varphi(a+b-x)\right]dx \quad (2.31)$$

$$+\int_{a}^{b}\frac{g'(x)}{\left[g(x)-g(a)\right]^{\alpha}}\left[\varphi(x)+\varphi(a+b-x)\right]dx\right]$$

$$\leq\frac{\varphi(a)+\varphi(b)}{2}.$$

By using the inequality (2.31), we obtain the following inequalities:

$$\frac{\Gamma(\alpha+1)}{4\left[g(b)-g(a)\right]^{\alpha}}\left[\mathcal{J}_{a+;g}^{\alpha}H\left(b,c\right)+\mathcal{J}_{b-;g}^{\alpha}H\left(a,c\right)\right]\leq\frac{f(a,c)+f(b,c)}{2}, \quad (2.32)$$

$$\frac{\Gamma(\alpha+1)}{4\left[g(b)-g(a)\right]^{\alpha}}\left[\mathcal{J}_{a+;g}^{\alpha}H\left(b,d\right)+\mathcal{J}_{b-;g}^{\alpha}H\left(a,d\right)\right]\leq\frac{f(a,d)+f(b,d)}{2}, \quad (2.33)$$

$$\frac{\Gamma(\beta+1)}{4\left[w(d)-w(c)\right]^{\beta}}\left[\mathcal{J}_{c+;w}^{\beta}G\left(a,d\right)+\mathcal{J}_{d-;w}^{\beta}G\left(a,c\right)\right]\leq\frac{f(a,c)+f(a,d)}{2}$$
$$(2.34)$$

and

$$\frac{\Gamma(\beta+1)}{4\left[w(d)-w(c)\right]^{\beta}}\left[\mathcal{J}_{c+;w}^{\beta}G\left(b,d\right)+\mathcal{J}_{d-;w}^{\beta}G\left(b,c\right)\right]\leq\frac{f(b,c)+f(b,d)}{2}.$$
$$(2.35)$$

Combining the inequalities (2.32)–(2.35), we obtain the last inequality in (2.15).
This completes the proof completely. $\qquad\square$

Remark 2 If we choose $g(t)=t$ and $w(s)=s$ in Theorem 5, then we have the following inequalities for Riemann–Liouville fractional integrals:

$$f\left(\frac{a+b}{2},\frac{c+d}{2}\right)$$

$$\leq\frac{\Gamma(\alpha+1)}{4(b-a)^{\alpha}}\left[J_{a+}^{\alpha}f\left(b,\frac{c+d}{2}\right)+J_{b-}^{\alpha}f\left(a,\frac{c+d}{2}\right)\right]$$

$$+\frac{\Gamma(\beta+1)}{4(d-c)^{\beta}}\left[J_{c+}^{\beta}f\left(\frac{a+b}{2},d\right)+J_{d-}^{\beta}f\left(\frac{a+b}{2},c\right)\right]$$

$$\leq\frac{\Gamma(\alpha+1)\Gamma(\beta+1)}{4(b-a)^{\alpha}(d-c)^{\beta}}\left[J_{a+,c+}^{\alpha,\beta}f(b,d)+J_{a+,d-}^{\alpha,\beta}f(b,c)+J_{b-,c+}^{\alpha,\beta}f(a,d)+J_{b-,d-}^{\alpha,\beta}f(a,c)\right]$$

$$\leq\frac{\Gamma(\alpha+1)}{8(b-a)^{\alpha}}\left[J_{a+}^{\alpha}f(b,c)+J_{a+}^{\alpha}f(b,d)+J_{b-}^{\alpha}f(a,c)+J_{b-}^{\alpha}f(a,d)\right]$$

$$+\frac{\Gamma(\beta+1)}{8(d-c)^{\beta}}\left[J_{c+}^{\beta}f(a,d)+J_{c+}^{\beta}f(b,d)+J_{d-}^{\beta}f(a,c)+J_{d-}^{\beta}f(b,c)\right]$$

$$\leq\frac{f(a,c)+f(a,d)+f(b,c)+f(b,d)}{4},$$

which are proved by Sarikaya in [36].

Corollary 3 *Under assumption of Theorem 4 with $g(t) = \ln t$ and $w(s) = \ln s$, then we have the following inequalities for Hadamard fractional integrals*

$$
f\left(\frac{a+b}{2}, \frac{c+d}{2}\right)
$$

$$
\leq \frac{\Gamma(\alpha+1)}{8\left[\ln \frac{b}{a}\right]^{\alpha}} \left[\mathbf{J}_{a+}^{\alpha} H\left(b, \frac{c+d}{2}\right) + \mathbf{J}_{b-}^{\alpha} H\left(a, \frac{c+d}{2}\right)\right]
$$

$$
+ \frac{\Gamma(\beta+1)}{8\left[\ln \frac{d}{c}\right]^{\beta}} \left[\mathbf{J}_{c+}^{\beta} G\left(\frac{a+b}{2}, d\right) + \mathbf{J}_{d-}^{\beta} G\left(\frac{a+b}{2}, c\right)\right]
$$

$$
\leq \frac{\Gamma(\alpha+1)\Gamma(\beta+1)}{16\left[\ln \frac{b}{a}\right]^{\alpha}\left[\ln \frac{d}{c}\right]^{\beta}} \left[\mathbf{J}_{a+,c+}^{\alpha,\beta} F(b,d) + \mathbf{J}_{a+,d-}^{\alpha,\beta} F(b,c) + \mathbf{J}_{b-,c+}^{\alpha,\beta} F(a,d) + \mathbf{J}_{b-,d-}^{\alpha,\beta} F(a,c)\right]
$$

$$
\leq \frac{\Gamma(\alpha+1)}{16\left[\ln \frac{b}{a}\right]^{\alpha}} \left[\mathbf{J}_{a+}^{\alpha} H(b,c) + \mathbf{J}_{a+}^{\alpha} H(b,d) + \mathbf{J}_{b-}^{\alpha} H(a,c) + \mathbf{J}_{b-}^{\alpha} H(a,d)\right]
$$

$$
+ \frac{\Gamma(\beta+1)}{16\left[\ln \frac{d}{c}\right]^{\beta}} \left[\mathbf{J}_{c+}^{\beta} G(a,d) + \mathbf{J}_{c+}^{\beta} G(b,d) + \mathbf{J}_{d-}^{\beta} G(a,c) + \mathbf{J}_{d-}^{\beta} G(b,c)\right]
$$

$$
\leq \frac{f(a,c) + f(a,d) + f(b,c) + f(b,d)}{4}.
$$

Corollary 4 *Under assumption of Theorem 4 with $g(t) = \frac{t^{\rho}}{\rho}$ and $w(s) = \frac{s^{\sigma}}{\sigma}$, then we have the following inequalities for Katugampola fractional integrals:*

$$
f\left(\frac{a+b}{2}, \frac{c+d}{2}\right)
$$

$$
\leq \frac{\Gamma(\alpha+1)\rho^{\alpha}}{8\left[b^{\rho} - a^{\rho}\right]^{\alpha}} \left[{}^{\rho}\mathcal{I}_{a+}^{\alpha} H\left(b, \frac{c+d}{2}\right) + {}^{\rho}\mathcal{I}_{b-}^{\alpha} H\left(a, \frac{c+d}{2}\right)\right]
$$

$$
+ \frac{\Gamma(\beta+1)\sigma^{\beta}}{8\left[d^{\sigma} - c^{\sigma}\right]^{\beta}} \left[{}^{\sigma}\mathcal{I}_{c+}^{\beta} G\left(\frac{a+b}{2}, d\right) + {}^{\sigma}\mathcal{I}_{d-}^{\beta} G\left(\frac{a+b}{2}, c\right)\right]
$$

$$
\leq \frac{\Gamma(\alpha+1)\Gamma(\beta+1)\rho^{\alpha}\sigma^{\beta}}{16\left[b^{\rho} - a^{\rho}\right]^{\alpha}\left[d^{\sigma} - c^{\sigma}\right]^{\beta}} \left[{}^{\rho,\sigma}\mathcal{I}_{a+,c+}^{\alpha,\beta} F(b,d) + {}^{\rho,\sigma}\mathcal{I}_{a+,d-}^{\alpha,\beta} F(b,c) + {}^{\rho,\sigma}\mathcal{I}_{b-,c+}^{\alpha,\beta} F(a,d) + {}^{\rho,\sigma}\mathcal{I}_{b-,d-}^{\alpha,\beta} F(a,c)\right]
$$

$$
\leq \frac{\Gamma(\alpha+1)\rho^{\alpha}\sigma^{\beta}}{16\left[b^{\rho} - a^{\rho}\right]^{\alpha}} \left[{}^{\rho}\mathcal{I}_{a+}^{\alpha} H(b,c) + {}^{\rho}\mathcal{I}_{a+}^{\alpha} H(b,d) + {}^{\rho}\mathcal{I}_{b-}^{\alpha} H(a,c) + {}^{\rho}\mathcal{I}_{b-}^{\alpha} H(a,d)\right]
$$

$$
+ \frac{\Gamma(\beta+1)\sigma^{\beta}}{16\left[d^{\sigma} - c^{\sigma}\right]^{\beta}} \left[{}^{\sigma}\mathcal{I}_{c+}^{\beta} G(a,d) + {}^{\sigma}\mathcal{I}_{c+}^{\beta} G(b,d) + {}^{\sigma}\mathcal{I}_{d-}^{\beta} G(a,c) + {}^{\sigma}\mathcal{I}_{d-}^{\beta} G(b,c)\right]
$$

$$
\leq \frac{f(a,c) + f(a,d) + f(b,c) + f(b,d)}{4}.
$$

References

1. Akkurt, A., Sarikaya, M.Z., Budak, H., Yildirim, H.: On the Hadamard's type inequalities for co-ordinated convex functions via fractional integrals. J. King Saud Univ. Sci. **29**, 380–387 (2017)
2. Alomari, M., Darus, M.: The Hadamards inequality for s-convex function of 2-variables on the coordinates. Int. J. Math. Anal. **2**(13), 629–638 (2008)

3. Azpeitia, A.G.: Convex functions and the Hadamard inequality. Rev. Colombiana Math. **28**, 7–12 (1994)
4. Bakula, M.K.: An improvement of the Hermite-Hadamard inequality for functions convex on the coordinates. Aust. J. Math. Anal. Appl. **11**(1), 1–7 (2014)
5. Belarbi, S., Dahmani, Z.: On some new fractional integral inequalities. J. Ineq. Pure and Appl. Math. **10**(3) (2009), Art. 86
6. Budak, H., Sarikaya, M.Z.: Hermite-Hadamard type inequalities for s-convex mappings via fractional integrals of a function with respect to another function. Fasciculi Mathematici **27**, 25–36 (2016)
7. Chen, F.X.: On Hermite-Hadamard type inequalities for *s*-convex functions on the coordinates via Riemann-Liouville fractional integrals. J. Appl. Math. **2014**, Article ID 248710, 8 p., 2014
8. Dahmani, Z.: New inequalities in fractional integrals. Int. J. Nonlinear Sci. **9**(4), 493–497 (2010)
9. Dahmani, Z.: On Minkowski and Hermite-Hadamard integral inequalities via fractional integration. Ann. Funct. Anal. **1**(1), 51–58 (2010)
10. Dahmani, Z., Tabharit, L., Taf, S.: Some fractional integral inequalities. Nonl. Sci. Lett. A **1**(2), 155–160 (2010)
11. Deng, J., Wang, J.: Fractional Hermite-Hadamard inequalities for (α, m)-logarithmically convex functions. J. Inequal. Appl. **2013**, Article ID 364 (2013)
12. Dragomir, S.S.: On Hadamards inequality for convex functions on the co-ordinates in a rectangle from the plane. Taiwan. J. Math. **4**, 775–788 (2001)
13. Dragomir, S.S., Pearce, C.E.M.: Selected Topics on Hermite-Hadamard Inequalities and Applications. Victoria University, RGMIA Monographs (2000)
14. Dragomir, S.S.: Inequalities of Hermite-Hadamard type for *h*-convex functions on linear spaces. Proyecciones J. Math. **37**(4), 343–341 (2015)
15. Dragomir, S.S.: Two mappings in connection to Hadamard's inequalities. J. Math. Anal. Appl. **167**, 49–56 (1992)
16. Dragomir, S.S., Pecaric, J., Persson, L.E.: Some inequalities of Hadamard type. Soochow J. Math. **21**, 335–341 (1995)
17. Farid, G., Rehman, A.U., Zahra, M.: On Hadamard inequalities for *k*-fractional integrals. Nonlinear Funct. Anal. Appl. **21**(3), 463–478 (2016)
18. Farissi, A.E.: Simple proof and re nement of Hermite-Hadamard inequality. J. Math. Inequal. **4**, 365–369 (2010)
19. Gorenflo, R., Mainardi, F.: Fractional Calculus: Integral and Differential Equations of Fractional Order, pp. 223–276. Springer, Wien (1997)
20. Iqbal, M., Qaisar, S., Muddassar, M.: A short note on integral inequality of type Hermite-Hadamard through convexity. J. Comput. Anal. Appl. **21**(5), 946–953 (2016)
21. Jleli, M., Samet, B.: On Hermite-Hadamard type inequalities via fractional integrals of a function with respect to another function. J. Nonlinear Sci. Appl. **9**, 1252–1260 (2016)
22. Katugampola, U.N.: New approach to a generalized fractional integrals. Appl. Math. Comput. **218**(4), 860–865 (2011)
23. Kilbas, A.A., Srivastava, H.M., Trujillo, J.J.: Theory and Applications of Fractional Differential Equations, North-Holland Mathematics Studies, 204. Elsevier Sci. B.V, Amsterdam (2006)
24. Latif, M.A.: Hermite-Hadamard type inequalities for *GA*-convex functions on the co-ordinates with applications. Proc. Pak. Acad. Sci. **52**(4), 367–379 (2015)
25. Miller, S., Ross, B.: An introduction to the Fractional Calculus and Fractional Differential Equations, p. 2. Wiley, USA (1993)
26. Ozdemir, M.E., Yildiz, C., Akdemir, A.O.: On the co-ordinated convex functions Appl. Math. Inf. Sci. **8**(3), 1085–1091 (2014)
27. Pavic, Z.: Improvements of the Hermite-Hadamard inequality. J. Inequal. Appl. **2015**, 222 (2015)
28. Pečarić, J.E., Proschan, F., Tong, Y.L.: Convex Functions. Academic Press, Boston, Partial Orderings and Statistical Applications (1992)

29. Peng, S., Wei, W., Wang, J.-R.: On the Hermite-Hadamard inequalities for convex functions via Hadamard fractional integrals. Facta Universitatis Series Math. Inf. **29**(1), 55–75 (2014)
30. Podlubni, I.: Fractional Differential Equations. Academic Press, San Diego (1999)
31. Sarikaya, M.Z., Budak, H.: Generalized Hermite-Hadamard type integral inequalities for fractional integrals. Filomat **30**(5), 1315–1326 (2016)
32. Sarikaya, M.Z., Set, E., Yaldiz, H., Basak, N.: Hermite -Hadamard's inequalities for fractional integrals and related fractional inequalities. Math. Comput. Modell. **57**, 2403–2407 (2013). https://doi.org/10.1016/j.mcm.2011.12.048
33. Sarikaya, M.Z., Filiz, H., Kiris, M.E.: On some generalized integral inequalities for Riemann-Liouville fractional integrals. Filomat **29**(6), 1307–1314 (2015). https://doi.org/10.2298/FIL1506307S
34. Sarikaya, M.Z., Budak, H.: Generalized Hermite-Hadamard type integral inequalities for functions whose 3 rd derivatives are s-convex. Tbilisi Math. J. **7**(2), 41–49 (2014)
35. Sarikaya, M.Z., Kiris, M.E.: Some new inequalities of Hermite-Hadamard type for $s-$convex functions. Miskolc Math. Notes **16**(1), 491–501 (2015)
36. Sarikaya, M.Z.: On the Hermite-Hadamard-type inequalities for co-ordinated convex function via fractional integrals. Integral Trans. Spec. Funct. **25**(2), 134–147 (2014)
37. Sarikaya, M.Z., Ogunmez, H.: On new inequalities via Riemann-Liouville fractional integration. Abstr. Appl. Anal. **2012** (2012), Article ID 428983, 10 p
38. Sarikaya, M.Z., Budak, H.: Some Hermite-Hadamard type integral inequalities for twice differentiable mappings via fractional integrals. Facta Universitatis Ser: Math. Inform. **29**(4), 371–384 (2014)
39. Set, E., Sarikaya, M.Z., Ozdemir, M.E., Yildirim, H.: The Hermite-Hadamard's inequality for some convex functions via fractional integrals and related results. JAMSI **10**(2), 69–83 (2014)
40. Tseng, K.L., Hwang, S.R.: New Hermite-Hadamard inequalities and their applications. Filomat **30**(14), 3667–3680 (2016)
41. Wang, D.Y., Tseng, K.L., Yang, G.S.: Some Hadamard's inequalities for co-ordinated convex functions in a rectangle from the plane. Taiwan. J. Math. **11**, 63–73 (2007)
42. Wang, J.R., Li, X., Zhu, C.: Refinements of Hermite-Hadamard type inequalities involving fractional integrals. Bull. Belg. Math. Soc. Simon Stevin **20**, 655–666 (2013)
43. Wang, J.R., Zhu, C., Zhou, Y.: New generalized Hermite-Hadamard type inequalities and applications to special means. J. Inequal. Appl. **2013**, 325 (2013)
44. Xi, B.Y., Hua, J., Qi, F.: Hermite-Hadamard type inequalities for extended s-convex functions on the co-ordinates in a rectangle. J. Appl. Anal. **20**(1), 1–17 (2014)
45. Yaldiz, H., Sarıkaya, M.Z., Dahmani, Z.: On the Hermite-Hadamard-Fejer-type inequalities for co-ordinated convex functions via fractional integrals. Int. J. Optim. Control: Theor. Appl. (IJOCTA) **7**(2), 205–215 (2017)
46. Yaldiz, H.: On Hermite-Hadamard type inequalities via Katugampola fractional integrals. ResearchGate paper, Available online at: https://www.researchgate.net/publication/321947179
47. Yang, G.S., Tseng, K.L.: On certain integral inequalities related to Hermite-Hadamard inequalities. J. Math. Anal. Appl. **239**, 180–187 (1999)
48. Yang, G.S., Hong, M.C.: A note on Hadamard's inequality. Tamkang J. Math. **28**, 33–37 (1997)
49. Yıldırım, M.E., Akkurt, A., Yıldırım, H.: Hermite-Hadamard type inequalities for co-ordinated $(\alpha_1, m_1) - (\alpha_2, m_2)$-convex functions via fractional integrals. Contemp. Anal. Appl. Math. **4**(1), 48–63 (2016)
50. Zhang, Y., Wang, Y.: On some new Hermite-Hadamard inequalities involving Riemann-Liouville fractional integrals. J. Inequal. Appl. **2013**, Article ID 220 (2013)
51. Zhang, Z., Wei, W., Wang, J.: Generalization of Hermite-Hadamard inequalities involving Hadamard fractional integrals. Filomat **29**(7), 1515–1524 (2015)
52. Zhang, Z., Wang, J.R., Deng, J.H.: Applying GG-convex function to Hermite-Hadamard inequalities involving Hadamard fractional integrals. Int. J. Math. Math. Sci. **2014**, Article ID 136035, 20 p